ACVP & ASVCP Concurrent Annual Meeting 2017

AA005033

Annual Meeting was combined between the American College of Veterinary Pathologists (ACVP) and American Society for Veterinary Clinical Pathology (ASVCP)

Vancouver, Canada
4 – 8 November 2017

ISBN: 978-1-5108-5329-4

Printed from e-media with permission by:

Curran Associates, Inc.
57 Morehouse Lane
Red Hook, NY 12571

Some format issues inherent in the e-media version may also appear in this print version.

Printed by Curran Associates, Inc. (2018)

For permission requests, please contact American College of Veterinary Pathologists (ACVP) at the address below.

American College of Veterinary Pathologists (ACVP)
2424 American Lane
Madison, WI 53704
USA

Phone: (608) 443-2466
Fax: (608) 443-2478

info@acvp.org

Additional copies of this publication are available from:

Curran Associates, Inc.
57 Morehouse Lane
Red Hook, NY 12571 USA
Phone: 845-758-0400
Fax: 845-758-2633
Email: curran@proceedings.com
Web: www.proceedings.com

TABLE OF CONTENTS

NATURAL DISEASE FOCUSED SCIENTIFIC SESSION I

NATURAL DISEASE FOCUSED SCIENTIFIC SESSION II

NATURAL DISEASE FOCUSED GROUP POSTER SESSION

CLINICAL PATHOLOGY FOCUSED SCIENTIFIC SESSION I

CLINICAL PATHOLOGY FOCUSED SCIENTIFIC SESSION II

EXPERIMENTAL DISEASE FOCUSED SCIENTIFIC SESSION I

EXPERIMENTAL DISEASE FOCUSED GROUP POSTER SESSION

DIAGNOSTIC PATHOLOGY FOCUSED SCIENTIFIC SESSION I

DIAGNOSTIC PATHOLOGY FOCUSED SCIENTIFIC SESSION II

DIAGNOSTIC PATHOLOGY FOCUSED GROUP POSTER SESSION

EDUCATION FOCUSED SCIENTIFIC SESSION

VETERINARY STUDENT POSTERS

ACVP LATE-BREAKING POSTER

Industrial and Toxicologic Pathology Focused Scientific Session I

November 5, 2017 | 8:00 AM – 12:00 PM
Session Chair: G. Palanisamy
Committee Members: K. Janardhan (Co-chair), T. Crabbs (Past Chair), G. Cain, C. Colleton, K. Helke, R. Johnson, and K. Knostman

November 5, 2017
9:00 AM – 9:15 AM

COMPREHENSIVE HISTOLOGIC STUDY OF THE REPRODUCTIVE SYSTEM OF HONEY BEE QUEEN

Ivanna V. Kozii, Sarah C. Wood, Roman V. Koziy, Claire Janse van Rensburg, Jessica Morris, Sophie Derveau, Igor Moshynskyy, Maud Ferrari, Ahmad Al-Dissi, Elemir Simko

Background: Increased mortality in honey bees coincided with widespread use of neonicotinoid insecticides in agriculture. However, discrepancies between field and laboratory experiments and incomplete understanding of toxicopathologic effects of neonicotinoids on honey bees reveal a gap in the risk assessment of pesticides for honey bees.

Mammalian risk assessment relies on histopathology as a “gold standard” for safety evaluation. Histologic assessment of gonadotoxicity, teratogenicity, neurotoxicity and cancerogenicity for determination of a safe dose range of a tested drug or chemical is sensitive and reproducible. However, comparable toxicopathologic approaches using histopathology have not been developed for honey bees.

Objective: Our objective was to characterize histologic morphology of the reproductive tract of honey bee queens for the future risk assessment of pesticides using gonadotoxicity in honey bee queens.

Methods: One-year old, genetically diverse, naturally mated honey bee queens (n=16) were harvested from local research apiaries. Serial H&E stained tissue sections from paraffin embedded blocks (every 100μm) were obtained for microscopic characterization.

Results: Following reproductive system segments were described: 1) ovaries, comprised of ovarioles showing sequential oocyte maturation, 2) tubular passage system formed by the calyx, lateral and median oviducts, genital chamber, and bursa copulatrix with bursal pouches, 3) spermatheca - a spherical organ for sperm storage during the entire life of queen (2-4 years), with spermathecal glands and duct.

This histologic study will provide a baseline for future investigation of gonadotoxicity of neonicotinoids in exposed honey bee queens and determination of a safe dose range for the most commonly used neonicotinoids in agriculture.

November 5, 2017
9:15 AM – 9:30 AM

CHARACTERIZATION OF REPRODUCTIVE TOXICITY OF IN UTERO EXPOSURE TO POLYCYCLIC AROMATIC HYDROCARBONS IN WILD-TYPE AND CYPB1 NULL MICE

Camila Dores, Erin Madeen, Tod Harper, Lyndsey Shorey, Frank Gonzalez, Emily Ho, Roderick Dashwood, David Williams, Ulrike Luderer, Christiane Loehr

Polycyclic aromatic hydrocarbons (PAHs) are a group of environmental compounds derived from incomplete combustion of organic matter and fossil fuel. A subset of activated PAHs, including PAH dibenzo[def,p]chrysene (DBC), are genotoxic and can cross the placental barrier. PAH are metabolized and activated via Cytochrome P450 1b1 (Cyp1b1) which is highly expressed in Leydig cells. We hypothesized that maternal gavage of mice with DBC (15, 12, 6.5, and 0 mg/kg) on gestation day 17 would cause reproductive toxicity in male offspring and that toxicity would require activation by Cyp1b1. Two groups of mice were used in this study, wild-type and Cyp1b1 knockout. At 21 days and ten months old, Cyp1b1 wild-type mice transplacentally exposed to DBC had reduced testes size ($p<0.001$), reduced numbers of seminiferous tubules ($p<0.001$), severe depletion of germ cells in the majority of tubules ($P < 0.001$). Immunohistochemistry against undifferentiated germ cell and meiotic markers (UCHL-1 and SCP-3) in the testes of 21 days-old animals demonstrated that Cyp1b1 wild-type mice treated with DBC had impaired spermatogenesis with reduced numbers of undifferentiated spermatogonia overlaying the basement membrane and had meiotic arrest with no round spermatids observed. Animals treated with 15mg/kg DBC had the most severe impairment. Cyp1b1 null offspring exposed in utero to DBC suffered no significant reproductive toxicity at 21 days or 10 months of age. Together these data indicate that a single in utero exposure to DBC at GD17 results in severe reproductive toxicity in male offspring and that these effects are dependent upon Cyp1b1 bioactivation.

November 5, 2017
9:30 AM – 9:45 AM

A PRECLINICAL SAFETY ASSESSMENT OF TWO CHIMERIC ANTIGEN RECEPTOR T (CAR T) CELL THERAPIES

James B. Rottman, Ken Ganley, Brenna Daly, Holly Horton, Kevin Friedman, Molly R. Perkins, Shannon Grande, Christopher J. Horvath

Chimeric antigen receptor T cells (CAR T) are emerging as promising cancer immunotherapeutics. Because CAR T cells cannot distinguish between target expression on normal versus neoplastic tissues, it is important to select target tumor antigens that are uniquely expressed by the neoplasm to avoid normal tissue ("on-target/off-tumor") toxicity. We developed a preclinical safety assessment paradigm involving bioinformatics, in situ hybridization (ISH), IHC and flow cytometry to assess novel targets for potential CAR T cell toxicity. We demonstrated that in cynomolgus macaques and mice, target X was expressed in the lungs and other tissues, whereas target Y expression was limited to lymphoid tissues. We generated CAR T cells to human X and Y, and demonstrated that anti-X, but not anti-Y CAR T cells also recognized the homologous mouse molecule. We subsequently used ISH and IHC to study anti-X and anti-Y CAR T cell activation and trafficking in NSG mice bearing X+Y+ xenografts. Anti-X CAR T cells accumulated in lung where they caused interstitial pneumonia, whereas anti-Y CAR T cells did not. Lung anti-X CAR T cells expressed the T cell activation / exhaustion marker PD-1. Animals treated with anti-X CAR T cells showed sparse T cell infiltration and poor control of X+Y+ xenografts. By contrast, animals treated with anti-Y CAR T cells showed robust T cell infiltration and xenograft clearance within 12 days. This preclinical safety assessment paradigm may be useful to de-risk potential CAR targets and model CAR T cell toxicity.

November 5, 2017
9:45 AM – 10:00 AM

VACCINATION OF MICE WITH VC2, A NOVEL MUTANT STRAIN OF HERPES SIMPLEX VIRUS 1, PROTECTS AGAINST OCULAR HERPESVIRAL INFECTION

Shan K. Naidu, Brent Stanfield, Nithya Jambunathan, Nagarjuna Cheemarla, Vladimir Chouljenko, Renee T. Carter, Ingeborg Langohr, Konstantin G. Kousoulas

Introduction: Recurrent activation of human simplex virus-1 (HSV-1) results in cold sores, periodic viral shedding in tears and herpetic stromal keratitis in humans. We evaluated the efficacy of a novel mutant herpes virus VC2 as vaccine against mouse ocular herpes models, and characterized the immunological correlates of protection.

Experimental Design: Mice were vaccinated twice at 21-day intervals by intramuscular injection, prior to ocular HSV-1 challenge. Cornea, lymph nodes, spleen and serum were analyzed on 5, 10 and 15 days post-infection. VC2 vaccine group was compared with live-attenuated HSV-1 (F) viral vaccine, and naïve group.

Methods: Immunohistochemistry and confocal microscopy were used to assess the features of keratitis. Serum antibody levels were measured against HSV-1 by ELISA and serum neutralization assay. Flow cytometry of lymph nodes and spleen, and cytokines levels in cornea and serum were assessed.

Results: VC2 vaccine reduced the lethality rate and protected mice from developing clinical keratitis upon ocular HSV-1 challenge. VC2 produced a prominent IgG2a type response while mice vaccinated with HSV-1(F) produced a significant IgG2b type response. Mice vaccinated with live attenuated HSV-1(F), unlike VC2, showed elevated levels of total systemic CD4+ T cells and persistence of ocular disease.

Conclusion: Immunization with our novel mutant strain VC2 provided complete protection against ocular herpes infection in mice, and elicits altered humoral and cell immune response compared to the live-attenuated wild type HSV-1 strain. Impact statement: Based upon our evaluation in mouse models, VC2 vaccine has a strong potential to protect against ocular herpes infection in humans.

November 5, 2017
11:30 AM – 12:00 PM

ANATOMIC PATHOLOGY FINDINGS IN THE SKIN AND LUNG OF RATS AND DOGS RECEIVING HIGH DOSES OF A POTENT ACETYL-COA CARBOXYLASE INHIBITOR ARE CONSISTENT WITH A FATTY ACID DEFICIENCY SYNDROME

Frank J. Geoly, Kathleen E. Biddle, Theodore J. Schmahai, Ahmed Shoieb, Ingrid D. Pardo, Kathryn E. Gropp, Karamjeet Pandher, Daniel J. Lettiere, William J. Reagan, Christopher J. Somps, Jay H. Fortner, Lawrence W. Updyke, Gregg D. Cappon, William P. Esler

Acetyl-CoA carboxylase (ACC) is a biotin carboxylase that catalyzes the initial and rate-limiting step in the endogenous de novo synthesis of fatty acids. Pharmacologic inhibition of ACC by a potent, investigational, small molecule ACC inhibitor impedes fatty acid synthesis in a dose-dependent manner. In general toxicology studies in rats and dogs, the molecule was administered at various doses for a minimum of two weeks, and the principal target organs were the lung and the skin. In rats, the lungs had features of interstitial injury, with random multifocal and coalescing areas of alveolar septal thickening, hyperplasia of Type II pneumocytes and bronchiolar epithelium, accumulations of alveolar macrophages, and eosinophil infiltrates. The rat skin lesions were generally on the head, muzzle, and feet and microscopically were characterized by epidermal hyperplasia, hyperkeratosis, atrophy of sebaceous glands, and secondary dermatitis. In dogs, lesions were present in the skin of the eyelids, specifically Meibomian glands, and the lung. Histologically, there was Meibomian gland atrophy in the eyelids with conjunctival inflammation secondary to decreased tear film quality, and lung lesions that were qualitatively similar to those in the rat though of lesser frequency and severity. Lesions in the skin and the lung are hypothesized to be the result of a fatty acid deficiency syndrome caused by high doses of the ACC inhibitor, since fatty acids are necessary for pulmonary surfactant phospholipid production and are also necessary for the production of cutaneous sebum, tear film meibum, and for normal epidermal barrier function.

Industrial and Toxicologic Pathology Focused Scientific Session II

November 7, 2017 | 1:30 PM – 5:00 PM
Session Chair: G. Palanisamy
Committee Members: K. Janardhan (Co-chair), T. Crabbs (Past Chair), G. Cain, C. Colleton, K. Helke, R. Johnson, and K. Knostman

November 7, 2017
2:30 PM – 3:00 PM

AN EPITHELIOTROPIC POLYOMAVIRUS PATHOGENIC IN IMMUNODEFICIENT RATS

Patricia A. Pesavento, Cynthia Beech-Williford, Shari Hamilton, Beth Bauer, Beatrix Kapusinszky, Tung Phan, Eric Delwart, Robert Livingston, Susan Cushing, Rie Watanabe, Matthew Myles

We have identified and characterized a novel polyomavirusin a colony of immune-deficient rats. Rats presented with signs of anorexia, emaciation, hunched posture, and/or dyspnea. Histological examination of multiple tissues demonstrated multifocalepithelial degeneration/necrosis and intranuclear inclusions located, most notably, within epithelial cells of the respiratory tract, salivary glands, lacrimal glands, uterus, and male secondary sex glands. A combination of serologic and molecular tests for common respiratory viral pathogens of rats were negative. Unbiased viral metagenomic sequencing of pooled tissues identified a novel virus of the family Polyomaviridae (RatPyV2), nearly simultaneously discovered in a colony of immune suppressed rats at Washington University in St. Louis. RatPyV2 phylogenetically clusters within the Wuki clade of the Betapolyomavirus genus. In situ hybridization analyses demonstrate viral nucleic acid within multiple tissues with cell targets including the epithelial lining of airways, apocrine and mucous glands, and epithelial cells lining both female and male reproductive tracts. Quantitative PCR supported the ISH findings with the highest viral loads in the respiratory and reproductive tracts and glandular tissues. Polyomavirus associated disease was experimentally reproduced in a second cohort of immune- deficient nude rats that were co-housed with the naturally infected rats. Population surveys found antibodies to RatPyV in 32% of rats used for biomedical research. This spontaneous and naturally occurring polyomavirus infection provides important information on viral pathogenesis and cell tropism, potential vertical mechanisms of transmission, anda potential system for studies on viral persistence of epitheliotropic polyomavirus.

November 7, 2017
4:30 PM – 5:00 PM

PATHOLOGICAL CHARACTERIZATION OF DRUG INDUCED CENTRAL NERVOUS SYSTEM (CNS) TOXICITY IN RATS ORALLY ADMINISTERED SMALL MOLECULE THERAPEUTIC CANDIDATE COMPOUNDS

Rie Kikkawa, Patrick J. Devine, Helen Gu, Alexandre Catoire, Robert Johnson, Karyn Colman

We have experienced test article-related central nervous system (CNS) toxicity in rats. Test article (Compound X) is a small molecule, whose target is expressed in certain hematopoietic cells and upregulated in inflamed conditions. The target is known to have no or little expression in the normal CNS or peripheral nervous system.

After nine consecutive days of oral dosing, animals developed neurological signs including paralysis in the hind limbs. Microscopically, compound-related changes were observed in the brain and spinal cord. These changes were characterized by neuronal degeneration/vacuolation, necrosis and/or loss accompanied by gliosis that obliterated the normal CNS structures. The lesions tended to be localized in the gray matter, including the parietal cortex, hippocampus and amygdala of the cerebrum; nuclei of the brain stem; granular layer of the cerebellum; and ventral horn of the spinal cord. Quantitative Whole-Body Autoradiography (QWBA) result indicated that compound-related radioactivity was moderately distributed through CNS. However the cause of rats' high susceptibility to the compound was not determined.

We concluded that this CNS toxicity case was likely to be associated with chemical structure, rather than pharmacology-related on-target effects, based on the fact that two other in-house compounds of the same chemical series reproduced the observed CNS toxicity and that there has been no CNS toxicity case reported with different chemical series compounds for this target, either internally or externally. The excitotoxicity and limbic encephalitis, which has been reported with the same chemical series compounds clinically, that mimicked our CNS toxicity may indicate possible mechanisms of toxicity.

Industrial and Toxicologic Pathology Focused Group Poster Session

T-01: REGIONAL RENAL TUBULAR EPITHELIAL HYPERTROPHY FOLLOWING ORAL ADMINISTRATION OF A SMALL MOLECULE RECEPTOR INHIBITOR

Maureen T. O'Brien, Kathleen Szabo

Background: Regional renal tubular epithelial hypertrophy (RTEH) is an uncommon change characterized by an increase in epithelial cell size but not an increase in cell number that is limited to a specific anatomic/functional region.

Methods: Fifteen female C57Bl/6 mice were dosed orally with the test article (a small molecule kinase inhibitor) or vehicle for 14 days at doses of 0, 20, or 50 mg/kg. Euthanasia (via CO2) occurred 1 hour following the last dose. Body, kidney, and liver weights were recorded. Selected tissues were collected and fixed with 10% buffered formalin. Histology slides were routinely prepared with hematoxylin and eosin staining. Microscopic evaluation was conducted on a limited tissue set.

Results: There were no mortalities or clinical abnormalities during the study. The primary microscopic finding, observed in all animals in the 50 mg/kg dose group and one animal in the 20 mg/kg dose group, was locally extensive RTEH hypertrophy limited to the inner stripe of the outer medulla (ISOM) and inner medulla (IM). Hypertrophic RTEH demonstrated increased cytoplasmic basophilia and the cytoplasm often had a slightly foamy appearance. Additionally, two animals in the 20 mg/kg dose group had RTEH within the ISOM and IM that had foamy, slightly basophilic cytoplasm without hypertrophy.

Conclusions: This case highlights an uncommonandsubtlelocalized renal epithelial change observed during a preclinical toxicology study. Additionally, this case provides the opportunity to review causes of RTEH and its relevance to preclinical safety assessment.

T-02: NOVEL OFF-TARGET EFFECT ON URINE CONCENTRATING ABILITY IN AN EXPLORATORY RAT TOXICITY STUDY

Bruce E. LeRoy, Mike C. Foley, Michael R. Logan

Objective: An investigative toxicity study was performed to enhance mechanistic understanding of toxic effects identified in rats following oral administration of an experimental compound during a dose-range finding (DRF) toxicity study.

Methods: Sprague-Dawley rats were administered the test item by oral gavage once daily for 14 days (DRF study) or 7 days (investigative study). The investigative study also included a 7-day recovery period. Study animals were placed in metabolic cages overnight for urine collections. Endpoints included urinalysis, urine volume, water intake, urine and serum osmolality, and serum arginine vasopressin concentration.

Results: at Day 14 of the DRF, urine specific gravity (USPG) from animals in the mid- and high-dose groups was markedly dilute (mean USPG of 1.004 vs. 1.032) compared to urine from controls. In the investigative study, markedly dilute urine was present following 7 days of administration of test item, but returned to pre-treatment levels following a 7-day recovery period. Water intake and urine volume were increased approximately 2- to 4-fold compared with controls. Urine osmolality results revealed meaningful differences between treated and control animals. Serum arginine vasopressin levels were increased following 7 days of administration of the test item compared to controls. An *in vitro* primary pharmacology receptor-binding assay demonstrated significant vasopressin receptor 2b antagonism by the experimental molecule.

Conclusions: The mechanism underlying the production of increased amounts of dilute urine observed during the dose-ranging toxicity study was impaired renal water resorption due to V2b arginine vasopressin receptor blockade caused by increased levels of the experimental molecule.

T-03: AUTOFLUORESCENCE FOR DIGITAL HISTOLOGY AND REFINED LIPOFUSCIN ANALYSIS IN AGING AND DIET RESTRICTION OR SUPPLEMENTATION STUDIES

Raoul V. Kuiper, Daniela C.F. Salvatori, Martijn E.T. Dollé, Erwin Reiling, Harry van Steeg, Jan H.J. Hoeijmakers

Objective: We explored the potential of multispectral imaging to advance the use of autofluorescence for quantitative analysis of tissue sections.

Method: For this evaluation we used end of life progeroid *Ercc1*$^{\Delta/-}$ mice that were kept as controls, or subject to 30% diet restriction, or rapamycin (14ppm) supplemented diet. Livers were collected in neutral buffered 4% formaldehyde, routinely processed and paraffin embedded. 4 µm sections were coverslipped without histochemical staining, and multispectral images were acquired using fluorescence microscopy with 20x objective. The signal was divided in a general autofluorescence designated as background, and a specific signal identifying autofluorescent pigment spots consistent with lipofuscin, which was extracted and digitally analyzed for spot number and intensity.

Results: Lipofuscin showed a consistent and specific spectrum which could be reproducibly extracted from the full autofluorescent signal, while background fluorescence provided structural information for evaluation of micro-anatomical distribution. We detected no differences in total lipofuscin signal (total spot area x intensity) between the groups, notwithstanding advanced age of the dietary restriction group. In rapamycin treated mice, spots were more frequent but less bright than in controls, while at similar age. This shift could indicate an effect of rapamycin on further lipofuscin processing or hepatocyte turnover, with less concentration in Kupffer cells and macrophages.

Conclusions: Spectral analysis of unstained histological sections provided useful structural information and enabled robust extraction of specific lipofuscin autofluorescence for reproducible digital analysis. The method refined manual scoring and indicated a distribution shift in rapamycin treated *Ercc1*$^{\Delta/-}$ mice, supporting hypothesis generating potential.

T-05: VASOACTIVE EFFECTS OF ACUTE ERGOT EXPOSURE IN SHEEP

Rossalin Yonpiam, Jair Gobbett, Ashok Jadhav, Kaushik Desai, Barry Blakley, Ahmad Al-Dissi

Introduction: Chronic exposure to ergot alkaloids is thought to cause arterial vasoconstriction via the activation of adrenergic and serotonergic receptors. The acute vascular effects and their mechanisms remain unknown.

Objective: To examine the effects of ergot alkaloids on the vasculature after single oral dose exposure and the mechanism of these effects.

Methods: Using a stomach tube, exposure group received a single dose of 600 µg/kg BW (total of 6 alkaloids in ground sclerotia) dissolved in water while the control group received a placebo (n= 6/group). Six hours post exposure animals were euthanized, and the dorsal pedal artery from each was collected and mounted in a tissue bath to determine the contractile response to phenylephrine (alpha1- adrenergic agonist) in the presence and absence of terazosin (alpha1- blocker) as well as serotonin.

Results: Acute exposure resulted in a significant increase in phenylephrine contractile response ($p = 0.0462$) while serotonin contractile response remained the same. Terazosin resulted in a dose dependent decrease in phenylephrine contractile response in both groups. Surprisingly, terazosin had more pronounced effect on the exposure group compared to control ($p < 0.05$).

Conclusion: Similar to chronic exposure, acute exposure to ergot alkaloids results in increased vascular sensitivity to phenylephrine but not serotonin. Terazosin is significantly more potent in blocking phenylephrine contraction in sheep exposed to ergot alkaloids.

T-06: IN SEARCH OF THE MOST SUITABLE RODENT MODEL AND INDICATORS OF TOPICAL GLUCOCORTICOID-INDUCED TOXICITY

Ana I. Blanco, Marta Calbet, Mercè Pont, Núria Godessart, Amadeu Gavaldà

Background: Long-term administration of topical glucocorticoids (GCs) is well-known for the risks of toxicity related to their potency and percutaneous absorption. Various strategies are being explored to improve the safety profile while maintaining the therapeutic benefits.

Objective: A reliable and predictive animal model is essential to investigate if amelioration of local and systemic adverse effects is to be achievedin preclinicalstudies.

Methods: Balb/c mice, Wistar rats and Sprague Dawley Hairless rats (Rj:SDH-Dsg4) received a singledaily topical applicationduring 4 to 9 consecutive daysof different marketed GCs on thedorsal thoracolumbarregion.

Results: Reduced thickness of the epidermis, in addition to atrophy of the dermis and adnexa were observed in all species/strains. The conventional haired mice and rats used intoxicological studies were very sensitive to cutaneous atrophy even with low potency GCs. However, subtle differences in atrophogenic potential could only be determined in hairless rats with a much thicker epidermis.Similarly, a good correlation was observed between the severity of systemic effects in relation to the GCs potencies and type of formulation in this strain of rat. The most suitable findings to compare compounds were the following: growth inhibition in juvenile animals; lower relative thymic weight as well as cortical lymphocytolysis; adrenocortical atrophy by histomorphometric analysis; and clinical pathological changes consistent with a GC leukogram, as well as metabolic alterations (e.g. hyperglycemia, hypercholesterolemia, hypertriglyceridemia and hyperproteinemia).

Conclusion: The hairless rat is considered the best rodent model to investigate GCs-induced toxicity in discovery studies.

T-07: INSIGHTS INTO THE EFFECT OF SUBTLE CHEMICAL MODIFICATION OF ARYL CHLOROETHYLUREA (CEU) AS SMALL MOLECULES TARGETING EITHER Β-TUBULIN OR PROHIBITIN AND THIOREDOXIN-1

Jessica S. Fortin, Alexandre Patenaude, Bernadette Bouchon, René C.-Gaudreault

Background: Chloroethylureas (CEU) are protein alkylating agents displaying potent antineoplastic properties that covalently bind to β-tubulin and affect microtubule polymerization dynamics. A different CEU subset has been shown to induce cell growth inhibition without alkylating β-tubulin.

Objective: Our research focus is to understand the mechanisms underlying the antiproliferative activity of that new class of compounds.

Methods: Proteins from B16 and MDA-MB-231 cells incubated with [(14)C-urea]-CEU-25 and [(125)I]-CEU-98 were separated using 2D-electrophoresis followed by MALDI-TOF identification of modified proteins. Protein expression and distribution were investigated by Western blot analyses and immunocytochemistry. Cell cycle analyses were obtained by flow cytometry.

Results: CEU-22 and its bioisosteric derivative CEU-98 are original CEU prototypes that covalently bind to β-tubulin via an ester linkage on Glu198. The alkylation leads to microtubule depolymerization phenotype, cell cycle arrest in G_2/M and inhibition of cell proliferation *in vitro*. A newly isolated subset of CEUs, exemplified by the prototypical CEU-25, alkylates prohibitin (PHB) on Asp40 and thioredoxin isoform-1 (TRX1). CEU-25 arrests cells predominantly in G_1 phase and inhibits Trx-1 and PHB nuclear translocation.

Conclusions: The intracellular proteins alkylated by the new CEU subset were identified as the PHB and TRX1. Different protein target profiles explain the G_1 cell cycle arrest. Our research emphasizes that a subtle chemical modification might lead to drastic change of protein target. Our finding might help to design new potent anticancer drugs that will target specific and lethal biological pathway essential to tumor growth.

T-08: COMPARATIVE CHRONIC TOXICITY OF THREE NEONICOTINOIDS ON NEW ZEALAND PACKAGED HONEY BEES

Sarah C. Wood, Ivanna V. Kozii, Roman V. Koziy, Claire Janse van Rensburg, Jessica Morris, Igor Moshynskyy, Sophie Derveau, Gary Wobeser, Tasha Epp, Karsten Liber, Elemir Simko

Background: Thiamethoxam, clothianidin, and imidacloprid are the most commonly used neonicotinoid insecticides on the Canadian prairies. There is widespread contamination of nectar and pollen with neonicotinoids, at concentrations which are sublethal for honey bees (Apis mellifera).

Objective: We compared the effects of chronic, sublethal exposure to the three most commonly used neonicotinoids on honey bee colonies established from New Zealand packaged bees using colony weight gain, brood area, and population size as measures of colony performance.

Methods: From May 7 to July 29, 2016 (12 weeks), sixty-eight colonies received weekly feedings of sugar syrup and pollen patties containing 0 nM, 20 nM (median environmental dose), or 80 nM (high environmental dose) of one of three neonicotinoids (thiamethoxam, clothianidin, and imidacloprid). Colonies were weighed at three week intervals. Brood area and population size were determined from digital images of colonies at week 12. Statistical analyses were performed by ANOVA and mixed models.

Results: There was a significant negative effect ($p<0.01$) on colony weight gain (honey production) after 9 and 12 weeks of exposure to 80 nM of thiamethoxam, clothianidin, or imidacloprid and on bee cluster size ($p<0.05$) after 12 weeks.

Conclusions: Chronic exposure of honey bees to high environmental doses of neonicotinoids has negative effects on honey production. Veterinary histopathology is an alternative tool for pesticide risk assessment in honey bees.

T-10: TOXIC ACUTE RENAL INJURY IN CATS: HISTOLOGIC LESIONS AND BIOMARKERS

Liam Broughton-Neiswanger, Nicolas Villarino, Pablo Piñeiro

Background: Specific mechanism of toxin-induced AKI in cats remains to be characterized. Defining the effects of nephrotoxins in feline kidneys will help identify key events in the progression of AKI to CKD. The aim of this study was to determine histologic changes caused by a nephrotoxin and to perform untargeted metabolomics to discover biomarker candidates for detection of AKI in cats.

Methods: A randomized blind placebo control prospective design was implemented. A proprietary nephrotoxic compound was administered to n=4 domestic short hair female cats. Age-matched control cats (n=4) were administered saline. Fixed samples of both kidneys were collected for histopathologic analysis using a semi-quantitative 0-4 scale and apoptosis immunohistochemistry. Higher scores represent more severe renal damage. In addition, we assessed the effect of administration of a nephrotoxin on more than 1400 low molecular weight substances in the renal cortex, using a metabolomics approach.

Results: All cats treated with the nephrotoxic compound had severe histologic changes associated with AKI including cortical tubular damage, with average histologic scores of 4, as compared to <1 in control cats ($p<0.05$). Apoptosis was detected in cortical tubular epithelial cells in 25% of examined fields in nephrotoxin-treated cats as compared to <5% in controls ($p<0.01$). Untargeted metabolomics revealed a decrease ($p<0.05$) in the levels of glycerophosphocholines (35:3), (36:2), and (38:3), sphingomyelin (d39:1), glycerol, and palmitoleic acid in the renal cortices of nephrotoxin-treated cats.

Conclusions: The results from this study identify key tissue and metabolic events that may be involved in the progression of AKI to CKD.

T-11: INCREASED GLIAL FIBRILLARY ACIDIC PROTEIN IMMUNOHISTOCHEMICAL STAINING OF THE BASAL GANGLIA IN THE BRAIN OF RHESUS MONKEYS TREATED WITH THE D2-5HT2 RECEPTOR ANTAGONIST, RISPERIDONE

Sabu Kuruvilla, Richard Briscoe, Guillermo Fernandez, Frédéric Poignant, James Marr

Objective: The objective was to characterize the brain histomorphology associated with administration of the dopamine type-2 and serotonin type-2 receptor antagonist, risperidone (an atypical antipsychotic), in rhesus monkeys.

Methods: Monkeys received daily oral doses of vehicle or risperidone (n=2/sex/group/time point) at 0.5 mg/kg/day (a pharmacologically active dose) for 3 weeks or 3 months, followed by a 3-month recovery period.

Results and Discussion: Histomorphological evaluation of monkey brains after staining with hematoxylin and eosin, double staining with Luxol fast blue and cresyl violet, or immunostaining with ionized calcium binding adapter molecule 1 showed no notable differences between control and treated animals. Evaluation of brain after immunostaining of glial fibrillary acidic protein (GFAP) showed increased staining of astrocytes in the putamen of monkeys dosed for 3 weeks and 3 months. The staining intensity was similar at 3 weeks and 3 months. This change was not accompanied with alterations in the distribution of astrocytes and was not consistent with an astroglial response to neuronal injury or loss. No increase in GFAP staining was observed in monkeys after a 3-month treatment-free recovery period. The observation in this study was similar to previous observations in the brain following administration of risperidone to rats or humans.

Conclusions: Risperidone-induced reversible increase in GFAP staining of astrocytes in the putamen of the brain in rhesus monkeys was considered an adaptive, non-adverse response, and was observed in the absence of any neuronal injury or loss. This is the first report describing this observation in non-human primates treated with risperidone.

T-12: INVESTIGATION OF KIDNEY FINDINGS IN A 4-WEEK CYNOMOLGUS MONKEY SAFETY STUDY

Karyn Colman, Sarah Tannehill-Gregg, Jonathan Moggs, Li Li

A novel small molecule therapeutic being developed for the treatment of solid tumors demonstrated an apparent effect on the kidneys in one preclinical safety study. In the pivotal 4-week monkey study some treated animals had evidence of a renal change that was not observed in any control animals (crystalline deposits with an inflammatory response in the kidney cortex in mid & high dose groups at the end of treatment, and following recovery in the high dose group). There was no clear dose-response in incidence or severity of the finding, however. No similar changes were noted in any rat studies, or in the follow-up 13-week monkey study. Evaluation of the literature and publicly available information showed that this observation is rarely reported as a background observation in safety studies nor are the crystals well-characterized. A report from another company showed identical changes in control monkeys and anecdotal evidence suggests it can be seen sporadically and in identified cohorts of animals. This may indicate some lithogenic environmental influence such as diet or water composition/consumption. This poster shows work done to characterize the crystalline deposits and evaluate any potential relationship to treatment. It was concluded that the kidney observation was spontaneous in nature based on our investigations that identified the composition of the crystals, and the data from control monkeys at another facility, indicating this observation can be seen spontaneously while not previously described as a background lesion in toxicology studies in the literature, as far as we could determine.

T-13: COMPREHENSIVE HISTOLOGIC STUDY OF THE REPRODUCTIVE SYSTEM OF HONEY BEE QUEEN

Ivanna V. Kozii, Sarah C. Wood, Roman V. Koziy, Claire Janse van Rensburg, Jessica Morris, Sophie Derveau, Igor Moshynskyy, Maud Ferrari, Ahmad Al-Dissi, Elemir Simko

Background: Increased mortality in honey bees coincided with widespread use of neonicotinoid insecticides in agriculture. However, discrepancies between field and laboratory experiments and incomplete understanding of toxicopathologic effects of neonicotinoids on honey bees reveal a gap in the risk assessment of pesticides for honey bees.

Mammalian risk assessment relies on histopathology as a “gold standard” for safety evaluation. Histologic assessment of gonadotoxicity, teratogenicity, neurotoxicity and cancerogenicity for determination of a safe dose range of a tested drug or chemical is sensitive and reproducible. However, comparable toxicopathologic approaches using histopathology have not been developed for honey bees.

Objective: Our objective was to characterize histologic morphology of the reproductive tract of honey bee queens for the future risk assessment of pesticides using gonadotoxicity in honey bee queens.

Methods: One-year old, genetically diverse, naturally mated honey bee queens (n=16) were harvested from local research apiaries. Serial H&E stained tissue sections from paraffin embedded blocks (every 100μm) were obtained for microscopic characterization.

Results: Following reproductive system segments were described: 1) ovaries, comprised of ovarioles showing sequential oocyte maturation, 2) tubular passage system formed by the calyx, lateral and median oviducts, genital chamber, and bursa copulatrix with bursal pouches, 3) spermatheca - a spherical organ for sperm storage during the entire life of queen (2-4 years), with spermathecal glands and duct.

This histologic study will provide a baseline for future investigation of gonadotoxicity of neonicotinoids in exposed honey bee queens and determination of a safe dose range for the most commonly used neonicotinoids in agriculture.

T-14: CHARACTERIZATION OF REPRODUCTIVE TOXICITY OF IN UTERO EXPOSURE TO POLYCYCLIC AROMATIC HYDROCARBONS IN WILD-TYPE AND CYPB1 NULL MICE

Camila Dores, Erin Madeen, Tod Harper, Lyndsey Shorey, Frank Gonzalez, Emily Ho, Roderick Dashwood, David Williams, Ulrike Luderer, Christiane Loehr

Polycyclic aromatic hydrocarbons (PAHs) are a group of environmental compounds derived from incomplete combustion of organic matter and fossil fuel. A subset of activated PAHs, including PAH dibenzo[def,p]chrysene (DBC), are genotoxic and can cross the placental barrier. PAH are metabolized and activated via Cytochrome P450 1b1 (Cyp1b1) which is highly expressed in Leydig cells. We hypothesized that maternal gavage of mice with DBC (15, 12, 6.5, and 0 mg/kg) on gestation day 17 would cause reproductive toxicity in male offspring and that toxicity would require activation by Cyp1b1. Two groups of mice were used in this study, wild-type and Cyp1b1 knockout. At 21 days and ten months old, Cyp1b1 wild-type mice transplacentally exposed to DBC had reduced testes size ($p<0.001$), reduced numbers of seminiferous tubules ($p<0.001$), severe depletion of germ cells in the majority of tubules ($P < 0.001$). Immunohistochemistry against undifferentiated germ cell and meiotic markers (UCHL-1 and SCP-3) in the testes of 21 days-old animals demonstrated that Cyp1b1 wild-type mice treated with DBC had impaired spermatogenesis with reduced numbers of undifferentiated spermatogonia overlaying the basement membrane and had meiotic arrest with no round spermatids observed. Animals treated with 15mg/kg DBC had the most severe impairment. Cyp1b1 null offspring exposed in utero to DBC suffered no significant reproductive toxicity at 21 days or 10 months of age. Together these data indicate that a single in utero exposure to DBC at GD17 results in severe reproductive toxicity in male offspring and that these effects are dependent upon Cyp1b1 bioactivation.

T-15: VACCINATION OF MICE WITH VC2, A NOVEL MUTANT STRAIN OF HERPES SIMPLEX VIRUS 1, PROTECTS AGAINST OCULAR HERPESVIRAL INFECTION

Shan K. Naidu, Brent Stanfield, Nithya Jambunathan, Nagarjuna Cheemarla, Vladimir Chouljenko, Renee T. Carter, Ingeborg Langohr, Konstantin G. Kousoulas

Introduction: Recurrent activation of human simplex virus-1 (HSV-1) results in cold sores, periodic viral shedding in tears and herpetic stromal keratitis in humans. We evaluated the efficacy of a novel mutant herpes virus VC2 as vaccine against mouse ocular herpes models, and characterized the immunological correlates of protection.

Experimental Design: Mice were vaccinated twice at 21-day intervals by intramuscular injection, prior to ocular HSV-1 challenge. Cornea, lymph nodes, spleen and serum were analyzed on 5, 10 and 15 days post-infection. VC2 vaccine group was compared with live-attenuated HSV-1 (F) viral vaccine, and naïve group.

Methods: Immunohistochemistry and confocal microscopy were used to assess the features of keratitis. Serum antibody levels were measured against HSV-1 by ELISA and serum neutralization assay. Flow cytometry of lymph nodes and spleen, and cytokines levels in cornea and serum were assessed.

Results: VC2 vaccine reduced the lethality rate and protected mice from developing clinical keratitis upon ocular HSV-1 challenge. VC2 produced a prominent IgG2a type response while mice vaccinated with HSV-1(F) produced a significant IgG2b type response. Mice vaccinated with live attenuated HSV-1(F), unlike VC2, showed elevated levels of total systemic CD4+ T cells and persistence of ocular disease.

Conclusion: Immunization with our novel mutant strain VC2 provided complete protection against ocular herpes infection in mice, and elicits altered humoral and cell immune response compared to the live-attenuated wild type HSV-1 strain. Impact statement: Based upon our evaluation in mouse models, VC2 vaccine has a strong potential to protect against ocular herpes infection in humans.

Natural Disease Focused Scientific Session I

November 5, 2017 | 8:00 AM – 12:00 PM
Session Chair: F.A. Uzal
Committee Members: G.J. Haldorson (Past Chair), D.R. Rissi (Co-Chair), Pompei G.F. Bolfa, J.B. Engiles, A.E. Pillatzki, H. Fenton

November 5, 2017
8:45 AM – 9:00 AM

PANCREATITIS, PANNICULITIS, AND POLYARTHRITIS (PPP) SYNDROME IN DOGS

Ashley Talley, Jennifer Luff, Heather Shive, Keith Linder

Panniculitis and polyarthritis are rare complications of pancreatic disease and, when concurrent, the condition is called pancreatitis, panniculitis, and polyarthritis (PPP)

syndrome. Circulating pancreatic enzymes are thought to cause necrosis and saponification of adipose tissue with suppurative steatitis. Necrosis of periarticular adipose tissue causes polyarthritis via release of free fatty acids into the synovial fluid and saponification of periarticular intraosseous adipose tissue can also contribute to arthritis. In humans and animals, causes are acute or chronic pancreatitis, pancreatic neoplasia, and pancreatic duct obstruction. Only 10 cases of pancreatic panniculitis are reported in dogs, a subset of which had clinical lameness (6/10) or intraosseous saponification (2/10). Here we retrospectively describe the clinicopathologic findings in five dogs with pancreatic panniculitis and document the occurrence of polyarthritis. Pancreatic neoplasia and pancreatitis were causes in 2 and 3 dogs, respectively. Serum lipases were markedly elevated (5/5), ranging from 18,380 to 95,520 IU/L. Panniculitis was multifocal on the trunk (2/5) and limbs (4/5) and draining tracts were uncommon (1/5). All dogs had clinical lameness and, at post mortem exam, joint lesions included synovial effusion (2/5), fibrinous to purulent synovitis (2/5), saponification of patellar adipose tissue (2/5), and periarticular intraosseous saponification (2/5). One (2/5) or multiple (2/5) joints were affected and stifles were the most commonly affected joint (4/5). Panniculitis and polyarthritis may significantly contribute to morbidity and mortality of patients with pancreatic disease, but are likely underdiagnosed. Therefore, a thorough examination should be performed in all patients with pancreatic disease and severe enzyme elevation.

November 5, 2017
9:00 AM – 9:15 AM

AMDOPARVOVIRUS INFECTION IN RED PANDAS (AILURUS FULGENS)

Charles E. Alex, Steven V. Kubiski, Linlin Li, Reza Sadeghi, Raymund F. Wack, Megan A. McCarthy, Joseph B. Pesavento, Eric Delwart, Patricia A. Pesavento

Aleutian mink disease virus is the type species in the genus *Amdoparvovirus*, and in mink and other *Mustelidae* can cause either subclinical disease or fatal chronic immune stimulation and immune complex disease. We describe a novel amdoparvovirus in the endangered red panda (*Ailurus fulgens*), discovered using viral metagenomics. We analyzed the prevalence, tissue distribution, and disease association by PCR, in situ hybridization, electron microscopy, and histology in a group of six red pandas from a single zoological collection. The study incorporates a 6-week fecal shedding survey and analysis of tissues from four necropsied animals over a twelve-year span. The tentatively named red panda amdoparvovirus (RpAPV) was detected in the feces in all six animals, and in tissues of all animals tested (4/4). In one necropsied, geriatric animal, infection was associated with pyogranulomatous peritonitis, pancreatitis, and myocarditis. Two other necropsied animals had detectable low level viral nucleic acid in lymph nodes and both oral and intestinal epithelium. Full-length RpAPV strains from two animals had 12% sequence divergence, demonstrating genetic diversity even among in-contact animals. RpAPV is a persistent infection in this cohort of red pandas, and has variable clinical expression.

November 5, 2017
9:15 AM – 9:30 AM

NONHUMAN PRIMATE YELLOW FEVER OUTBREAK IN ESPIRITO SANTO, BRAZIL, 2017: PATHOLOGICAL AND MOLECULAR INVESTIGATIONS

Natália C.C.A. Fernandes, Mariana Sequetin, Juliana Mariotti Guerra, Rodrigo A. Ressio, Cinthya Cirqueira, Silvia Iglezias, Julia de Carvalho, Emerson Luiz Lima Araujo, José Luiz Catão-Dias, Josué Díaz-Delgado

Background: In January 2017, the Yellow Fever (YF) outbreak spread to Espirito Santo state (Brazil), an area with no record of YF virus circulation in the last 50 years and low human immunization coverage.

Objective: To report the epidemiology and diagnostics in 22 YF-positive nonhuman primates (NHPs) deceased early in the ongoing outbreak.

Methods: Necropsies were performed on 22 NHPs: two were howler monkeys (*Alouatta* sp.), and 20 were New World primates (NWPs) not further identified. Selected tissues were collected for histopathology and immunohistochemistry (n= 22), and polymerase chain reaction (PCR) analysis (n= 11).

Results: Histologically, all animals had zonal bridging or massive liver necrosis with apoptotic bodies, variable steatosis and pleocellular, mainly lymphocytic and histiocytic, inflammatory infiltrates, accompanied or not by hemorrhage. Few atypical cases displayed unusual histological features in the liver. Additional consistent findings were: splenic lymphoid depletion and follicular necrosis/lymphocytolysis; acute renal tubular necrosis; and multisystemic hemorrhage. All animals had positive granular cytoplasmic immunolabeling for YF-virus antigen, more intense in degenerating and remaining hepatocytes, while necrotic hepatocytes consistently lacked immunoreactivity. All PCR-tested animals were positive with quantification values ranging from 11 to 26.

Conclusions: NWPs were effective YF sentinels and enabled rapid government response, prompting medical deployment such as, vaccination campaigns, and vector control measurements. Immunohistochemical detection of YF-virus antigen remains a highly reliable diagnostic tool, allowing for detection of atypical presentations and diagnosis when molecular analysis is not an option.

Update: The occurrence of YF in southern Brazil is 14% (01/01/17 to 06/28/17).

November 5, 2017
9:30 AM – 9:45 AM

HOOKWORM CLEARANCE IN SOUTH AMERICAN FUR SEAL PUPS (ARCTOCEPHALUS AUSTRALIS): MECHANISMS AND ROLE IN PUP SURVIVAL

Mauricio Seguel, Felipe Montalva, Diego Perez-Venegas, Josefina Gutierrez, Victor Alvarado, Elizabeth Howerth, Nicole Gottdenker

Background: Hookworms (*Uncinaria sp.*) are a major cause of pup mortality in many otariid populations. A unique feature of otariid hookworm infection is the clearance of these nematodes when pups are between 2 to 6 months-old.

Objective: Determine how hookworms are cleared from the intestine, and determine the importance of this process in fur seal pup survival.

Methods: Between 2014 and 2017 hookworm infection and immune function were closely monitored in a total of 490 South American fur seal (*Arctocephalus australis*) pups at Guafo Island, Northern Chilean Patagonia.

Results: Most pups (402, 82%) were able to clear hookworms between 3 to 6 weeks after infection, and 18% (n=88) of pups failed to clear infection and died due to hookworm disease. Hookworm clearance was associated with increased numbers of lymphocytes and basophils, and presence of parasite specific IgG in the blood (GLMs, $P<0.001$). These pups also had more reactive T-lymphocytes and higher numbers of T-lymphocytes in the intestine (Mann-Whitney U-test, $P<0.001$). Parasite specific IgG band to the intestinal brush border of the hookworms, an important structure for blood digestion and absorption. In generalized linear models the length of hookworm infection was the most significant factor predicting fur seal pup mortality (GLM Binomial, $P=0.004$).

Conclusions: Fur seal pups clear hookworm infection through an immune -mediated mechanism, where T-lymphocytes, basophils and parasite-specific IgG are key players. The capacity of a pup to mount the immune response required for parasite clearance significantly impacts its chances of hookworm elimination and survival.

November 5, 2017
9:45 AM – 10:00 AM

EQUINE PAPILLOMAVIRUS TYPE 2 RNA IN SITU HYBRIDIZATION AS A DIAGNOSTIC TOOL FOR EQUINE PENILE LESIONS

Cameron G. Knight, Sarah Greenwood, Claudia Klein, Samuel Sharpe, Garrett L. Wachoski-Dark, Bruce Wobeser

Background: The equine penis is predisposed to development of neoplasms. This has recently been associated with infection by equine papillomavirus type 2 (EcPV-2).

Objectives: To use RNA in situ hybridization (RISH) to evaluate certain equine epithelial lesions for the presence of EcPV-2 and to compare RISH and PCR results. To determine whether distribution of RISH signal patterns provides additional diagnostic information.

Methods: Equine tissues were collected from 28 genital lesions and 9 non-genital papillomas or squamous cell carcinomas (SCCs). Each sample received: independent diagnosis by five pathologists; consensus PCR testing for papillomaviral DNA; and RISH for EcPV-2 presence and distribution.

Results: Among pathologists there were frequent differences in diagnoses of penile lesions, but not of vulvar or non-genital lesions. Papillomaviral DNA sequences were detected by PCR in 24/28 genital lesions and 7/9 non-genital lesions. RISH for EcPV-2 was positive in 24/28 genital lesions, corresponding exactly with PCR results. RISH for EcPV-2 was negative for all 9 non-genital lesions, regardless of PCR results. RISH signal patterns differed among penile papillomas, in situ carcinomas, and SCCs.

Conclusions: Histologic differentiation of male genital papillomas, carcinomas in situ, and SCCs may be difficult in some cases, possibly because diagnostic criteria are not standardized. RISH may be useful as an adjunct test for deciding diagnosis in EcPV-2 induced penile lesions. RISH is a sensitive and specific test for EcPV-2 infection, with no cross-reactivity for other equine papillomaviruses. EcPV-2 infection in non-genital locations is rare.

November 5, 2017
10:30 AM – 10:45 AM

RESPIRATORY DISEASE IN PYTHONS EXPERIMENTALLY INFECTED WITH PYTHON NIDOVIRUS AND OTHER UPDATES ON THIS EMERGING VETERINARY DISEASE

Laura L. Hoon-Hanks, Marylee L. Layton, Robert J. Ossiboff, Edward J. Dubovi, Mark D. Stenglein

Introduction: A severe respiratory illness of pythons has been observed by veterinarians since the 1990s. Circumstantial evidence has linked a novel python nidovirus to the disease. We conducted an experimental infection in ball pythons (*Python regius*) to test the hypothesis that ball python nidovirus (BPNV) infection would yield clinical signs and histologic lesions consistent with respiratory disease. Additionally, we have investigated natural disease in mixed snake colonies, providing further perspectives on this emerging veterinary disease.

Methods: Five juvenile ball pythons were inoculated: three with BPNV-infected medium and two with uninfected medium. Antemortem swabs were performed weekly and tested for BPNV RNA by PCR. Euthanasia and postmortem examination were performed on infected snakes at 5 weeks, 10 weeks, and 12 weeks post-inoculation (PI) based on clinical signs.

Results: The most significant lesions in the infected snakes included chronic-active catarrhal rhinitis, stomatitis, tracheitis, and esophagitis with variable epithelial proliferation and an interstitial and proliferative pneumonia. Infectious virus was recovered from swabs and tissues from infected snakes. Control snakes remained negative throughout the experiment and did not show clinical signs or share histologic lesions.

Conclusion: Our findings establish a causal relationship between BPNV infection and respiratory disease in ball pythons. BPNV is part of an expanding group of related viruses that have been associated with respiratory disease in reptiles and mammals. This work, as well as our additional studies, have revealed insights into clinical course, possible routes of transmission, useful diagnostics, viral genomic diversity, species specificity, and disease epidemiology of the python nidovirus

November 5, 2017
10:45 AM – 11:00 AM

CYSTIC AND MYXOMATOUS LESIONS OF THE FELINE SYNOVIUM

Linden E. Craig, Paula M. Krimer

Cystic and myxomatous lesions from synovial joints of 14 cats were examined by light microscopy and immunohistochemistry. The lesions consisted of fluid-filled cysts lined by synoviocytes, more solid foci of stellate cells in a myxomatous matrix, or a combination of the two. Mitoses and other features of malignancy were rare to nonexistent. Immunohistochemistry for IBA-1 was positive in some or all of the cells in all 14 cases. Nine of the 14 cases were positive for CD18. All cases were consistently vimentin positive. The elbow was the most commonly affected joint (11/14) and degenerative joint disease was a frequent concurrent and bilateral condition (9/10 cats with orthopedic evaluation or radiographs). The median age of onset was 14 years. In 11 of 12 cats with follow-up information, the lesion gradually increased in size over a period of years; in one cat the lesion did not recur following biopsy. Treatments included aspiration of fluid, radiation, debulking, and amputation, but most cats were not treated other than biopsy. None of the cats died or were euthanized because of this lesion. We propose that synovial cysts arise in some joints with degenerative disease, and have the potential to transform to a solid tumor with benign behavior, synovial myxoma.

Natural Disease Focused Scientific Session II

November 5, 2017 | 1:30 PM – 5:00 PM
Session Chair: F.A. Uzal
Committee Members: G.J. Haldorson (Past Chair), D.R. Rissi (Co-Chair), Pompei G.F. Bolfa, J.B. Engiles, A.E. Pillatzki, H. Fenton

November 5, 2017
2:15 PM – 2:30 PM

DEMONSTRATION OF OVHV2, THE AGENT OF SHEEP ASSOCIATED MALIGNANT CATARRHAL FEVER, IN FORMALIN-FIXED TISSUES

Patricia A. Pesavento, Donal O'Tool, Cristina Cunha, Hong Li

Background: Malignant catarrhal fever (MCF) viruses are gammaherpesviruses that asymptomatically infect one host species, typically ungulates such as sheep or wildebeest. They cause fatal disease in poorly adapted species that serve as end-stage hosts. Domestic sheep are the natural reservoir for ovine herpesvirus 2 (OvHV-2). Sheep are the source of outbreaks of sheep-associated MCF in susceptible species, particularly domestic cattle, American bison, and exotic ungulates in zoological collections. Sheep-associated MCF is a common cause of acute MCF in farmed American bison due to high disease susceptibility. Histologic hallmarks of acute infection are lymphoproliferation, mucosal necrosis-apoptosis, and systemic arteritis-phlebitis. Although never propagated *in vitro*, the primary genomic structure of OvHV-2 has been characterized. Lesions of sheep-associated MCF are well defined in experimentally-infected bison, cattle, pigs, rabbits, and sheep.

Objective: An unanswered question is the pathogenesis of MCF. Addressing this question was hampered by an inability to detect OvHV-2 in fixed tissue. This results in multiple difficult-to-test hypotheses about the basis for lesions, such as immune dysregulation, altered expression of inflammatory cytokines, and/or excessive production of cytotoxic T cells. *In situ* hybridization (ISH), in combination with quantitative PCR and immunohistochemistry for lymphocyte subset, should help clarify the role of OvHV-2 in nascent lesions.

Results: For the first time, we demonstrate viral nucleic acid of OvHV-2 in nuclei of inflammatory cells in naturally-infected (bison, cattle, sheep) and experimentally-infected (rabbits, pigs, bison) host species. Tissue distribution and kinetics of dissemination correlate well with severity of inflammation and with quantitative PCR-estimated viral loads.

November 5, 2017
2:30 PM – 2:45 PM

BACTERIAL BIOGEOGRAPHY IN THE COLON OF DOGS WITH CHRONIC ENTEROPATHY

Paula R. Giaretta, Jan S. Suchodolski, Joerg M. Steiner, Jonathan A. Lidbury, Raquel R. Rech

Background: Canine chronic enteropathy (CE) is characterized by gastrointestinal signs that persist for more than three weeks and is diagnosed by exclusion of other possible causes. Lymphoplasmacytic inflammation is the most common lesion in dogs with CE. The intestinal microbiota is believed to play a role in the pathogenesis of CE in dogs and humans.

Objective: The objective of this study was to characterize the spatial distribution and quantify selected bacterial groups in the colonic mucosa of dogs with CE and control dogs.

Methods: Formalin-fixed paraffin-embedded samples of colon from seven dogs with CE and seven control dogs were used. Bacteria were investigated using fluorescence in situ hybridization (FISH) with a eubacterial probe (EUB338) and specific probes for *Helicobacter* spp., *Escherichia coli/Shigella, Faecalibacterium prausnitzii,* and *Akkermansia muciniphila*. Ten fields with labeled bacteria on the surface and in the crypts were photographed. Bacterial quantification was performed using ImageJ software.

Results: On the colonic surface, dogs with CE had higher numbers of total bacteria (P=0.0409) and *Escherichia coli/Shigella* (P=0.0021) than control dogs. Also, the number of *Helicobacter* spp. was decreased on the colonic surface (P=0.0407) and in the crypts (P=0.0095) of CE dogs. The number of total bacteria (P=0.038) and *Akkermansia muciniphila* (P=0.037) was decreased within the crypts of CE dogs. No differences in the number of *Faecalibacterium prausnitzii* between the two groups were detected.

Conclusion: In summary, the composition, number, and distribution of the colonic microbiota in dogs with CE differ from control dogs.

November 5, 2017
2:45 PM – 3:00 PM

HISTOPATHOLOGICAL CHARACTERIZATION AND PROPOSED GRADING SYSTEM FOR EQUINE ARYTENOID CHONDROSIS/CHONDROPATHY

Pompei Bolfa, Michelle Dennis, Bernard Grevemeyer, Lusan Dellagrotte, Marta Cercone, Norm Ducharme

Background: The etiology of equine arytenoid chondrosis (chondropathy) is not clear. Trauma or infection/inflammation of the corniculate process and arytenoid cartilage from mucosal damage have been suggested.

Objective: Our objective was to develop a grading system that would correlate the size of the affected arytenoid cartilage with the degree of cartilage loss. This would facilitate a less invasive arytenoidectomy and optimization of the current therapy with restoring of a good athletic capacity of the horses.

Methods: 19 diseases arytenoids from horses that had previous antibiotic therapy and 4 controls were sectioned and measured (area) at the same level, caudal to the corniculate process. Based on their size, they were divided into control, grade I, grade II and grade III. Al tissues were stained with HE and Alcian blue and evaluated histologically.

Results: 10 arytenoids (52.6%) were included in grade I (less severe), 5 (26.3%) in grade II and 4 (21.1%) in grade III. There was a positive correlation between size and hyaline cartilage loss. The lesions were composed of a center area of fibrous connective tissue, and for most one or more sinus tracts communicating with the mucosa. Other histological features that were variably present were granulation tissue (N=5), thrombosis (N=8), distortion of arterial architecture (N=5), acute or chronic inflammation (N=14) with multinucleated macrophages (N=2), presence of fibrocartilage (N=7), bone metaplasia (N=2) and involvement of nearby skeletal muscle (N=15).

Conclusions: We propose that size is an accurate predictor of disease severity in equine arytenoid chondropathy helping surgical intervention strategies.

November 5, 2017
3:30 PM – 3:45 PM

THYROIDITIS IN FREE-LIVING WOLVERINES (GULO GULO) FROM NORTHERN ALASKA, 2014-2017

David S. Rotstein, Qaiyaan Harcharek, Raphaela Stimmelmayr

Background: Understanding the natural history and disease of free-living species can be hindered by sample collection and sample sizes. For this reason, population trends may be revealed over time.

Objective: To present a population-based endocrine system trend in examined wolverine.

Methods: Eleven previously frozen carcasses of subsistence (hunter trapped) harvested adult wolverines submitted to the Department of Wildlife Management, North Slope from 2014 to 2017 were necropsied and limited organs were collected based on availability including thyroid glands, adrenal glands, lung, heart, lymph nodes, kidneys, spleen, and gonads. 73% (8/11) were male and 27% (3/11) were female.

Results: Body conditions included emaciated (1/11), thin (3/11), good (2/11), and robust (5/11). Lymphoplasmacytic thyroiditis was observed in 55% (6/11) include 2 females and 4 males. Follicular atrophy was observed in 50% (3/6) and follicular hyperplasia in one animal. Alopecia was not reported. There was minimal testicular activity characterized by approximately 10% of tubules with small numbers of spermatazoa. Two males had a lymphoplasmacytic epididymitis and one male had a Leydig (interstitial) tumor. Sarcocysts were present within skeletal muscle (36%; 4/11) and diaphragm (9%; 1/11) animals. There was no inflammation.

Conclusions: Although thyroid dysfunction could not be confirmed due to the lack of available blood for a full thyroid panel, the lesions identified within the thyroid glands and gonads in this population of wolverines merits further exploration to better characterize these lesions, as well as potentially identify a cause (*e.g.* genetic, environmental or other).

November 5, 2017
3:45 PM – 4:00 PM

EXPRESSION OF NANOG, NESTIN, OCT3-4 AND SOX2 IN CANINE PROSTATE CANCER

Michelle M. Story, Hsi-Yu Lin, Ashna Mawjii Alladin, Carlos Eduardo Fonseca-Alves, Renee Laufer-Amorim, Valeria Grieco, Brett Stringer, Rodney Straw, Chiara Palmieri

Background: Cancer stem cells are considered to play a role in human prostate cancer

Objective: Our aim was to use immunohistochemistry to evaluate if cells resembling cancer stem cells are present in canine prostate cancer (PC) by comparing the expression of Nanog, Nestin, Oct3/4 and Sox2 between canine PC and benign prostatic hyperplasia (BPH).

Methods: Immunohistochemistry was performed on formalin-fixed, paraffin-embedded canine prostate samples. Differences in the percentage of positive cells for each protein were calculated with the ANOVA test followed by the post hoc Bonferroi-Holm test. $P<0.05$ was considered significant.

Results: Ten cases of BPH and 19 cases of PC were analyzed. Cytoplasmic expression of Nanog was greater in PC (89.05% ± 21.21 positive cells) than in BPH (21.60% ± 25.81 positive cells). Nuclear expression of Nanog was greater in PC (24.78% ± 32.66 positive cells) than in BPH (12.90% ± 17.85 positive cells). Sixteen PCs expressed cytoplasmic Nestin (79.42% ± 36.00 positive cells) and 4 expressed nuclear Nestin (6.47% ± 17.42 positive cells). Nestin was expressed in the hyperplastic epithelial cells without any nuclear distribution in 2 cases of BPH. Cytoplasmic and nuclear expression of Oct3/4 and cytoplasmic expression of Sox2 were similar in PC and BPH. Nuclear Sox2 expression was greater in PC (70.32% ± 26.35) than BPH (30.19% ± 30.33). Differences in cytoplasmic Nanog, cytoplasmic Nestin and nuclear Sox2 between BPH and PC were statistically significant.

Conclusions: Alterations of the expression of Nanog, Nestin and Sox2, but not Oct3/4, may be associated with canine PC growth.

November 5, 2017
4:00 PM – 4:15 PM

ELEVATED GAMMA-GLUTAMYL TRANSFERASE ACTIVITY IN RACING THOROUGHBREDS AND ITS ASSOCIATION WITH EQUINE HEPACIVIRUS, EQUINE PEGIVIRUS, AND THEILER'S DISEASE-ASSOCIATED VIRUS INFECTION

Joshua D. Ramsay, Ryan Evanoff, Robert H. Mealey

Background and Objective: In racing Thoroughbreds elevated serum gamma-glutamyl transferase (GGT) activity is positively correlated with cumulative days in training and, when extremely elevated, is associated with decreased racing performance. The cause of this condition remains unknown, but the recent discovery of three viruses in association with equine liver disease has raised the possibility of a viral etiology. The objective of this study was to test the hypothesis that elevated GGT in racing Thoroughbreds is positively associated with equine hepacivirus (EHCV), equine pegivirus (EPgV), and/or Theiler's disease associated virus (TDAV) infection.

Methods: Pre-race blood samples from 800 Thoroughbreds at the Santa Anita Park and Los Alamitos Race Course were tested for GGT activity (IU/L), and for the presence of EHCV, EPgV, and TDAV RNA by end-point RT-PCR.

Results: Serum GGT activity above the reference range (0-40 IU/L) was detected in 19.3percent of horses tested. From the effected group, 10.7percent were EPgV positive, 2.3percent were EHCV positive, and none were TDAV positive. The relative risk of being infected with EPgV and having elevated GGT was 0.61(P = 0.026). The relative risk for the other two viruses did not approach statistical significance.

Conclusions: The data generated from this study does not support a causal relationship between EHCV, EPgV, or TDAV infection and elevated GGT in racing Thoroughbreds. In fact, the data suggests that EPgV infection may have protective effects and warrants further investigation into the underlying mechanism.

Natural Disease Focused Group Poster Session

N-01: NOCARDIA PAUCIVORANS ASSOCIATED NECROTISING PNEUMONIA IN A DISPLAY BELUGA

Stephen Raverty, Martin Haulena, Erin Zabek, Heindrich Snyman

A 20 year old male display beluga presented with acute death and few premonitory signs. The animal was in fair to moderate body and good post mortem condition. The most significant necropsy finding was focally extensive abscessation in the caudal third level of the left lobe. The abscess thickly spanned the ventral third portion of the lung and protruded above the parenchyma and on cut surface centrally contained abundant brown red mucoid necrotic debris and was bound by granulomatous infiltrate and encapsulated. There was marked regional lymphadenopathy and the parenchyma was mottled grey black. Microscopically, there was marked necrotising, fibrinosuppurative to pyogranulomatous pneumonia with multifocal dystrophic mineral deposition and alveolar histiocytosis. There was reactive change with the regional lymph nodes, lobular collapse in the liver, and lymphoid hyperplasia in lymphoglanular elements the larynx. Special stains disclosed finely beaded to filamentous agyrophilic intralesional bacilli. Apparently normal and abscessed lung tissue were negative for influenza virus, Mollicutes, consensus Mycobacterium, and consensus herpesvirus and positive for Nocardia consensus. DNA sequencing identified Nocardia paucivorans which was also isolated from the lung. No other significant aerobic or anaerobic bacteria were recovered from the lung or hilar and prescapular lymph nodes. Although Nocardia has previously been identified in stranded and display cetaceans, N paucivorans has only recently been reported in humans with pneumonia, meningoencephalitis or occasionally, septicemia. The advent of advanced molecular techniques has enhanced our ability to detect unusual or novel pathogens which valuable insights into potential pathogen source and control or preventative studies.

N-02: CHARACTERIZATION OF A NOVEL MYOPATHY IN DOGS

Hayley Hunt, Nick J. Cave, Brett D. Gartrell, Jenni Petersen, Wendi D. Roe

Background: A novel myopathy, colloquially referred to as ‘Go Slow,’ was first reported in hunting and working farm dogs in New Zealand in 2000. Dogs presented with muscle tremors, weakness and collapse, followed by a prolonged period of exercise intolerance.

Objective: This prospective case series describes the epidemiological, histological and electron microscopic features of ‘Go Slow’ in dogs.

Methods: Cases were recruited through pig hunting clubs, veterinarians and diagnostic laboratories in New Zealand from 2014-2017. Dogs eligible for inclusion had clinical signs, serum biochemistry and/or skeletal muscle histology findings supportive of a myopathy, with no etiology determined.

Results: A total of 86 cases were recruited, and skeletal muscle histology was performed in 18 of these. Pig hunting dogs were predominantly affected (58 cases), but cases also occurred in pet dogs (16) and working farm dogs (12). The onset of clinical signs was associated with confirmed consumption of wild pork in 88% of cases. Acutely, serum creatine kinase, aspartate aminotransferase and alanine aminotransferase were increased, with degeneration of myofibres in the absence of inflammation on histology. In chronic cases, serum biochemistry was frequently normal, but histology showed muscle regeneration, fibre splitting and myofibre loss. Electron microscopy of skeletal muscle in 4 cases showed enlarged mitochondria with disrupted cristae, and accumulation of lipid and glycogen within the mitochondrial matrix.

Conclusions: ‘Go Slow’ is a myopathy in dogs that affects skeletal muscle mitochondria. The disease is associated with the consumption of wild pork, and further work is required to identify possible causative compound(s).

N-03: MALIGNANT CATARRHAL FEVER ASSOCIATED WITH A NOVEL GAMMAHERPESVIRUS IN RED-FLANKED DUIKERS

Francisco R. Carvallo, Janet D. Moore, Kenneth Jackson, Akinyi Nyaoke, Lisa Naples, Patricia Pesavento

Malignant catarrhal fever (MCF) is a widely distributed, often fatal lymphoproliferative viral disease predominantly affecting domestic and wild ruminant species, caused by a group of viruses from the gender Macavirus, sub family *Gammaherpesvirinae,* family herpesviridae. MCF is reported as endemic in animal parks, because MCFV are present in various adapted, host ruminant species, and exposure to non-adapted species occurs in mixed ruminant collections. Under these circumstances, spread of disease is facilitated by direct contact or vector transmission. Two adult red-flanked duikers from a zoological collection in southern California were submitted for necropsy in a period of a week, each with a one day history of fever, depression, tremors and death. At necropsy, generalized lymphoadenomegaly and hepato-splenomegaly, transmural urinary bladder hemorrhages and fibrin in body cavities were noted. Microscopically, infiltrates of atypical lymphoid cells were identified in multiple organs. Vascular involvement, a hallmark histologic lesion for most MCF cases, was present but limited to thromboses in several organs of one animal, and vasculitis in the brain, lung and liver of the other. PCR for common MCFV, including OvHV2, CaHV2, and AlHV 1 and 2 was negative. Using a degenerate primer strategy that would be inclusive for viruses of the MCF complex, we amplified and sequenced a segment of a novel gammaherpesvirus most closely related by phylogeny to caprine herpesvirus 2.

N-04: INTRA-OSSEUS EPIDERMAL INCLUSION CYSTS IN TWO BROWN KIWI (APTERYX MANTELLI)

Stuart Hunter

Two brown kiwi(*Apteryx mantelli*), separated geographically and spatially, presented to awildlife hospital after being found caught in and rescued from leg hold traps (designed to capture pest species such as possum). Both birds had an open, simple and slightly oblique, mid-diaphysealfracture of the right tarsometatarsal bone. After initial stabilisation both birds underwent surgical fracture repair with external fixation. Unfortunately one kiwi died from an unrelated chronic respiratory condition while the second kiwi was euthanased due to chronic malunion of the fracture site. Post mortem sectioning of both affected tarsomatatarsi revealed a fairly discrete, roughly rectangular, dull yellow mass effect occupying the medullary cavity of the mid-diaphysis. Histology of both bones showedreplacement of the medullary cavity with granulation and fibrous tissue surrounding a large central core of eosinophilic and pyknotic debris admixed with irregular fragments of necrotic lamellar (cortical) bone. The interface between the necrotic core and surrounding granulation tissue was composed of stratified, keratinizing sqamous epithelium or a thin layer of multinucleated giant cells and epithelioid macrophages. In both cases a diagnosis of an intra-osseous epidermal inclusion cyst was made.The etiology of intraosseous epidermal inclusion cysts are uncertain. The two main hypotheses include a congenital etiology with intraosseous inclusion of embryonal epithelial tissue while the second is the result of traumatic implantation of epidermal fragments into the bone by any type of injury.Given the history of previous traumatic injury to these birds, the latter hypothesis appears the more likely.

N-05: 'WOBBLY HEDGEHOG SYNDROME': A NEUROANATOMICAL AND ULTRASTRUCTURAL STUDY

Josué Díaz-Delgado, Derick Whitley, Ralf Storts, Jill Heatley, Sharman Hoppes, Brian Porter

Background: Wobbly hedgehog syndrome (WHS) is a leading cause of neurologic disease in African pygmy hedgehogs (APH; *Atelerix albiventris*).

Objective: To describe detailed neuroanatomical and ultrastructural features of WHS in a cohort of 12 pet APH.

Methods: Twelve APHs were submitted for necropsy to the Department of Veterinary Pathobiology at Texas A&M University between 2010 and 2016. Signalments and clinical histories were analyzed. Complete necropsies were performed and histopathological examinations were conducted on selected tissue sections, along with special histochemistry (Holmes and Luxol fast blue stains), electron microscopy, and immunohistochemistry (polyclonal anti-Iba [ionized calcium-binding adapter molecule]-1 antibody; polyclonal anti-GFAP [glial fibrillary acidic protein] antibody). For analysis of neuroanatomical lesion distribution, a template including 25 coronal brain sections was used.

Results: Histologically, lesions consisted of varying degrees of white matter spongiosus, typically bilateral and symmetrical, accompanied by axonal degeneration and loss, reactive microgliosis and astrocytosis, and neuronal degeneration and loss. Lesions were most severe in the cerebellum and medulla oblongata, cervical and thoracic spinal cord, corona radiata, corpus callosum and striatum, internal capsule, and mesencephalon. Ultrastructurally, the lesions consisted of myelin sheath splitting initiated at the intraperiod line with subsequent disruption, dilatation, rhexis, and phagocytosis.

Conclusions: Based on these results, WHS is best described as a 'spongy myelinopathy' with widespread central nervous system involvement.

N-06: EFFECTS OF INNATE IMMUNE STIMULATION ON NATURALLY OCCURRING RESPIRATORY DISEASE IN BEEF CALVES

Laura L. Bassel, Joanne Hewson, Shayan Sharif, Alaina Macdonald, Carmon Co, Laura Siracusa, Ksenia Vulikh, Jeff L. Caswell

Stress and virus-induced immunosuppression is considered a major risk factor for the development of bovine respiratory disease (BRD). We hypothesized that stimulation of innate immune responses on arrival to a feedlot could overcome this immunosuppression and decrease the prevalence and severity of BRD. Sixty calves at high risk of disease were temporally divided into 4 cohorts and randomly assigned to receive aerosolized E. coli and S. aureus lysate or phosphate-buffered saline control solution delivered via nebulization. Body weight, rectal temperature, serum haptoglobin, fibrinogen, and results of targeted lung ultrasounds were recorded at baseline and regular intervals up to 1 month after arrival. Animals exhibiting clinical signs were treated according to farm protocols. Animals dying during the study period received a full postmortem examination. Pilot studies established that aerosolization of bacterial lysate was well tolerated and resulted in transient increases in temperature, respiratory rate and neutrophils in bronchoalveolar fluid. Unexpectedly, the mortality rate attributed to *M. bovis* was 20% (6/30) in calves receiving the bacterial lysate compared to 3% (1/30) in control calves. Significant respiratory disease was seen in 70% of immunostimulated calves versus 53% of control calves. Calves receiving the immunostimmulant had lower weight gains at one month after arrival versus control calves. Inflammatory markers and lung ultrasound scores were similar between the two groups. Although stimulation of innate immune responses was unsuccessful in preventing BRD, the results of this study suggest a potential relationship between pulmonary inflammation and M. bovis pneumonia.

N-07: HISTOLOGIC CHARACTERISTICS OF CUTANEOUS DERMATOPHYTOSIS IN PERSIAN CATS

Alexandra N. Myers, Joanne Mansell, Aline Rodrigues Hoffmann

Background: Persian cats are clinically more susceptible to chronic and severe dermatophytosis than other breeds. This susceptibility is presumed to be genetic and is likely associated with a defect in the innate or adaptive immune system.

Objective: The objective of this study is to identify any differences in the histologic lesions between Persians and other breeds of cats with dermatophytosis. We hypothesize that Persian cats will exhibit a decreased inflammatory response compared with other breeds.

Methods: The Texas A&M University Dermatopathology Specialty Service archives were searched for cases of cutaneous dermatophytosis between the years of 1999 and 2017, and the hematoxylin and eosin stained slides from all of these cases were pulled for review. Histologic features, including folliculitis and perifolliculitis, epidermal hyperplasia, hyperkeratosis, and numbers of affected follicles, were scored and compared between Persian and non-Persian cats.

Results: Persian cats with dermatophytosis exhibited primarily minimal to mild folliculitis and perifolliculitis whereas other breeds of cats presented more often with moderate to marked inflammation. No differences were observed for the other investigated parameters.

Conclusions: The lesions of cutaneous dermatophytosis in Persian cats are characterized by milder folliculitis and perifolliculitis compared with other cat breeds. These findings suggest that Persian cats may mount a less robust immune response to dermatophytes than other breeds, potentially due to a genetic mutation in the innate or adaptive immune pathways. Future work should investigate the types of immune cells present in the skin and if a possible genetic mutation is contributing to an altered immune response.

N-08: DIAGNOSTIC UTILITY OF CYTOKERATIN 5 FOR THE IDENTIFICATION OF PROLIFERATIVE INFLAMMATORY ATROPHY IN CANINE PROSTATES

Michelle M. Story, Fabian Z. X. Lean, Syeda Hasina Akter, Valeria Grieco, Angelo Michael De Marzo, Chiara Palmieri

Background: It has been postulated that proliferative inflammatory atrophy (PIA) is associated with the development of prostate cancer in men. PIA lesions consist of atrophic but proliferative epithelia and show increased expression of the basal cell marker cytokeratin 5 (CK5). Lesions resembling PIA have been seen in canine prostates.

Objective: The aim of this study was to determine if CK5 immunohistochemistry could be used to assist in the identification of PIA in canine prostates.

Methods: Hematoxylin and eosin (H&E) stained slides from 87 formalin-fixed, paraffin-embedded samples of canine prostate were first classified as normal, benign prostatic hyperplasia (BPH), prostatic carcinoma (PC) and prostatitis, and were then examined for lesions resembling PIA. Immunohistochemistry for CK5 was subsequently performed.

Results: PIA was observed in 24 out of 87 cases (27.6%) when only H&E stained slides were examined, whereas CK5 positive PIA lesions were detected in 43 out of 87 cases (49.4%). This meant that 19 out of 43 cases (44.2%) with CK5 positive PIA lesions were considered to be negative for PIA on assessment of the H&E stained slides. CK5 positive PIA lesions were detected in 2 of 6 normal prostates, 20 of 30 cases of BPH, 10 of 34 cases of PC, 8 of 10 cases with mixed lesions (BPH and PC) and 3 of 7 cases of prostatitis.

Conclusion: Our study demonstrates that lesions resembling PIA can be found in the canine prostate, and that CK5 immunohistochemistry may be of significant diagnostic value in the identification of these lesions.

N-09: OUTBREAK OF CANINE PARVOVIRUS TYPE 2B AND CLOSTRIDIUM DIFFICILE ENTEROTOXEMIA INFECTION IN ASIAN SMALL-CLAWED OTTERS (AMBLONYX CINEREUS)

Tatiane Terumi Negrao Watanabe, Edward J. Dubovi, Dawn E. Evans, Ingeborg M. Langohr, Fabio Del Piero

Background: Canine parvovirus (CPV) type 2 is a potential cause of gastrointestinal disease in otters. Over one to four days, five of seven 6-month- to 2-year-old Asian small-clawed otters (*Amblonyx cinereus*) from the same colony presented comatose and acutely dehydrated. Four of the five affected otters died while one survived after supportive care.

Methods: Three otters were submitted for post-mortem examination.

Results: Grossly, the small intestine was diffusely reddened and contained red to brown malodorous watery digesta without formed feces (3/3). Histologically, there was frequent loss of luminal enterocytes and necrosis of crypt epithelial cells. Denuded villi were often lined by mixed bacterial colonies. Splenic follicles had lymphoid depletion (2/3). Immunofluorescence assay (IFA) revealed CPV antigen in the mesenteric lymph nodes (3/3), small intestine (2/3), and spleen (1/3). Immunohistochemistry revealed CPV antigen in the enterocytes, lymphocytes and dendritic cells of the Peyer's patches and spleen (3/3), and glossal epithelial cells (1/2). PCR and DNA sequencing identified CPV-2b as the cause (2/3). No canine distemper virus was detected by IFA (3/3). No parasites were identified through fecal floatation exams (3/3). *Streptococcus* spp. were cultured from the intestine (3/3). *Clostridium difficile* producing A and/or B toxins was identified in the intestinal content by ELISA (3/3).

Conclusions: We describe herein the clinical, pathologic and ancillary test findings of a CPV-2b outbreak in otters complicated by *Clostridium difficile* toxins exacerbating the viral necrosis. The source of the viral infection remains unknown although otters or other carnivores may have been the source.

N-10: ENDOMETRIAL ADENOCARCINOMA IN TWO ASIAN ELEPHANTS (ELEPHAS MAXIMUS)

Jessica K. Wong, Allan P. Pessier

Background: A high prevalence of cystic endometrial hyperplasia (CEH) and endometrial polyps that may affect fertility hasbeen observed in captive North American elephants. In these cases, repeated reproductive cycles combined with nulliparity and aging is the suspected pathogenesis. In humans, endometrioid-type adenocarcinomas are among the most common gynecological cancers and develop from within premalignant endometrial proliferative lesions (endometrial intraepithelial neoplasia, or EIN). The occurrence of EIN is associated with prolonged estrogen stimulation unopposed by progesterone. To date, the development of uterine carcinomas in elephants is largely unexplored.

Objective: To expand the knowledge of endometrial proliferative lesion pathogenesis in Asian elephants (*Elephas maximus*).

Methods: A retrospective review of Asian elephant necropsies at the San Diego Zoo was performed. Tissues were characterized by routine histology and immunohistochemistry, including pancytokeratin (CK), vimentin (VIM), and estrogen receptor (ER).

Results: Two aging elephants with endometrial neoplasms were identified. Both cases exhibited diffuse CEH and endometrial polyps. In Case 1, two polyps contained adenocarcinoma *in situ*. In Case 2, a papillary adenocarcinoma expanded the endocervix and uterine body with abdominal carcinomatosis and regional lymph node metastasis. In both cases, neoplastic cells stained positively for CK and ER and variably for VIM.

Conclusion: This report expands the spectrum of reported proliferative endometrial lesions in Asian elephants. The observations of adenocarcinoma*in situ* arising within endometrial polyps in Case 1 and positive ER staining of neoplastic cells in both cases suggest development of these lesions may have similarities to the pathogenesis of endometrioid adenocarcinomas in humans.

N-11: PROTEIN BIOMARKERS IN SERUM AND URINE FOR DIAGNOSIS OF CANINE HIP DYSPLASIA

Chantelle C. Bozynski, Carin E. Ahner, Aaron M. Stoker, Emily Leary, James L. Cook

Background: Canine hip dysplasia (CHD) is one of the most commonly diagnosed orthopedic disorders in dogs. According to the Orthopedic Foundation for Animals (OFA) database, the prevalence of CHD can be as high as 74% in certain breeds. Dysplastic dogs are often diagnosed in the irreversible phase (i.e., presence of osteoarthritis), which limits treatment options. Proteins in bodily fluids have been used as biomarkers for early diagnosis of several other disorders, providing potential for prevention and improving efficacy of treatment.

Objective: To identify protein biomarkers in serum/urine that differentiate dysplastic from normal dogs.

Methods: Whole blood and urine were collected from seventy-four client-owned dogs (normal hips [n=49] and dysplastic hips cohorts [n=25] based on OFA grading criteria). Seventeen biomarkers reflecting direct and indirect measures of joint health were examined. Comparisons between cohorts were performed using two sample *t*-tests with significance set at $p<0.05$. Combinations of biomarkers with the greatest ability to determine group membership (dysplastic or normal) were analyzed using receiver operator characteristic (ROC) curves.

Results/Conclusions: Dogs with dysplastic hips had significantly different urine concentrations of cartilage degradation products, intercellular signaling molecule, bone synthesis protein, and proteinase inhibitor when compared to normal dogs. ROC curve analyses established a panel of 4 biomarkers (CTX-I and CTX-II in urine; PIICP and MMP-9 in serum) with high discriminatory capability (area under the curve = 0.89). Protein biomarkers in serum/urine have the potential to provide for early/accurate diagnosis of CHD, and ongoing biomarker panel analyses in puppies will validate this method for clinical use.

N-12: NEOPLASIA IN STRANDED CETACEANS FROM NORTH CAROLINA AND VIRGINIA, 2011-2017

David S. Rotstein, William McLellan, Craig Harms, Victoria Thayer, Susan Barco, Kristy Volker, Alexander Costidis, D. Ann Pabst

Background: Marine mammal strandings aid in monitoring for diseases, population health assessments, and understanding the natural history of coastal and pelagic oceans. This is achieved through a thorough biometrics data and necropsy.

Objective: To determine prevalence of neoplasia in cetaceans stranded along the North Carolina and Virginia coast.

Methods: Necropsy records for stranded odontocetes and mysticetes evaluated by a single pathologist from 2011 to 2017 were reviewed for cases of neoplasia including adenomatous hyperplasia. Overall and species-specific prevalence rates were determined.

Results: A total of 233 records representing 19 species were reviewed. Neoplasia was present in 2.1% (5/233). Four odontocete species had neoplasia including *Tursiops truncatus*, *Delphinus delphis*, *Grampus griseus*, and *Lagenorhynchus albirostris* with an intraspecies prevalence of 1.2% (1/163), 5.9% (2/17), 50% (1/2), and 100% (1/1), respectively. No neoplasia was observed in the mysticetes. Neoplasms were observed in the endocrine system (80%, 4/5), hematopoietic system (20%, 1/5), and reproductive system (20%, 1/5). In the endocrine system, there were cases of thyroid gland adenomatous hyperplasia and medullary carcinoma, pheochromocytoma, and schwannoma. In the reproductive system, there was a seminoma. In the hematopoietic system, there was an immunoblastic lymphosarcoma, which was the only metastatic tumor.

Conclusions: Endocrine system neoplasia was the most prevalent. Metastasis was not common. The prevalence of neoplasia is low in the stranded population, but the accuracy is limited to stranded animals and cannot account for off-shore mortalities, grossly observable tumors, decomposed animals not sampled, and unknown total population numbers.

N-13: IDIOPATHIC HYPERTROPHIC PACHYMENINGITIS IN A DOG

Timothy W. Carlson, Aditya M. Davé, Erik J. Olson

A 6-year-old, spayed female, Greyhound dog was presented for acute blindness and inability to close the jaw. Subsequently, the dog developed tremors, panting, weakness, incoordination, and brief periods of stupor. There was lymphocytic pleocytosis of cerebrospinal fluid (CSF), and magnetic resonance imaging (MRI) revealed diffuse, symmetric, and marked pachymeningeal thickening and enhancement with multifocal dilation of the subarachnoid space. At necropsy examination, the dura mater within the cranial vault was diffusely thickened, along with less extensive involvement of the dura mater of the spinal cord. Histologically, the cerebral dura mater was expanded by a diffuse, densely cellular infiltrate composed of numerous plasma cells, macrophages, spindle cells, and fewer lymphocytes admixed with a large amount of collagenous matrix. This infiltrate similarly expanded the meninges of the optic nerves, as well as the leptomeninges and the dura mater of the spinal cord to approximately the level of the third lumbar vertebra. The MRI, CSF, and necropsy findings were consistent with previous case reports of idiopathic hypertrophic pachymeningitis, as described in dogs and humans.

Idiopathic hypertrophic pachymeningitis has been rarely reported in dogs. Although the total number of reported canine cases is low, Greyhounds appear to be over-represented. The pathogenesis is not known, but is thought to be related to an autoimmune inflammatory response localized to the dura mater. Previous case reports in dogs and humans include minimal to no histologic detail and differ in the degree of involvement of the leptomeninges and dura mater overlying the spinal cord.

N-14: SARCOCYSTIS SP. ASSOCIATED POLYMYOSITIS IN STRANDED CALIFORNIA SEA LIONS (ZALOPHUS CALIFORNIANUS)

Mauricio Seguel, Kathleen Colegrove, Cara Field, Elizabeth Howerth, Padraig Duignan

Background: Skeletal muscle disease is an uncommon but constant cause of stranding among pinnipeds, however little is known regarding the etiology and pathogenesis of muscle disease in marine mammals.

Objective: Determine the etiology and characterize the pathology of polymyositis in California sea lions (*Zalophus californianus*).

Methods: We retrospectively investigated cases of polymyositis at the Marine Mammal Center, Sausalito, California, from 2001 to 2016.

Results: A total of 133 polymyositis cases were found. All animals had *Sarcocystis sp.* cysts in striated muscle. In 96 sea lions myositis was considered incidental, and in 42 cases a major factor contributing to stranding and death. The skeletal muscles of these animals (42/133) had multiple areas of coagulative necrosis, regeneration, and chronically atrophy and fibrosis. Myofibers were surrounded by numerous T-lymphocytes that sometimes infiltrated the endomysium and sarcoplasm of intact myofibers. In some inflammatory foci there were rare 2-3 µm zoites and occasional immature protozoan cysts that had moderate immunolabelling with an anti-*Sarcocystis neurona* antibody. *S. neurona* antibody titers were higher in animals with severe myonecrosis when compared to animals with mild necrosis. In at least 10 cases with significant polymyositis the protozoan species in the skeletal muscle was molecularly identified as *S neurona*.

Conclusion: *S. neurona* is a major cause of polyphasic polymyositis in California sea lions. Pathologic findings suggest that immune mediated mechanisms drive muscle damage in animals with severe polymyositis.

N-16: PCR PREVALENCE AND RISK FACTOR ANALYSIS OF COXIELLA BURNETII IN CATS FROM CYPRUS

Charalampos Attipa, Chris Helps, Tom Chisnall, Severine Tasker, Mark Eisler

Background: *Coxiella burnetii*, causative agent of Q fever in humans, is an important zoonotic pathogen, especially in ruminants. In cats, infection with*C. burnetii*can cause abortions and is commonly subclinical, however, it has been proposed that cats can be source of infection for humans. No prevalence studies for*C. burnetii*have been performed in cats from Cyprus, which has a high seroprevalence of this pathogen in ruminants (goats up to 48.2%) and humans (52.7%).

Objective: Determine the PCR-prevalence of*C. burnetii*in Cypriot cats, and identify risk factors for this infection.

Methods: Samples of DNA (extracted from blood) were available from 172 cats of known signalment and lifestyle characteristics, and underwent*C. burnetii*PCR. Statistical analysis was performed to assess for any associations between signalment, lifestyle characteristics and other infectious agents (feline haemoplasmas, retroviruses,*Hepatozoon*spp.,*Leishmania*spp. and*Bartonella henselae*) and*C. burnetii*PCR positivity.*P*-values ≤ 0.05 was considered statistically significant.

Results: Of the 172 samples, 5 (2.9%) were*C. burnetii*PCR positive. Statistical analysis showed a significant association between*C. burnetii*infection and not using ectoparasiticides (*P*=0.05).*C. burnetii*infection positive status had a trend towards significance in non-healthy (*P*=0.19) and FIV seropositive (*P*=0.11) cats.

Conclusions: This is the first report of molecular detection of*C. burnetii*in Cypriot cats. The prevalence in cats is not as high as in other animal species from Cyprus, but it is noticeable and given the zoonotic nature of this pathogen, regular and effective ectoparasitic prevention is recommended for cats.

N-17: PLACENTA ACCRETA IN THREE AFRICAN GREEN MONKEYS (CHLOROCEBUS AETHIOPS SABAEUS)

Caralyn S. Labriola, Rachel N. Andrews, Gayathriy Balamayooran, Alexander Robert Hutchison, David L. Caudell, Matthew J. Jorgensen, James Mark Cline, Nancy D. Kock

Placenta accreta is the anomalous adherence of the placenta to the uterus, and has been extensively studied in women, as it is associated with failure of placental detachment during parturition, excessive hemorrhage, and peripartuirent mortality. This condition is currently estimated to occur in 1 of every 500-2500 pregnancies in developed countries, and has been recognized with increasing frequency, paralleling the increased incidence of Caesarian section in these countries. Caesarian section, placenta previa, increased maternal age and parity are the most important risk factors for the development of abnormally invasive placenta. Herein we describe cases from two 16 year-old and one 20 year-old multiparous African green monkeys (*Chlorocebus aethiops sabaeus*) diagnosed with abnormally adherent placenta. All three cases resulted in the death of both mother and fetus. The causes of maternal death were exsanguination following fetotomy, clinical euthanasia following cardiac arrest with placenta previa, and exsanguination with abortion. The two 16 year-old animals had a retained placenta in a previous pregnancy, one that resulted in abortion. None of the animals had a history of Caesarian section. Placenta accreta, the lack of decidua with attachment of the chorionic plate to the myometrium, was confirmed on histologic examination. Abnormally invasive placenta has rarely been described in non-human primates, and appears to be a novel finding in African green monkeys.

Supported in part by P40 OD010965 and T32 OD010957.

N-18: OVARIAN ADENOCARCINOMA IN CAPTIVE NORTH AMERICAN JAGUARS (PANTHERA ONCA): INVESTIGATING THE GENETIC MECHANISMS OF TUMORIGENESIS

Sarah M. Corner, Maciej Parys, Anneke Moresco, Vilma Yuzbasiyan-Gurkan, Dalen Agnew

Background: Ovarian adenocarcinoma (OC) is a frequent cause of mortality in adult, female jaguars in zoological institutions. OC is extremely rare in other felids. Formalin-fixed, paraffin-embedded (FFPE) tissues from 56 female jaguars between 1988-2017 were collected and reproductive lesions characterized. 23 jaguars (40% of study population) had OC. Pedigree analysis of the jaguar population was suggestive of an autosomal dominant pattern of inheritance.

Objective: Our goal is to investigate the genetic pathogenesis by screening for germline mutations associated with jaguar OC in cancer genes with known involvement in human ovarian and breast cancer, such as BRCA1 and BRCA2.

Methods: Next generation sequencing (NGS) was performed to evaluate exon sequences of cancer genes (n=276) in 10 paired FFPE ovarian adenocarcinomas and normal tissues from each jaguar, 4 unaffected jaguars, and 4 domestic cat control tissues. Due to lack of a jaguar reference genome, primers were designed using the domestic cat and other felids from the genus *Panthera*. Sequencing was performed using the Illumina HiSeq 2500 System.

Results: In BRCA2, 111 germline single nucleotide variants were detected, and 64 were missense variants. One single nucleotide variant was predicted to be potentially deleterious using SIFT analysis, and is locatedin a region of exon 11 necessary for RAD51 binding in DNA repair.

Conclusions: Investigating the genetic pathogenesis of jaguar OC may enable the identification of mutation carriers for breeding decisions, and improve disease monitoring. Evaluation of variant allele frequency in the jaguar population and loss of heterozygosity in neoplastic tissue is underway.

N-19: AN UNUSUAL OCULAR PATHOLOGY IN A STRANDED KILLER WHALE

Stephen Raverty, Bruce Grahn, Paul Cottrell, Joe Gaydos

An adult male killer whale found floating dead in the Georgia Strait, British Columbia December 21, 2016. The animal was towed ashore and presented in fair post mortem and moderate body condition. Gross exam revealed a large hematoma in the right cervical region, which caused the fibroelastic sheath to bulge and on cut section, extended deep into the skeletal muscle. Blunt force trauma of undefined origin was a prime consideration and histopathology revealed lakes of eosinophilic material interpreted as autolyzed and hemolyzed red blood cells. Sections of the left eye were prepared, stained with hematoxylin and eosin and Schiff's periodic acid and forwarded to the Western College of Veterinary Medicine for consultation. The most salient histopathology included marked cataract formation with spherophakia, liquefaction of the lens cortex, extensive reduplication of the lens capsule and multiple scrolled anterior and posterior capsule remnants, consistent with a prior lens capsule rupture. Degenerate capsular remnants were surrounded by histiocytes, lymphocytes and plasma cells. Histiocytes were engorged with PAS positive material, which may represent lens capsule; however, there were clusters of acicular clefts which also suggest long past resolving hemorrhage. Although whales rely on sonar for predation and sound production for social interactions, the role of vision is not yet fully resolved. Most of the lesions in this animal would have preceded the final trauma by several weeks, given the fibrosis, advanced retinal atrophy and repair of Descemet's membrane all of which we estimated to be at least 3-4 weeks duration at a minimum.

N-20: DISEASE SURVEILLANCE IN SMALL POULTRY FLOCKS IN ONTARIO

Nancy Brochu, Michele Guerin, Csaba Varga, Marina Brash, Leonardo Susta

Rationale: Small poultry flocks may act as potential reservoirs of avian and zoonotic pathogens, due to limited biosecurity and contact with wild birds. The recent increase in the number of small flocks throughout Ontario calls for a better assessment of baseline disease prevalence in this sector.

Objective: A prospective surveillance study of small flock postmortem submissions to the Animal Health Laboratory was conducted over a 2-year period to determine the prevalence of infectious agents in small flocks throughout Ontario.

Methods: Upon the owner's consent and completion of a standardized questionnaire concerning husbandry and biosecurity, a full postmortem examination and pre-set array of tests for infectious agents were conducted.

Results: During the first year, 64 submissions (97 birds total) with associated questionnaires were received. Chickens were most common (85% of submissions), followed by turkeys (6%), game birds (5%), and ducks (3%). The most common causes of death were bacterial (42% of birds), viral (14%, including Marek's disease [50%]), neoplastic (10%), parasitic (7%), and idiopathic inflammatory (8%). Microbiological tests isolated *Campylobacter* spp., *Brachyspira* spp., *Mycoplasma synoviae*, *Mycoplasma* *gallisepticum*, and *Salmonella* spp. (*S.* Uganda and Anatum) in 36, 34, 27, 17, and 3% of submissions. Infectious bronchitis virus, fowl adenovirus, infectious laryngotracheitis virus, and reovirus were detected in 31, 26, 13 and 7% of submissions. Avian paramyxovirus serotype 1 was isolated from 2 chickens and avian influenza virus from one turkey (H10N8, LPAIV).

Conclusions: Results from this study will aid in the prevention and control of relevant diseases among Ontario's small poultry flocks.

N-21: A CASE OF CHRONIC WASTING DISEASE IN CAPTIVE RED DEER (CERVUS ELAPHUS) IN SOUTH KOREA, 2016

In-Soon Roh, Hyo-Jin Kim, Tae-Young Suh, Hae-Eun kang, Hyun-Joo Sohn

Background: Tissue samples(obex, retropharyngeal lymph node: RPLN, tonsil) from a 30-month-old, male red deer having a history of emaciation, ataxia and inability to stand due to unknown reasons were submiited to the CWD OIE Ref Lab, Animal and Plant Qurantine Agency(APQA) for CWD testing on Feb. 4, 2016.

Objective: Our objective was to report the case of a outbreak of CWD in capive red deer.

Methods: The samples(obex, RPLN, tonsil) were analyzed by using a commercial ELISA test(rapid test) for detection of PrP^{CWD} according to the manufacture's instructions. After an initial positive result, the presence of PrP^{CWD} was demontrated by a commercial available Western blot test, and immunohistochemical method using polyclonal antibody S1(made in APQA) at a dilution of 1:3000 for confirmation. These tissues were stained with hematoxylin-eosin(HE) to see spongiosis and gliosis.

Results: The red deer was found to be positive for the abnomal prion protein in the obex, RPLN and tonsil by the rapid test. Furthermore, all samples of the red deer demonstrated proteinase K-resistant three bands pattern between 17 and 29 kDa. The microscopic lesion indentified was spongiform encephalopathy in the obex. Diffuse patterns of PrP^{CWD} immunolabelling was presented in the obex region, espicially, in the dorsal motor nucleus the vagus nerve and solitary tract, and in the germinal center of lymphoid follicles in RPLN and tonsil.

Conclusions: The suspect case of CWD in a captive red deer was confirmed to be a new outbreak of CWD in South Korea, 2016.

N-22: PATHOLOGICAL INVESTIGATIONS IN AGED CAPTIVE BEARS (URSUS THIBETANUS FORMOSANUS AND URSUS ARCTOS ALASCENSIS)

Wen-Ta Li, Chien-Hao Chen, Victor Fei Pang, Chian-Ren Jeng, Chen-Hsuan Liu, Fun-In Wang, Yuan-Shyuee Lai, Hui-Wen Chang

Background: Pathological investigation is important for the diagnosis of diseases and improvement of the management in captive wild animals. Several non-infectious diseases, such as dental diseases, degenerative joint diseases, and neoplasms, have been reported in captive bears, but the information on the senile diseases, especially the cardiovascular disease, is rather limited.

Objective: Investigate the diseases in aged captive bears for improving the welfare of captive bears.

Materials and Methods: Four aged captive bears (>24 year-old), including two Formosan black bears, *Ursus thibetanus formosanus*; and two Alaskan brown bears, *Ursus arctos alascensis*, died spontaneously during Jun. 2015 and Jun. 2017. Necropsy was performed and representative tissue samples were collected and processed routinely for histopathological examination.

Results: Cardiovascular conditions, including arteriolosclerosis and arterial medial calcification and osseous metaplasia with myocardial fibrosis and hemorrhage, pharyngeal varices, non-traumatic brainstem hemorrhage or hematoma; and tumor/tumor-like conditions, including branchial cysts, pulmonary adenocarcinoma, adrenal cortical nodular hyperplasia/carcinoma or Brunner's gland hamartoma, were found in all bears examined. Other findings included chronic renal disease (3/4), pulmonary fibrosis (2/4), disseminated intravascular coagulation (2/4), pancreatitis (1/4), and suppurative uveitis (1/4).

Conclusion: Cardiovascular and tumor/tumor-like conditions were the most common diseases affecting the health of aged bears in captivity. The cardiovascular abnormalities seen in these bears are mainly aging and hypertension-associated. Therefore, early clinical detection and management of hypertension in captive bears would be extremely important in preventing the subsequent cardiovascular and/or other related disease development.

N-23: PRELIMINARY EVALUATION OF AN AUTOLOGOUS DENDRITIC CELL VACCINE UTILIZING NANOPARTICLE TECHNOLOGY FOR THE TREATMENT OF CANINE MALIGNANT MELANOMA

Meaghan V. Eren, Julianne K Hwang, Cleverson D. Souza

Background: Canine malignant melanoma (CMM) carries a poor prognosis due to its minimal responsiveness to treatment.

Objectives: To generate an autologous dendritic cell (DC) vaccine using poly-lactic-co-glycolic nanoparticles (PLGA-NPs) containing antigens from a whole tumor lysate, and evaluate its safety following intradermal administration.

Methods: A study was performed in 5 dogs with CMM. Protein was collected from 1 g of tumor tissue using repetitive freeze-thaw cycles, and concentration was measured using spectrophotometric methods. Protein was diluted, and encapsulated within PLGA-NPs. Encapsulation efficiency (EE) and protein release from the PLGA-NPs were determined, along with zeta potential and dynamic light scattering (DLS). PLGA-NPs were cultured with autologous DCs, and surface marker expression was assessed using flow cytometry. Local and systemic reactions, outlined by the National Cancer Institute's Common Terminology Criteria for Adverse Events, were recorded.

Results: An average of 69.2 mg/mL of protein was obtained, with prominent bands observed at 15, 45 and 70 kDa as determined by SDS-PAGE. EE ranged from 44.9-63.2%, and 75.2-93.7% of proteins were released over 21 days, with peak release on day 4. Zeta potential was negative for all PLGA-NPs (-12.5 mV to -15.6 mV), indicating increased stability, and DLS confirmed minimal aggregation. Flow cytometry of the DCs following incubation with PLGA-NPs revealed increased expression of MHC I, MHC II and CD80. No significant vaccine reactions were observed.

Conclusion: PLGA-NPs can be used to successfully encapsulate tumor antigens and mature dendritic cells from dogs with CMM in vitro. Future studies are warranted to assess specific immune responses.

N-24: DEVELOPMENT OF A LARGE ANIMAL MODEL OF FAMILIAL LEFT VENTRICULAR HYPERTROPHY AND SUDDEN CARDIAC DEATH IN THE RHESUS MACAQUE

Yu Ueda, Joshua A. Stern, Ross P. Tarara, Christina Cruzen, Sree Kanthaswamy, Muthuswamy Raveendran, R. Alan Harris, Jeffery Rodgers, Jeffery A. Roberts, J. Rachel Reader

Funding: CNPRC base grant NIH-P510D011107

Background: Hypertrophic cardiomyopathy (HCM) affects 1 in 500 humans and can causesudden death. Little progress in prevention or treatment of HCM may be due to the lack of a suitable large animal model of HCM. Left ventricular hypertrophy (LVH) with sudden death has been identified in rhesus macaques through post-mortem and ante-mortem investigations.

Objective: To identify sarcomeric mutations that may explain a rhesus macaque model of HCM through extensive phenotyping and whole exome sequencing (WES).

Methods: LVH was diagnosed at necropsy when the ratio of LV diameter to the LV lumen diameter was greater than 3 in specific transverse sections. DNA from 65 LVH-affected rhesus macaques was used to investigate genetic factors contributing to disease. WES was performed and the data mined for sarcomeric variants. 133 randomly selected rhesus macaques were used as a control population for association analysis.

Results: WES identified multiple variants in sarcomeric genes. 46 of 65 LVH-affected rhesus macaques (71%) carried a missense mutation in the sarcomeric gene myosin heavy chain 7 (MYH7). This frequency is dramatically elevated relative to controls ($p < 2 \times 10^{-11}$). Three animals also exhibit a missense mutation in the cardiac myosin binding protein C gene. All variant genotypes were heterozygous, consistent with observations that dominant MYH7 mutations can cause human HCM.

Conclusions: A nonhuman primate model of HCM is being developed. These animals share clinical, pathologic and genetic similarities with human HCM and represent an exciting area for future research into this devastating disease.

N-25: SAFETY OF RECOMBINANT MYXOMA VIRUS THERAPY IN CANINE CANCER PATIENTS

Katelyn M. Polemi, Amy L. MacNeill

Background: Oncolytic virotherapy is a viable treatment option for cancer. However, many oncolytic viruses that are efficacious in murine models of cancer are ineffective in humans. More predictive cancer models are being explored to further advance the field of oncolytic virotherapy. Over six-million dogs are diagnosed with cancer in the United States yearly. Many of these patients fail to respond to current treatment strategies. The outcomes of oncolytic virus treatment of dogs with spontaneous tumors may be better predictors of human cancer response to treatment and may improve treatment options for dogs.

Objective: Evaluate the safety of intratumoral injection of myxoma virus lacking the serp2 gene (MYXVΔserp2) in dogs with subcutaneous tumors.

Methods: MYXVΔserp2 was injected intratumorally in dogs with spontaneous soft tissue sarcoma. Tumor volume, tumor biopsies, and blood, urine, feces, saliva, and swabs from the injection site of the virus were collected at several time points following the injection of MYXVΔserp2 to evaluate organ function, immune response, and virus distribution.

Results: To date, no adverse effects have been observed in any canine cancer patients following the MYXVΔserp2 therapy. Results of complete blood profiles, serum chemistry analyses, and urinalyses were within reference intervals. Virus was detected by PCR in some biopsied tumors, but virus shedding was not observed. Anti-MYXV antibodies were detected in 1/8 dogs.

Conclusions: These studies provide needed safety information to advance clinical trials using MYXVΔserp2 to treat dogs with cancer. We believe oncolytic virotherapeutics that are efficacious in dogs will be effective in human cancer patients.

N-26: COMPARISON OF SERUM MICRORNA BETWEEN DOGS WITH AND WITHOUT CANINE MAMMARY CARCINOMA BY DEEP SEQUENCING AND COMPARISON TO HISTOPATHOLOGIC CHARACTERISTICS

Eric J. Fish, Gisela Martinez-Romero, Patricia Delnnocentes, Jey W. Koehler, Nripesh Prasad, Kristopher J. Irizarry, Annette N. Smith, Richard Curtis Bird

Background: Canine mammary carcinoma prognosis depends on factors such as histologic subtype, grade, and presence of lymphatic invasion, which require invasive biopsies. Serum microRNA (miRNA/miRs) have generated interest as minimally invasive cancer biomarkers.

Objective: To compare the serum microRNA profile between dogs with and without mammary carcinoma, and correlate expression of specific miRs to histologic subtypes and grades.

Methods: RNA was extracted from serum and submitted for miRNA deep-sequencing (Illumina HiSeq2500). Tumor biopsy samples were blindly assessed for histologic subtype, grade, and presence of lymphatic invasion. Bioinformatic and univariate statistical analyses comparing miRNA between groups were performed.

Results: 10 healthy females (5 spayed, 5 intact) and 10 female dogs with mammary carcinoma were included. Mammary carcinoma samples varied by histologic subtype and grade (n=4 Grade I; n=3 Grade II; n=3 Grade III). 6/10 had lymphatic invasion. 452 unique miRNA were identified in the serum deep-sequencing dataset. Comparing the neoplastic group to controls, 65 serum miRs were significantly upregulated >1.5-fold, while 12 miRs were significantly down-regulated >1.5-fold. A signature of upregulated miRNA predicted to regulate key hormone and tumor suppressor pathways (previously identified in canine mammary tumor cell culture exosomes) was identified, including miR-18a, miR-19b, miR-29b/c, miR-34c, miR-181c, miR-215 and miR-345. Of these, serum miR-18a was significantly higher in dogs with lymphatic invasion (p=0.0141) and Grade III tumors (p=0.0255).

Conclusions: Results suggest serum microRNA are differentially expressed in dogs with mammary carcinoma. Expression of serum miR-18a was significantly higher in dogs with tumor lymphatic invasion and Grade III mammary carcinomas.

N-27: METASTATIC ANGIOLEIOMYOSARCOMA INDUCED BY A FOREIGN BODY IN A KANGAROO (MACROPUS GIGANTEUS)

Laura Christina Setyo, Mikel Sabater, Alex Young, Barry H Rickman

An adult female Eastern Grey kangaroo (*Macropus giganteus*) was examined for lameness due to a mass located at the right tibiotarsal joint. Radiographs revealed a metal arrowhead foreign body within the mass alongside a pathologic fracture. Grossly, at necropsy, there was a multilobulated and haemorrhagic mass with intralesional metal fragments, associated osteolysis of the tibiotarsus and pulmonary metastatic nodules. Histologically the neoplasm was composed of plump elongated and fusiform cells that often formed concentrically around small vessels. The pulmonary nodules were composed of similar neoplastic cells, necrosis and haemorrhage. Neoplastic cells were positive immunohistochemically for vimentin and smooth muscle actin. The histological features, immunohistochemical profiles and behavior of this tumor support a diagnosis of a primary angioleiomyosarcoma with lung metastasis associated with a metal foreign body.

N-28: SEVERE GENERALIZED OSTEOLYTIC AND OSTEOPROLIFERATIVE HEREDITARY DISEASE OF YOUNG WISTAR RATS

Inbal E. Biton, Ori J. Brenner, Alon Harmelin

Background: We identified a spontaneous bone disease of striking severity which affects young outbred Wistar rats. A breeding colony was established at the Weizmann Institute to study this condition.

Objective: Our objective was to characterize the pathologic and radiographic findings of this novel disease.

Methods: Full necropsy and microscopic analysis were performed on 42 sick rats, of which 36 were also studied with CT.

Results: The disease affected the appendicular and axial skeleton of male and females pups. Its incidence suggested autosomal recessive inheritance. Certain bones were preferentially involved, but as the disease progressed, it became widespread, often symmetrical and caused severe swelling. Splenic necrosis was a common finding in advanced cases. Microscopically, there was extensive destruction of bone, both normal pre-existing and newly formed woven bone, by numerous large osteoclasts. Affected bones were converted into mesenchymal tissue and woven bone. Many cases had neutrophilic cellulitis, arthritis and tenosynovitis of variable severity.

Conclusions: The microscopic findings suggest a genetic defect that causes excessive osteoclastic activity and sterile inflammation. Identification of the underlying abnormality could shed light on the control of osteoclast function.

N-30: STREPTOCOCCAL LYMPHADENITIS IN AN IMPORTED COLONY OF EGYPTIAN FRUIT BATS (ROUSETTUS AEGYPTIACUS)

Joy M. Gary, Christie Ferrecchia, Michael Garner, Amy Denison, Jana Ritter

Several bats in a colony of wild-caught Egyptian Fruit Bats (*Rousettus aegyptiacus*) imported to the US from Israel developed cervical, axillary, or inguinal lymphadenitis in quarantine. Seven of the bats died or were euthanized, while the remainder of the colony was treated by lancing visible abscesses and subsequent topical/systemic treatment or by prophylactyic antibiotic treatment. Cultures of the lymph node lesions in these bats revealed a variety of bacteria, including *Streptococcus equi subsp zooepidemicus*, other beta and alpha hemolytic streptococci, *Staphylococcus aureus, Enterococcus spp.*, *Raoultella ornithinolytica,* and *E. coli*. Of the seven bats that died, three had histologic evidence of a suppurative and necrotizing lymphadenitis and cellulitis. Other histologic findings in these 3 bats included increased circulating neutrophils and alveolar hemorrhage in the lung (1 bat) and a cerebral abscess containing gram-positive cocci (1 bat), suggesting septicemia. Abundant gram-positive cocci were seen within the lymph node lesions; these cocci labeled by an immunohistochemical assay that detects group C streptococci. *Streptococcus spp*-specific PCR assays performed on extracts from the formalin-fixed, paraffin-embedded lymph nodes from two bats, and from the brain abscess of the third bat were positive. Sequencing of the PCR product revealed *Streptococcus equi*, with no further speciation possible, for one bat, and *Streptococcus spp*, with no further species identification was possible, in the other two bats. *Streptococcus equi*-associated suppurative cervical lymphadenitis is reminiscent of similar conditions in horses ("strangles") and guinea pigs ("lumps"), and to the authors' knowledge, has not been previously reported in bats.

N-32: MOLECULAR CARCINOGENESIS IN EQUINE PENILE PAPILLOMAS AND SQUAMOUS CELL CARCINOMAS: RELATIONSHIP OF COX-2, E-CADHERIN, PTEN AND 14-3-3SIGMA WITH GRADE AND TUMOR PROGRESSION

Alejandro Suárez-Bonnet, Claire Elizabeth Willis, Kenneth C. Smith, Timothy S. Mair, Simon Lawrence Priestnall

Background: Squamous cell carcinoma (SCC) is the commonest equine penile neoplasm however, information on molecular carcinogenesis and the expression of tumor suppressors and oncoproteins is limited.

Objective: To describe, for the first time, the expression of E-cadherin, PTEN and 14-3-3σ in equine PP and SCC, and their possible association with histological grade and COX-2 expression.

Methods: Penile papilloma or SCC cases were retrieved from the pathology archives of the RVC and University of Bristol. Mitotic index (MI), inflammation, presence of vascular invasion and histologic grade were evaluated. Immunolabelling of serial sections for COX-2, 14-3-3σ, E-cadherin and PTEN was performed, and associations between protein expression and histopathological features were analyzed.

Results: 45 cases, 11 PP and 34 SCC (14 well- and 20 poorly-differentiated), were included. 85% cases were COX-2-positive, with no significant difference between PP and SCC. E-cadherin was down-regulated and aberrant nuclear expression of 14-3-3σ was detected in 64% and 79% SCC respectively. Cytoplasmic translocation of E-cadherin was strongly correlated with nuclear expression of 14-3-3σ ($p = 0.01$). E-cadherin expression correlated with grade ($p = 0.01$), and MI was positively correlated with COX-2 expression ($p = 0.03$) and grade ($p = 0.03$). Cases of SCC exhibited loss of PTEN expression in comparison with normal skin and PP.

Conclusions: High MI, loss of E-cadherin and PTEN and 14-3-3σ nuclear translocation are associated with increased malignancy in equine penile SCC. These markers have potential prognostic and therapeutic application. In contrast, COX-2 expression in equine penile neoplasia is not statistically associated with malignancy.

N-33: METALLOTHIONEIN EXPRESSION IS RELATED TO KI-67 IMMUNOREACTIVITY WITHIN BILE DUCT EPITHELIUM AND PARENCHYMAL INFLAMMATORY CELLS IN EQUINE LIVER DISEASE

Jolanda N. C. Verhoef, Andrew L. Allen, John C. S. Harding, Ahmad N. Al-Dissi

Background: Chronic liver disease is an important cause of illness in horses and may result in mortality in up to 25% of affected animals. Metallothionein (MT) is a highly conserved intracellular protein with a high binding affinity for divalent cations. It has been shown to play a significant role in inflammation and cellular regeneration.

Objective: To evaluate the role of MT in horses affected by chronic liver disease.

Methods. The relationship between hepatocyte MT expression, assessed by immunohistochemistry (IHC), and the degree of inflammation, fibrosis and bile duct proliferation was evaluated in 77 selected cases. Additionally, proliferation of hepatocytes, bile duct epithelium, and inflammatory cells was determined using IHC for Ki-67, a protein expressed during all active stages of the cell cycle. Inflammation and fibrosis were given scores from 0 to 3 depending on severity, and bile duct proliferation was assessed by counting bile duct profiles.

Results: Increased MT expression was observed in 73 of 77 (94.8%) cases. Ki-67 expression was seen in resident Kupffer cells (42/77 cases, 54.6%), lymphocytes (39/77 cases, 50.7%), bile duct epithelium (10/77 cases, 13.0%), and hepatocytes (8/77 cases, 10.4%). Median MT expression was higher in cases containing lymphocytic infiltrates compared to cases with no lymphocytic infiltrate ($P < 0.05$). Strong Ki-67 expression was found in all lymphocytic inflammatory foci. MT expression was also significantly associated with Ki-67 staining in bile duct epithelium and Kupffer cells.

Conclusion: These results suggest a putative role for MT during liver inflammation and proliferation of bile duct epithelia in horses.

N-34: CUTANEOUS MAST CELL TUMOR IN A 3 MONTH OLD PUPPY WITH PROLONG FOLLOW

Charalampos Attipa, Hannah Wong, Christos Yiapanis, Annelize Ide, Matti Kiupel, Ana Lara

A three-month-old, intact male French bulldog was presented with a cutaneous mass caudal to the right ear. Histopathological examination following surgical excision revealed an intermediate grade (Patnaik system) or low grade (Kiupel system) mast cell tumor (MCT) with 3 mitoses in 10 high power fields. The MCT was prognosticated by deteriming c-KIT mutations in exons 8 and 11, the KIT expression pattern and cell proliferation based on the Ki67 index and the number of AgNORs.

PCR did not detect mutations in c-KIT in exons 8 and 11. Immunohistochemistry (IHC) for KIT showed stippled cytoplasmic labeling and loss of perimembrane labeling in more than 10% of neoplastic cells, consistent with a KIT pattern 2There was strong nuclear labelling for Ki67 in an average of 19.6 cells per microscopic grid, below the previously published threshold value of 23 that indicates aggressive biological behavior. There was an average of 3.2 AgNORs per nucleus in neoplastic cells. The combined AgNORxKi67 score was 62.72, above the threshold value of 54 that indicates aggressive biological behavior.

Physical examinations were performed every six months and the dog had no MCT recurrence or metastasis 5 years post-surgery.

While the combined AgNORxKi67 score indicated a more aggressive biological behavior, neither histological grade, mitotic count or mutation status supported the likelihood of systemic disease. This case documents that complete surgical removal of low grade MCTs without mutations in exon 11 of c-KIT results in complete cure.

N-35: IMMUNOHISTOCHEMICAL CHARACTERIZATION OF THE INFLAMMATORY CELL POPULATION ASSOCIATED WITH FELINE GLIOMA

Dan R. Rissi, Brian F. Porter, Christen E. Boudreau, Paula M. Krimer, Andrew D. Miller

Background: The relationship between inflammation and tumor behavior and morphology has been characterized in human glioma, but this information is scarce in veterinary medicine.

Objective: The objective of this study is to characterize the inflammatory response in 18 feline gliomas and its association with tumor morphology and type.

Methods: Cases were retrospectively searched from 3 institutions. Tissue sections were submitted to immunohistochemistry (IHC) for CD3, CD20, Iba1, MAC387, and factor VIII–related antigen. The number of stained cells for each antibody was counted in ten 400 x fields and a cumulative value for each antibody was generated. Spearman's rank correlations and Mann-Whitney tests with Bonferroni's correction were evaluated for associations between IHC results and to compare IHC values in astrocytomas and oligodendrogliomas.

Results: Intratumoral and peritumoral CD3+ T-lymphocytes were observed in all cases. Perivascular CD20+ B-lymphocytes were detected in 12 cases and occurred within tumors, around tumors, and near areas of leptomeningeal spread. MAC387 immunoreactivity highlighted intravascular monocytes in 9 cases. Intratumoral and peritumoral Iba1 immunoreactivity was observed in all cases, with increased overall intensity around areas of necrosis and leptomeningeal spread. Intratumoral and peritumoral factor VIII-related antigen immunoreactivity was also detected in all cases and was concentrated in areas of microvascular proliferation and necrotic foci. No significant associations were found between IHC values. Average factor-VIII reactivity was higher in astrocytomas than oligodendrogliomas ($p=0.006$).

Conclusions: Although inflammation was detected within feline gliomas, no significant associations were found between the type of inflammation and tumor morphology and diagnosis.

N-36: HISTOPATHOLOGY AND IMMUNOHISTOCHEMICAL EVALUATION OF MYOGLOBIN, HSP70 AND S100 IN KIDNEY OF LIVE-STRANDED ODONTOCETES OF BRAZIL

Angélica María Sánchez-Sarmiento, Rodrigo Albergaria Réssio, Natalia A. C. Fernandes, Juliana M. Guerra, Paloma Martins Nascimento, Vitor Luz Carvalho, Carlos Sacristán, Kátia R. Groch, Samira Costa-Silva, Omar Gonzales-Viera, Ana C.O. Meirelles, José L. Catão-Dias

Background: Live-stranded cetaceans are highly susceptible to the capture myopathy, characterized by muscular damage and acute renal failure associated to myoglobin toxicity.

Objective: To analyze renal lesions related to capture myopathy by histopathology and injury/stress related immunomarkers (IHC): Myoglobin (MYO), HSP70 and S100, in odontocete cetaceans.

Methods: Kidney samples from 30 live-stranded cetaceans that subsequently died were evaluated by means of histopathology and IHC using polyclonal antibodies: HSP70 (1:200), MYO (1:45.000), S100 (1:50.000) + avidin biotin complex (HSP70) or horseradish peroxidase polymer (MYO, S100). IHC expression of MYO {positivity, score (% labelled area*Intensity) and pattern (absent, cellular, casts, mixed); HSP70 positivity; S100 positivity; acute tubular necrosis-ATN and protein casts; medical treatment (yes/not), survival time (T1=≤4h; T2=4-36h; T3=≥36h), family, sex and age class were annotated. Differences among parameters were assessed (Chi-squared & Fisher's test=$p \leq 0.05$).

Results: Overall, 56.7% (17/30) specimens presented ATN and 63% (19/30) presented protein casts. All immunomarkers were overexpressed: MYO (73.3%; 22/30), HSP70 (82.1%; 24/29) and S100 (83.3%; 25/30), with some significant differences. MYO expression was higher in adults, and in specimens treated (93.3%; 14/15) *vs.* not treated (50%; 7/14). Treated also had higher MYO score and differences on MYO pattern. MYO expressed more frequently in T2 (100%; 12/12) *vs.* T3 (80%; 4/5) and T1 (41.7%; 5/12) with higher MYO score and differences on MYO pattern.

Conclusions: MYO, HSP70 and S100 are useful kidney injury/stress related markers for live-stranded odontocetes and may differ according age class, time between stranding and death and if the individual was subjected to treatment.

N-37: CLINICAL AND HISTOPATHOLOGIC CHARACTERISTICS OF NON-OCULAR MELANOCYTIC NEOPLASMS IN CATS WITH LONG-TERM OUTCOME

Rachel Pittaway, Melanie Dobromylskyj, Kerstin Erles, Simon L. Priestnall

Background: Non-ocular melanocytic neoplasms are considered rare in cats, however are frequently reported in the literature and exhibit a wide range of behaviors, from benign to highly malignant. Unlike in dogs, there are no defined prognostic features for these tumors in the cat making prognostication challenging.

Objective: To describe signalment and histopathologic characteristics of feline non-ocular melanocytic neoplasms, and to evaluate these features alongside outcome data to identify potential prognostic features.

Methods: Signalment and location is described for 328 feline non-ocular melanocytic neoplasms from two UK diagnostic pathology laboratories (2006-2017). Where available, histologic characteristics including histologic pattern, mitotic index, degree of pigmentation, cellular pleomorphism and evidence of junctional activity were evaluated (n=142). Preliminary outcome data was obtained from primary practitioners (n=45).

Results: The majority were domestic shorthair cats with no sex predilection. Median age was 11 years. The head, most notably the pinna and oral cavity, was the site most commonly affected. The most frequent histologic pattern was spindloid (59%). Median survival was 150d (range 8-1720d). Forty-seven percent of cats were alive at 6 months with 27% surviving at least one year. Mitotic count and degree of pigmentation were significantly associated with survival ($p<0.01$).

Conclusions: Melanocytic neoplasia should be considered an important differential diagnosis for cutaneous tumors in cats, particularly those arising on the head, and more specifically affecting the pinna or oral cavity. The clinical behavior and histologic appearance of feline non-ocular melanocytic neoplasms is highly variable and mitotic count and degree of pigmentation may be useful prognostic features.

N-38: PHYSEAL VARIABILITY AND SUBCLINICAL LESIONS IN TWO COMMERCIAL STRAINS OF YOUNG FEMALE TURKEYS

Laura R. Chen, Harold John Barnes, Oscar J. Fletcher, Luke B. Borst

Skeletal disease is an important contributor to morbidity, mortality, and economic losses in commercial turkeys. However, detailed descriptions with histomorphometry of the physis and metaphysis in modern, commercial turkeys is unavailable. Therefore the thicknesses of the cartilage zones in the physis and primary spongiosa of clinically normal birds were measured and compared between 2 modern commercial turkey strains. The free thoracic vertebrae (FTV), proximal femur (PF), and proximal tibiotarsus (PTT) from clinically healthy turkey hens from 2 commercial strains (A, B) were collected at 1, 3, 5, and 7 weeks of age. At each time point, 6 birds per strain were evaluated histologically. In the PF and PTT, length of the zone of hypertrophy and primary spongiosa varied among individuals without consistent significant differences between the two strains. However, in both PF and PTT, the zone of proliferation in strain A was significantly thicker than strain B at weeks 1 and 7 ($p<0.05$) as well as in week 5 in the PTT. In the FTV, measurements were less variable and no significant differences were observed between the 2 strains. During the course of evaluation, striking, but apparently subclinicallesions were observed. These included robust osteoclast aggregates subjacent to the periosteum at the metaphysis of the PF and PTT and osteochondrosis-spectrum lesions in the FTV. The physeal morphometrics and subclinical lesions described herein are useful as references for diagnostic and experimental evaluation of turkey skeletal disease. Further research is warranted into the clinical significance of the described subclinical lesions.

N-39: DEVELOPMENTAL LUNG DISEASE LEADING TO EARLY DEATH IN NORWICH TERRIERS

Linda Huang, Csaba Galambos, Steven Abman, Kurt Williams

Background: Developmental lung disease (DLD) has recently been identified in puppies, though to date, there has been no recognized association of DLD with specific breeds of dogs. Although unexplained neonatal mortality is common in Norwich Terriers (NT), the underlying cause is unknown. Objective: We investigated whether DLD is present in NT neonates that died unexpectedly.

Methods: The lungs of 9 deceased NT puppies, <1 month of age, were collected; an 8 day old mixed-breed puppy with no lung pathology served as control. As part of a complete necropsy, the lungs were inflated with formalin through tracheal cannulation. In addition to routine staining with H+E, CD31 immunohistochemistry for endothelial cells was performed to highlight the pulmonary vasculature.

Results: Mild thoracic and abdominal effusion and diffusely mottled lungs were grossly evident in all of the NT puppies. Microscopically, similar pathology was present within all lung lobes of each of the NT puppies. Findings included reduced but thickened alveolar septae and decreased acinar development. Persistent double capillary layers along the alveolar septa were seen in some areas, while others showed decreased capillary numbers. In some lung regions disorganized and congested capillaries were present, often associated with dilated congested small veins. The media of pulmonary arteries and arterioles had markedly thickened smooth muscle and many large veins had increased adventitial collagen. Rare bronchopulmonary "shunt" vessels were identified.

Conclusions: Diffuse developmental lung disease causes early death in Norwich Terrier puppies, suggesting an underlying genetic cause for DLD in the breed.

N-40: SARCOCYSTIS FALCATULA INFECTION IN TWO SOUTHERN ROCKHOPPER PENGUINS (EUDYPTES CHRYSOCOME)
Shannon G.M. Kirejczyk, Michael J. Yabsley, Julie Ter Beest, Zoli Gyimesi, Mary B. Ard, Rita McManamon

Background: Two captive Southern rockhopper penguins (*Eudyptes chrysocome*) exhibited anorexia, lethargy, dyspnea, and died within a 5-day period in April 2017. Birds were temporarily housed for 6 weeks, in an outdoor, completely meshed enclosure, with supplemental cooling when needed. Grossly, lungs were dark red, wet, and (for one bird) sank in formalin. Histopathology revealed necrotizing interstitial pneumonia, with myriad intravascular and interstitial protozoal schizonts and zoites. Zoites were also in skeletal muscle, heart and spleen. Current immunohistochemical reagents often exhibit partial cross-reactivity between closely-related apicomplexan parasites. Disseminated toxoplasmosis has been reported in two other penguin species, but not in rockhoppers. There are no peer-reviewed publications of *Sarcocystis falcatula* infection in penguins.

Objective: Our objective was definitive parasite identification, using immunohistochemical, ultrastructural and molecular tests.

Methods: Multiple tissues from both penguins were evaluated via light microscopy; immunohistochemistry and transmission electron microscopy were performed on lung; polymerase chain reaction (PCR) methodology was used on lung and spleen. Primers targeted partial cytochrome b gene (Cytb), the internal transcribed spacer (ITS) 1 region, partial RNA polymerase beta subunit gene (RpoB), and the partial surface antigen 2 (SAG2) gene. Amplicons were amplified and bidirectionally sequenced.Results: Protozoal life stages were strongly immunopositive for *Sarcocystis neurona* antibody. Cytb, RpoB and SAG2 gene sequences were 100% similar to *S. falcatula* samples. ITS-1 sequences were 98-99% similar to *S. falcatula*.Ultrastructure confirmed schizonts and zoites with morphological features of *Sarcocystis falcatula*.

Conclusion: *S. falcatula* infection was confirmed by PCR and sequence analysis in two Southern Rockhopper penguins.

N-41: COLLAGEN ARCHITECTURE OF THE PROSTATE IN NEUTERED AND INTACT CANINES
Hannah Ruetten, Sara Colopy, Ruth Sullivan, Kyle Wegner, Anoop Chandrashekar, Chad Vezina

Background: Benign prostatic hyperplasia (BPH) causes clinical complications in humans and dogs. Though BPH is well studied in humans, less is known about canine BPH, and underlying mechanisms are not fully known in either species. Fibrosis is emerging as a key contributor to voiding impairments experienced by human males with BPH. If prostatic fibrosis plays a comparable role in the dog, it will provide rationale for future clinical drug trials in canine targeting fibrotic pathways. Despite today's neutering trend, there exists a population of canine athletes and breeding stock that are intact.

These dogs inevitably develop BPH. Therefore, the objective of this study is to determine collagen architecture of intact and neutered canine prostate.

Methods: Whole canine prostates, from euthanized dogs of various age and neuter status, were obtained. Prostates were fixed in formalin, embedded in paraffin, and sectioned. Sections were stained with picrosirius red, a general collagen marker, and collagen fiber thickness, length, and density were analyzed. Hematoxylin and eosin-stained sections were also analyzed.

Results: Collagen fiber thickness, length, and density varied between neutered and intact animals. It also varied between region of the prostate: capsule, prostate ducts, and urethra.

Conclusions: The collagen architecture of the canine prostate varies between neutered and intact animals. Pending functional voiding analysis collagen may also contribute to symptoms experienced by canines with BPH.

Student Funding provided by: UW- Madison School of Veterinary Medicine

Research Funding provided by: NIH U54DK104310

N-42: SMALL RNA-SEQ EVALUATION OF MICRORNAS DURING DISEASE PROGRESSION IN KIDNEY BIOPSIES FROM DOGS WITH X-LINKED HEREDITARY NEPHROPATHY

Candice P. Chu, Mary B. Nabity

Background: Dogs with X-Linked Hereditary Nephropathy (XLHN) have a glomerular basement membrane defect leading to juvenile-onset renal failure. These dogs provide a good model of rapidly progressive chronic kidney disease (CKD).

Objective: To identify differentially expressed (DE) microRNAs (miRNAs) in the kidney, top-ranked miRNA target genes, and enriched biological processes and pathways involved in CKD progression.

Methods: Small RNA sequencing (small RNA-seq) was performed on total RNA isolated from kidney biopsies at three time points (T1, T2, and T3) from dogs with XLHN (n=6) and age-matched unaffected male littermates as controls (n=4). Small RNA-seq reads were mapped to the canine genome, read counts were normalized, and the Benjamini-Hochberg procedure was performed to obtain the false discovery rate (FDR). Target prediction and functional analyses determined the enriched biological processes and pathways for DE miRNAs.

Results: Up to 23 DE miRNAs were identified comparing affected versus control dogs at different time points (FDR < 0.05, fold change > 2). Most DE miRNAs were discovered at T1 and decreased over time. MiR-802, miR-146b, miR-147, miR-150, miR-21, miR-142, miR-31, and miR-29a/c were consistently upregulated in affected dogs in multiple comparisons. Top-ranked target genes of these miRNAs are involved in intracellular signal transduction, cellular protein modification, and the integrin signaling pathway. Using paired comparisons, miR-486, a known inhibitor of muscle wasting in CKD, was down-regulated in affected dogs versus up-regulated in controls from T2 to T3.

Conclusions: This study identifies miRNAs that might regulate critical pathways contributing to CKD progression in dogs with XLHN.

N-43: EFFECT ON CUTANEOUS ABSORPTION OF THE GREEN TREE FROG BY BATRACHOCHYTRIUM DENDROBATIDIS INFECTION

Sho Kadekaru, Yuki Shibata, Toshihiro Tokiwa, Masakazu Suzuki, Yumi Une

Background: Chytridiomycosis is caused by *Batrachochytrium dendrobatidis* (Bd). The pathogenesis is reported that causes plasma osmolality reduction, electrolyte depletion and cardiac arrest, but still unclear. Frog ventral skin maintain the body water balance by water absorption under the dehydration.

Objective: To determine the pathogenesis of chytridiomycosis focusing on cutaneous water absorption via aquaporins (AQPs).

Methods: Ten green tree frogs (*Litoria caerulea*) were divided into 2 groups (5 frogs each). Bd-exposed group was developed mild chytridiomycosis by immersing in water containing, 5,000, 6,000 and 1×10^6 zoospores per 1 ml for 24 hours and kept for 6 days in fresh water. Control group was routinely kept in fresh water for 7 days. Water permeability of the thigh and dorsal skins was measured with an Ussing chamber, and these skins were histopathologically analysed. The blood were biochemically tested. Bd infection was examined by histopatological analysis and/or nested PCR.

Results: In Bd-exposed group, 3 frogs sloughed skin excessively and all the frogs histopathologically showed slight thickening of the stratum corneum and slight epidermal hyperplasia. In control group, water transport was observed in the thigh skin, but not in the dorsal skin. However, water transport across the thigh skin was disrupted in Bd-exposed group. Hematological analysis showed plasma osmolality, Na, Cl, Mg and Ca ion levels decreased in Bd-exposed group. Bd was detected in 2 exposed frogs.

Conclusions: These results suggest that Bd infection causes disruption of water absorption across the ventral skin via AQPs and the plasma electrolyte abnormality.

N-44: NEWLY RECOGNIZED EQUUS CABALLUS PAPILLOMAVIRUS-8 (EcPV-8) ISOLATED FROM MULTIPLE VIRAL PLAQUES AND INVASIVE SQUAMOUS CELL CARCINOMA LESIONS IN A 16 YEAR OLD PALOMINO MIXED BREED GELDING

Stacy L. Rine, Jeanine Peters-Kennedy, Richard Hackett, Christian Lange

Most papillomaviruses are species-specific pathogens that cause a spectrum of epithelio-proliferative disorders, ranging from papillomas, *in situ* squamous cell carcinomas (SCC) to invasive SCC. Eight *Equus caballus* papillomaviruses (EcPV) have been described in horses. Horses infected with these viruses exhibit four distinct clinical presentations; classical equine viral papillomatosis, aural plaques, genital papillomas and the recently recognized EcPV-8 associated generalized papillomatosis. With the exception of EcPV-2, which has been reported to evolve into squamous cell carcinoma of the genitalia, most equine papillomaviruses are associated with benign viral papillomas and plaques. Here we describe a 16 year old quarter horse gelding with dozens of cutaneous viral plaques and invasive SCCs. Initially, one 2.5cm verrucous, non-painful, non-ulcerated mass was excised from the inguinal region four years ago. It was diagnosed as SCC and a viral plaque via histopathology. Two years later, the horse presented with dozens of similar masses in the inguinal region, sheath and penis. Masses were well-demarcated, frond-like, flesh colored to black and 1x1x1cm to 3x2x2cm. Six masses were excised. Cisplatin beads were implanted into incision sites and cisplatin emulsion was injected into 8 of the smaller masses that were not removed. The following year, the horse presented with a similar clinical scenario and received the same treatment regime. The fourth year, with a similar clinical presentation, the horse returned for mass removal and chemotherapy. Electrochemotherapy was performed to maximize the absorption of cisplatin into the cells. PCR for viral DNA identified the presence of the recently sequenced EcPV-8.

N-45: NASITREMA SP. - A FREQUENT CULPRIT IN MELON-HEADED WHALE (PEPONOCEPHALA ELECTRA) STRANDINGS IN TRINIDAD

Ayanna C.N. Phillips, Rod B. Suepaul

Background: At least thirty species of cetaceans are known to inhabit the waters of the Wider, of which nineteen species have been documented in the waters of Trinidad and Tobago. There have been five reports of stranded melon-headed whales (*Peponocephala electra*) during the period March 2013 to November 2015.

Methodology: Necropsy was performed and the helminths were preserved in 70% alcohol for identification. Tissue samples were fixed in 10% buffered formalin for 48 hours, embedded in paraffin and stained with haematoxylin and eosin for histological examination.

Results: Clinical signs in live specimens included; respiratory distress, cardiovascular collapse, disequilibrium and periodic muscle tremors. Specimens 1 and 4 were in generally poor body condition. On necropsy, there were multiple areas with a small volume of green tinged exudate on the meninges and trematode eggs were observed on wet mounts of smears in all cases. Adult trematodes were found in the cranial vault of 1 case. Severe unilateral hydrocephalus of the lateral ventricle was found in two of the cases. In the brain of three specimens, there were multiple small foci of necrosis with moderate pyogranulomatous inflammation and multiple triangular to oval-shaped trematode (*Nasitrema sp.*) eggs.

Conclusions: Eggs and/or adult trematodes (*Nasitrema sp.*) were identified in the brain of all stranded melon-headed whales examined on the island of Trinidad during the period March 2013 to November 2015. Findings strongly suggest that *Nasitrema sp.*-associated encephalitis is a common cause of melon-headed whale strandings in the south eastern Caribbean.

N-46: MULTIPLE CASES OF BOVINE ABORTION ASSOCIATED WITH LOCOWEED INGESTION IN NON-CLINICAL DAMS FROM SOUTHEASTERN COLORADO

Michael J. Betley, Bryan L. Stegelmeier, Gene Niles, Chad Frank

Chronic ingestion of certain species from the genera *Oxytropis*, *Astragalus*, *Ipomoea* and *Swainsonia*, colloquially known as locoweeds, is well known to induce locoism in grazing cattle. Poisoning in adult cattle is characterized by sensory deficits, ataxia, behavior changes, loss of condition and failure to grow. The toxic principle swainsonine is rapidly absorbed and distributed through the bloodstream to multiple organs where it inhibits lysosomal alpha-mannosidase and Golgi mannosidase II leading to lysosomal dysfunction and disruption of glycoprotein processing. Excessive mannose accumulation causes cytoplasmic vacuolation and cellular dysfunction. Chronic consumption of locoweeds is most commonly associated with disease in adult animals and abortion in affected, neurologic dams. However, we report multiple cases of bovine abortion associated with chronic locoweed ingestion in non-clinical dams from a herd in Southeastern Colorado. Furthermore, in the aborted animals, neurons from the cerebral cortex, brainstem, cerebellum and renal tubular epithelial cells display the characteristic cytoplasmic vacuolation observed in adult cases of locoism. Our findings expand the understanding of locoweed-induced disease in cattle and indicate that fetuses may be more sensitive to maternal locoweed ingestion than previously suspected.

N-47: GENE EXPRESSION PROFILING IDENTIFIES CANINE CD4+ T-CELL LYMPHOMA AS A NATURALLY OCCURRING MODEL OF AN AGGRESSIVE SUBSET OF HUMAN PERIPHERAL T-CELL LYMPHOMA-NOT OTHERWISE SPECIFIED

Lauren Harris, Kelly Hughes, E.J. Ehrhart, Janna Yoshimoto, Robert Burnett, Anne Avery

Background: Peripheral T-cell lymphoma-not otherwise specified (PTCL-NOS) is a heterogeneous group of neoplasms with poor responsiveness to therapy and grave clinical outcomes in both canine and human patients. Recent gene expression studies of human PTCL have identified prognostically significant subsets of PTCL-NOS with distinct biomarkers. Because PTCL-NOS is relatively uncommon in human patients, an appropriate model is needed to facilitate investigation of the underlying pathobiology, identify novel therapeutic targets, and carry out targeted clinical trials.

Objective: To determine if canine PTCL serves as naturally occurring model for human PTCL-NOS.

Methods: RNA-sequencing was performed on lymph node aspirates from 6 dogs diagnosed with CD4+ PTCL and sorted CD4+ T-cells from control dogs. Differential gene expression and pathway enrichment analysis was performed and results were compared to similar studies on human PTCL-NOS.

Results: Human and canine PTCL-NOS upregulate pathways involved in proliferation, cell adhesion, extracellular matrix modification, and signal transduction. Canine cases overexpress GATA3 and CCR4; two markers which have been identified in human cases of PTCL-NOS and characterize a distinct subset associated with worse clinical outcomes. GATA3 expression is corroborated by immunohistochemistry and may serve as a histologic marker to identify this entity. Overexpression of PGDFR and upregulation of pathways regulating IL-10 production are also conserved across both human and canine cases and have been suggested as potential therapeutic targets.

Conclusions: Canine CD4+ PTCL is molecularly similar to an aggressive subset of human PTCL-NOS and serves as a useful model to further investigate mechanisms underlying oncogenesis and identify novel therapeutic targets.

N-48: PRE-CHEMOTHERAPY TREATMENT OF PRIMARY AND METASTATIC CANINE APPENDICULAR OSTEOSARCOMA CELLS WITH AN AUTOPHAGY INHIBITOR

Geoffrey A. Wood, Courtney R. Schott

Progression to metastatic disease is the most common outcome in dogs diagnosed with appendicular osteosarcoma (OSA), despite aggressive chemotherapy post-amputation. Autophagy can allow cells to survive harsh conditions, and neoplastic cells may enter autophagy to survive chemotherapy. To investigate this, canine OSA cell lines were pre-treated with an autophagy inhibitor (spautin-1, a beclin-1 inhibitor) followed by doxorubicin, using low and high doses of each drug (LD and HD). We hypothesized that autophagy inhibition would enhance chemosensitivity in cell viability assays and limit colony formation in clonogenic assays. Six canine cell lines were used, including primary, metastatic, and control lines. In OSA lines, LD-spautin-1 decreased cell viability by up to 11% compared to no treatment. LD-spautin-1 pre-chemotherapy reduced cell viability by up to 12% or 8% compared to LD- or HD-doxorubicin alone, respectively. Compared to no treatment, HD-spautin-1 alone decreased cell viability by 18-38%. HD-spautin-1 pre-chemotherapy reduced the cell viability by 10-30% or 6-20% compared to LD- or HD-doxorubicin alone, respectively. The effects on colony formation were more variable. In some OSA lines, LD-spautin-1 increased the surviving fraction of colonies compared to no treatment and as a pre-chemotherapy treatment compared to doxorubicin alone. Overall, trends for HD-spautin-1 were more similar to those seen in the cell viability assays. Autophagy inhibition with spautin-1 prior to chemotherapy had an effect that was additive rather than synergistic, but clinically it could serve to lower the dose of chemotherapy used, and thus avoid toxicity, and/or increase the overall killing of metastatic OSA cells post-amputation.

N-49: MERKEL CELLS IN THE FOLLICLE SINUS COMPLEXES OF MUZZLE SKINS ARE USEFUL FOR ANTE AND POSTMORTEM DIAGNOSIS OF RABID DOGS
Nozomi Shiwa, Kazunori Kimitsuki, Chikage Nakajima, Daria L. Manalo, Satoshi Inoue, Chun-Ho Park

Background: Rabies is a highly fatal zoonotic disease. It is occurring worldwide and more than 55,000 people die of rabies every year. The direct fluorescent antibody test (dFAT) using the fresh brain tissues is the most commonly used method for detection of the rabies virus antigen in dogs. However, this method is laborious, time-consuming and there is a high risk of exposure to the rabies virus. In addition, the sensitivity of dFAT decreases once brain specimens are decomposed under warm temperature in the field.

Objective: We evaluated the diagnostic utility of follicle-sinus complexes (FSCs) in the muzzle skins of the rabid dogs by immunohistochemistry (IHC) and FAT.

Methods: Brain and muzzle skins were obtained from 250 rabies-suspected dogs (72 Euthanasia, 161 Found dead, 17 Unknown). FAT and IHC analyses confirmed the presence of viral antigen in the brain tissue in 234 of 250 dogs.

Results: Viral antigen was also demonstrated in the MCs (Merkel cells) of the muzzle skin in 234 of 250 dogs (100% specificity). Virus antigen was clearly detected in a part of the outer root sheath at the level of the ring sinus of FSCs and most of these cells were positive for MCs markers (CK 20, CAM5.2), respectively. Electron microscopically, virus particles and virus-associated structures were observed in the cytoplasm of MCs.

Conclusions: Our results suggest that MCs in FSCs are target for virus replication and the FSCs are very useful as alternative source for ante- and post-mortem diagnosis of rabies, especially in rabies-endemic developing countries.

N-50: EXPRESSION OF E-CADHERIN AND COX-2 IN AVIAN SQUAMOUS CELL PAPILLOMAS AND CARCINOMAS
Alwyn Llewelyn Jones, Alejandro Suárez-Bonnet, Gustavo Adolfo Ramirez, Simon Lawrence Priestnall

Background: The incidence of skin neoplasia in birds is very low; most reports describe squamous cell carcinomas (SCCs) affecting domestic fowl and psittacines. A paucity of information exists on potential oncogenic proteins involved in the development and progression of avian SCCs.

Objective: To examine histologically 32 avian papillomas and SCCs and evaluate the diagnostic utility of COX-2 and E-cadherin immunohistochemistry by semi-quantitative scoring.

Methods: Histological examination included mitotic index, nuclear polymorphism, inflammation and the presence of surface ulceration. Neoplasms were classified as papilloma, carcinoma *in-situ* and SCC (poorly or well-differentiated).

Immunohistochemical labelling for COX-2 and E-cadherin was performed. COX-2 total score (TS) was calculated as the product of the labelling intensity and distribution scores. E-cadherin labelling intensity and location (membranous or cytoplasmic) were evaluated.

Results: Tumors comprised masses from 30 psittacine and 2 non-psittacine birds; 10 males and 4 females (18 unspecified) with an age range from 3-35 years (21 unspecified). The majority of lesions were located around the uropygial gland. Diagnoses included 7 papillomas, 1 carcinoma *in-situ* and 24 SCCs; 12 well-differentiated and 12 poorly differentiated. SCCs had a higher mean COX-2 TS (5.3) compared to papillomas (1.1). 68% SCCs demonstrated reduced membranous E-cadherin labelling, compared with 14% papillomas, while 64% SCCs demonstrated cytoplasmic labelling versus 14% papillomas.

Conclusion: E-cadherin and COX-2 immunolabelling is useful in differentiating papillomas and SCCs in birds. Increased expression of COX-2 and reduced membranous and increased cytoplasmic expression of E-cadherin is shown, as in other species, to be associated with a more malignant phenotype.

N-51: NATURALLY OCCURRING OVINE HERPESVIRUS 2 (OVHV-2) INFECTION WITH MALIGNANT CATARRHAL FEVER (MCF)-LIKE DISEASE IN LAMBS (OVIS ARIES) WITH CONCURRENT WHITE MUSCLE DISEASE

Rahul B. Dange, Patricia Pesavento, Federico Giannitti, Santiago S. Diab

OvHV-2 is one of a growing cluster of ungulate gamma-herpesviruses that cause MCF. Sheep are adapted hosts for OvHV-2, and carry the virus asymptomatically. Non-adapted ungulate species (cattle, bison), are susceptible to MCF, so there is significant risk in mixing ungulate species at food animal operations or zoos. While it is generally assumed that OvHV-2 does not cause disease in sheep, very little is known about the cell and tissue targets of infection or pathogenesis in this species. Here, we describe the pathological findings in two unrelated lambs spontaneously infected with OvHV-2. Lamb A, an 8-month-old female, had a history of weakness and staggering when stressed. Autopsy revealed multifocal ulceration of the soft/hard palate and lip, and moderate lymphohistiocytic, plasmacytic and neutrophilic bronchointerstitial pneumonia with multifocal lymphocytic perivasculitis suggestive of a viral etiology. Lamb B, a 10-week-old, male Dorper lamb, died following a 3-day history of weakness, staggering and diarrhea. Autopsy findings included severe multifocally extensive ulceration of the soft palate, tongue, esophagus, rumen and small and large intestine. OvHV-2 was positive and abundant by quantitative PCR on splenic tissue from both animals. Distribution of virus was established by in situ hybridization using a probe designed to hybridize OvHV-2. Additionally, both lambs had skeletal and cardiac myofiber degeneration and necrosis, with fibroplasia and mineralization compatible with white muscle disease. Hepatic levels of selenium, vitamin E and copper were deficient in lamb A. We speculate that stressors or nutritional deficiencies may have predisposed these two lambs to develop MCF-like lesions.

N-52: RNA-SEQ TRANSCRIPTOME ANALYSIS OF FORMALIN-FIXED PARAFFIN EMBEDDED CANINE GLIOMA

Mason C. Jager, Jennifer K. Grenier, Erica A. Sloma, Andrew D. Miller

Gliomas are the second most commonly reported primary intracranial tumor in dogs and share some similarities with human gliomas including some histologic features and biological behavior. Because of the similarities between canine and human gliomas, dogs have been proposed as models and surrogates for studying and treating gliomas. However, compared to human glioma research, less is known about specific pathways and individual genes that are involved in the development and progression of canine gliomas. In addition, there is a paucity of suitable fresh material available for studies of differential expression (DE) of genes in cases of veterinary neoplasia compared to formalin-fixed, paraffin embedded tissue. We report here the use of formalin-fixed paraffin embedded (FFPE) specimens up to two years old to provide RNA suitable for transcriptome analysis using next-generation sequencing (NGS). RNA was extracted from FFPE of six canine gliomas that represented four low grade oligodendrogliomas, one high grade oligodendroglioma, and one low grade astrocytoma. RNA from the gliomas was compared to RNA extracted from normal brain tissue from the same animal and revealed 938 significantly differentially expressed genes. Multiple genes, including SOX6 and THY1, were validated using qPCR. Among the DE genes were oncogenes, tumor suppressors, transcription factors, and members of the WNT and hedgehog pathways. Our work demonstrates that RNA of sufficient quality can be extracted from FFPE samples to provide biologically relevant transcriptome analyses using RNA-seq. Furthermore, due to the abundance of FFPE tissue, this technique can provide transcriptome data on larger cohorts for increased statistical power.

N-54: CD31 REACTIVITY IN CANINE NORMAL TISSUES AND NEOPLASMS

José A. Ramos-Vara, Margaret A. Miller, Dee DuSold

CD31 (PECAM-1) is an adhesion molecule of leukocytes, platelets, and endothelial cells. Vascular tumors are typically positive for CD31. Detection of CD31 reactivity in canine renal cell carcinomas prompted evaluation of CD31 expression in 295 specimens including 131 normal tissues (adrenal gland, gallbladder, heart, intestine, kidney, liver, lung, lymph node, mammary gland, ovary, pancreas, parathyroid gland, pituitary gland, prostate, salivary gland, skin, spleen, stomach, testis, thymus, thyroid, tongue, tonsil, urinary bladder, and uterus) and 164 pathologic samples (hepatic nodular hyperplasia and regenerative nodules, adenomas, carcinoids, carcinomas, sarcomas, melanomas, lymphomas, plasmacytomas, cutaneous histiocytomas, gonadal tumors, thymomas, insulinomas, pheochromocytomas, and mesotheliomas). The CD31 mouse monoclonal antibody (clone JC/70A, Dako, Carpenteria, CA) was diluted 1/100 and incubated at room temperature for 60 min. Heat antigen retrieval (REVEAL, Biocare, Concord, CA) was used. Tissues were considered positive if >5% of nonendothelial cells were reactive. Besides endothelial cells (positive in all tissues examined), CD31 membranous immunoreactivity was detected in hepatocytes of 3/5 normal livers, all 6 hyperplastic nodules, and all 3 regenerative nodules. Hepatic sinusoidal endothelial cells were negative. In addition to endothelial tumors (hemangiomas and hemangiosarcomas), 1/4 apocrine carcinomas of the anal sac glands, 1/5 mammary carcinomas, 2/5 plasmacytomas, and 18/20 hepatocellular carcinomas had membranous labeling of variable intensity. The significance of CD31 labeling of neoplastic hepatocytes is unknown, but expression was typically stronger than in adjacent nonneoplastic hepatocytes. Human nonvascular tumors with CD31 expression include breast cancer, lymphoma, dendritic cell proliferations, and plasmacytomas, but not hepatocellular neoplasms.

N-55: BREED AND AGE-RELATED DIFFERENCES IN CELLULAR DETERMINANTS OF CANINE PARVOVIRUS-2 IN THE CANINE MYOCARDIUM

Laura McEndaffer, Jordan C. Ford, Alex Molesan, Kathleen M. Kelly

Canine parvovirus (CPV)-2 is well-known as a cause of myocarditis of very young puppies. Despite the assumption thatvaccination has eliminated parvoviral (PV) myocarditis in dogs, our research indicates that PV is an under-recognized cause of myocarditis and cardiac damage in young dogs. The mechanism of CPV-2 myocardial cell infection and injury has not been characterized. We hypothesized that differences in growth patterns and corresponding cardiac mitotic activity between different breed sizes (small, large, and giant breeds) may underlie cardiac susceptibility to PV. Retrospectively, we investigated the breed and age-related effects on cellular determinants of PV infection by evaluating expression of the viral receptor transferrin receptor type-1 (TfR-1)) and mitotic activity (Ki67) by immunohistochemistry in the hearts of prepartum late pregnancy (n=13), early (0d to 14 d; n=23), and late neonate canines (15d to 28d; n=18) identified from the NYS Animal Health Diagnostic Center pathology database. TfR-1 expression gradually increased in the canine myocardium from prepartum through the early and late neonatal period. Myocardial Tfr-1 expression was correlated to the post-partum age ($p=0.008$, $R= 0.74$ Spearman rank) in small breed dogs (n=12) but not in large breed dogs. Myocardial Ki67 signal was most abundant in the early neonatal period and gradually decreased into the late neonatal period compared to the prepartum; however, there were not significant differences in myocardial Ki67 expression between small and large breed dogs. These factors may contribute to thesusceptibilityof the canine myocardium to CPV-2 but other factors likely play a role.

N-56: NATURAL DISEASE AND ITS ASSOCIATION WITH ZOONOTIC PATHOGEN INFECTION IN WILD NORWAY RATS (RATTUS NORVEGICUS)

Jamie L. Rothenburger, Chelsea G. Himsworth, Nicole M. Nemeth, David L. Pearl, Piper M. Treuting, Claire M. Jardine

Background: Norway rats (*Rattus norvegicus*) are successful invasive species that inhabit cities worldwide. Despite their role in agriculture damage, transmission of zoonotic pathogens, and as urban pests, remarkably little is known about their diseases.

Objectives: 1) to establish the spectrum and characteristics of natural pathology in wild rats; 2) to examine associations between disease in rats and zoonotic pathogen carriage.

Methods: We described and summarized the macroscopic and microscopic lesions in 672 rats trapped in Vancouver, Canada.

Results: Grossly-evident lesions were present in 12% of rats and were significantly associated with *Leptospira* sp. but not *Bartonella* sp. infection. Major histological lesions in a subset (n=430) included interstitial nephritis (30%), *Trichosomoides crassicauda* (nematode) in the urinary bladder (31%), cardiomyopathy (32%), *Capillaria hepatic* infection (36%), esophageal and stomach *Eucoleus sp.* (nematode) infection associated with proliferative gastritis (60%) and respiratory tract inflammation (67%). Many rats had multiple lesions in a variety of systems. Of the rats with respiratory inflammation, 35% had one or more of the other five major lesions; six rats had all six lesions. There were frequent co-infections with bacterial and parasitic pathogens, including those that are potentially zoonotic. For instance, 11 rats were co-infected with *Leptospira* sp. and *Bartonella* sp.

Conclusions: Given the severity and frequency of lesions and infections, natural disease may be an important factor contributing to urban rat mortality. The results of this study enhance our understanding of spontaneous disease in laboratory rats, wild rat ecology and how co-infections and co-morbidities influence zoonotic pathogen ecology in rats.

N-57: PATHOLOGY OF A NATURALLY OCCURRING MODEL OF NEURONAL CEROID LIPOFUSCINOSIS (CLN7) IN JAPANESE MACAQUES (MACACA FUSCATA)

Anne D. Lewis, Rebecca M. Ducore, Robert Zweig, Betsy Ferguson, Samuel Peterson, Martha Neuringer, Laurie M. Renner, Trevor J. McGill, Jodi L. McBride, Jackie Domire, Lois M.A. Colgin

Neuronal ceroid lipofuscinosis (NCL), a neurodegenerative lysosomal storage disorder, has been identified in a colony of Japanese macaques. Affected animals developed clinical signs of ataxia, motor incoordination, intention tremor, and hypermetria between four and five years of age. Progression of clinical signs prompted euthanasia by six years of age. Pedigree analysis demonstrated an autosomal recessive pattern of inheritance. Whole exon sequencing identified a single base deletion resulting in frame shift mutation in the CLN7/MFSD8 gene; all affected individuals were homozygous for the frame shift mutation. Tissues from five affected animals were evaluated. Marked cerebellar and cerebral atrophy was evident on post mortem evaluation. The average brain weight of affected animals was 69.1 g $\pm$ 5.4 g compared to age-matched controls 97.8 g $\pm$ 4.3 g. In the cerebellum, there was widespread loss of cells in the Purkinje and granular cell layers with collapse of cerebellar microarchitecture. There was microarchitectural disruption and atrophy of the cerebral cortex and diffuse retinal atrophy. Astrogliosis and microgliosis were evident throughout the brain. Abundant pigmented, autofluorescent granular material distended the cytoplasm of neurons in the brain, spinal cord, ganglia, and retina and was present within the sarcoplasm of cardiomyocytes. Storage material was PAS positive, diastase resistant; Sudan black and luxol fast blue positive. Transmission electron microscopy demonstrated cytoplasmic granular osmophilic deposits (GROD) in neurons. The Japanese macaque model of NCL recapitulates key features of the human disease and will serve as a valuable resource for developing biomarkers of disease progression and evaluating promising therapeutic strategies.

N-58: PATHOLOGY OF NATURAL OLEANDER (NERIUM OLEANDER) POISONING IN LLAMAS (LAMA GLAMA) AND ALPACAS (VICUGNA PACOS)

Sebastian E. Carrasco, F. Charles Mohr, Francisco R. Carvallo, Birgit Puschner, Federico Giannitti, Santiago S. Diab

Oleander is an evergreen shrub commonly found in California. Toxic effects are a result from exposure to cardiac glycosides and have been reported in most domestic animals, including South American camelids. To date, there are limited data characterizing the pathology of oleander poisoning in llamas and alpacas. We reviewed gross and microscopic pathology records of 16 llamas and 9 alpacas admitted to CAHFS between 2003 and 2015 with confirmed diagnosis based on detection of oleandrin in gastrointestinal content using LC-MS/MS. Grossly, of the 16 llamas, 11 (69%) had pulmonary edema, 9 (56%) epi/endocardial hemorrhages, 6 (38%) hydropericardium, 6 (38%) small and/or large intestinal congestion and/or hemorrhagic intestinal contents, 4 (25%) hydrothorax, 2 (13%) ascites, and 2 (13%) renal hemorrhages. Of the 9 alpacas, 3 (33%) had pulmonary edema, 3 (33%) ascites, 2 (22%) subepi/subendocardial hemorrhages, 2 (22%) hydrothorax, and 1 (11%) hydropericardium, small intestinal congestion and hemorrhagic intestinal contents, or renal hemorrhages. Major microscopic lesions included swollen or fragmented, hypereosinophilic cardiomyocytes with subtle vacuolation and mineralization and/or loss of cross-striations in 15 llamas (94%) and 8 alpacas (89%), accompanied by neutrophilic and/or histiocytic infiltration in the myocardium of 12 llamas (75%) and 6 alpacas (67%). Congestion and multifocal hemorrhages in the small and large intestines were noted in 11 llamas (73%) and 4 alpacas (44%). Preliminary renal changes consisted of tubular degeneration/necrosis in 4 llamas (44%) and 4 alpacas (50%). In summary, common lesions in llamas and alpacas naturally intoxicated with oleander are myocardial degeneration/necrosis, pulmonary edema, and intestinal congestion/hemorrhage.

N-59: EQUINE CABALLUS PAPILLOMAVIRUS TYPE 2 (ECPV-2)- NOT JUST ANOGENITAL SQUAMOUS CELL CARCINOMAS

Bruce Wobeser, Sarah Greenwood

Equine caballus papillomavirus-type 2 (EcPV-2) has recently been identified as a potential cause of anogenital squamous cell carcinoma (SCC) in horses. Attempts to identify the presence of papillomaviral DNA within primary SCCs from other anatomical sites has been largely unrewarding. The purpose of this study is to determine the prevalence of papillomaviral DNA, within a variety of equine squamous cell carcinomas from a variety of anatomical locations. PV DNA was identified using conventional PCR, targeting two separate sites within the L1, (major capsid protein)gene. Amplicons were sequenced and identification determined by BLAST analysis. Anatomical locations included: penis, sheath, vulva, perianal, periocular (eyelids and nictitating membrane), ocular, trunk, limbs and other. The majority of PVs identified shared between 95-100 % identity to EcPV-2, although HPV types 4 and 150 were also identified in a small number of tumors. In addition to the anticipated anogenital tumors, EcPV-2 DNA was detected in ocular and periocular tumors, as well as those of other anatomical sites, with similar frequency.To our knowledge, this is the largest study assessing the presence of papillomavirus within equine squamous cell carcinomas from a variety of anatomical sites and is the first study to identify EcPV-2 within primary periocular and ocular equine SCC. Similarly, this is the first report of Human papillomaviruses (HPV) types 4 and 150 being identified within equine SCCs.

N-60: NEW INSIGHTS ON BRUCELLA IN CETACEANS OF BRAZIL: SEROLOGICAL, MOLECULAR, PATHOLOGICAL AND IMMUNOHISTOCHEMICAL INVESTIGATION

Angélica M. Sánchez-Sarmiento, Vitor L. Carvalho, Josué Díaz-Delgado, Rodrigo A. Réssio, Natalia A. C. Fernandes, Juliana M. Guerra, Carlos Sacristán, Kátia R. Groch, Natalia Silvestre-Perez, Eduardo Ferreira-Machado, Samira Costa-Silva, Pedro Navas, Ana C.O. Meirelles, Waleska Gravena, Cintia Favero, Juliana Marigo, Carolina P. Bertozzi, Adriana C. Colosio, Milton C.C. Marcondes, Marta Cremer, Nairléia dos Santos Silva, Jose Soares Ferreira, Eva Sierra, Antonio Fernández, José L. Catão-Dias

Background: Marine brucellosis is increasingly recognized in cetaceans worldwide. We recently reported the first case of cetacean brucellosis in the Southwest Atlantic Ocean, involving a Clymene dolphin (*Stenella clymene*).

Objective: To investigate the occurrence of *Brucella* in cetaceans of Brazil and to characterize *Brucella*-associated lesions in exposed or infected individuals.

Methods: We tested sera of 63 cetaceans (37 free-ranging; 26 dead-stranded or live-stranded that subsequently died, comprising 11 species) by Rose Bengal test [RBT], competitive-enzyme-linked immunosorbent assay [C-ELISA] and 2-mercaptoethanol [2-ME], this last in RBT- or C-ELISA-positive. Conventional and quantitative polymerase chain reaction [PCR/qPCR] were conducted on selected frozen tissues of 118 autopsied animals, some of them tested by serology. In those positive for at least one test, postmortem gross and histopathological data were analyzed, and immunohistochemistry (polyclonal anti-*B. abortus* antibody) was conducted on selected tissues.

Results: Thirteen animals (eight species) were positive in at least one test: 4.8% (3/63) to RBT; 15.9% (10/63) to C-ELISA; 10% (1/10) to 2-ME; and 3.4% (4/118) to PCR/qPCR. A variety of *Brucella*-type lesions (e.g., skeletal abscess, meningoencephalomyelitis, pneumonia, nephritis, multicentric lymphadenopathy) was observed in some of them. Three of 13 (23.1%) had cytoplasmic immunopositivity in phagocytic cells or extracellular in inflammatory foci.

Conclusions: We found evidences of exposure and infection by *Brucella* without *Brucella*-associated lesions, as well as cases suggestive of acute/chronic brucellosis.

This study represents the first large-scale survey of *Brucella* in cetaceans of Brazil, widening the spectrum of susceptible hosts and the geographic distribution range of this agent with zoonotic potential.

N-61: SEBACEOUS VIRAL PLAQUES ASSOCIATED WITH FELIS CATUS PAPILLOMAVIRUS TYPE 5 IN A CAT

John S. Munday, Neroli A. Thomson, Adrienne French, Keren E. Dittmer, Simon F. Hills, Rebecca E. Laurie

Feline viral plaques are pre-neoplastic lesions that are generally caused by *Felis catus* papillomavirus (FcaPV) type -2. A 15-year old cat developed multiple 1cm diameter crusting plaques on the face. Histology revealed a well-demarcated focus of mild epidermal hyperplasia that was typical for a feline viral plaque. However, markedly hyperplastic sebaceous glands were also present. Cytoplasmic vacuoles that contained grey-blue fibrillar material, consistent with papillomavirus-induced cytopathology, were prominent within the lesion. Due to the unusual histologically features the presence of papillomavirus DNA was investigated using multiple different consensus PCR primers. Only one sequence was amplified which was determined to be from a novel papillomavirus type. Using DNA extracted from a swab of a plaque, specific 'outward facing' primers amplified a 7600bp length of circular papillomaviral DNA. As this is the fifth papillomavirus from domestic cats, it was designated FcaPV-5. FcaPV-5 had the highest *ORF L1* similarity with FcaPV-3 and -4 and, due to the shared host species of all three papillomaviruses, and the similar lesions associated with FcaPV-3 and -5, it is proposed all three papillomaviruses are classified within a new genus. While papillomaviruses can asymptomatically infect skin, the presence of prominent papillomaviral changes within the lesions support a papillomaviral etiology. Additionally, marked p16 immunostaining, previously shown to be predictive of a papillomaviral etiology, was visible throughout the hyperplastic epidermis and sebaceous structures of the plaque. A subset of feline viral plaques exhibit prominent sebaceous differentiation and may be more likely to have a non-FcaPV-2 etiology.

N-62: HISTOLOGICAL EVALUATION OF OVERGROWN MOLAR TEETH IN PRAIRIE VOLES (MICROTUS OCHROGASTER)

Gayathriy Balamayooran, Caralyn Schwartz Labriola, David Caudell, Tyler Aycock, Nancy Donehue Kock

The dentition of prairie voles (PV) is comprised of one incisor and three molars in all four quadrants, all of which grow continually. The continual growth of teeth in rodents has been attributed to retention of the stellate reticulum from which stem cells are generated. Molar tooth overgrowth in PV has been associated with a spontaneous mutation in the regulatory signaling pathway in molar stem cells. We report seven cases of mandibular and maxillary molar overgrowth in laboratory PV, in which the apical surfaces of all had stellate reticulum surrounded by inner and outer enamel epithelium. All had overgrown first mandibular molars surrounded by thin cortical bone with marked remodeling. Two had oral abscesses with intralesional plant material and mandibular osteomyelitis with osteonecrosis. Five had apically overgrown maxillary molars which extended into the skull and compressed the nasal sinuses, although the brains were unaffected. The literature on PV oral and dental anatomy is sparse, and previous reports have not described the histological changes, which are documented in this report.

N-63: CLINICOPATHOLOGICAL EVALUATION OF DEFORMED WING VIRUS INFECTION IN HONEY BEES

Roman Koziy, Sarah Wood, Ivanna Kozii, Sophie Derveau, Claire Janse van Rensburg, Igor Moshynskyy, Jess Morris, Ihor Dvylyuk, Elemir Simko

Background: Varroa mite-mediated transmission of the deformed wing virus (DWV) is a major threat to honey bee health worldwide. The clinical symptoms of the infection include increased mortality, deformed wings, abdominal bloating and discoloration. However, the pathological basis of the DWV infection is not well understood and there is little known about the histopathological manifestation of this disease.

Objective: We investigated clinicopathological and histological aspects of DWV infection.

Methods: Combs with synchronized age of brood (larvae/pupae) were collected from five colonies, which had high prevalence of both mites and DWV infection. The emergence time was recorded and both affected and non-affected newly emerged bees were collected. Serial sections of the head were produced and analysed using Image-Pro Premier software. DWV infection was confirmed by RT-qPCR. Mann-Whitney U test and Student's t-test were used for data analysis.

Results: The emergence time for the bees with deformed wings was significantly longer ($p<0.05$) than for non-affected bees and some affected bees (~30%) died during the emergence process. All of the affected bees also demonstrated protruded proboscis during emergence. The total area of hypopharyngeal and mandibular glands in serial sections of the severely affected bees was 3.2 and 3.5 times lower than in clinically non-affected bees (both $p<0.001$).

Conclusion: Prolonged emergence time and protruded proboscis were observed as DWV infection symptoms. Furthermore, in DWV affected bees hypopharyngeal and mandibular glands were hypoplastic. Further histologic evaluation of the DWV infection will enhance our understanding of the pathogenesis of this important disease in honey bees.

N-64: PATHOLOGIC AND IMMUNOHISTOCHEMICAL CHARACTERIZATION OF PRIMARY NERVOUS SYSTEM LYMPHOMA IN CATS

Brittany J. McHale, Andrew D. Miller, Dan R. Rissi

Background: Lymphoma is the most common hematopoietic neoplasia of cats, but primary nervous system (NS) lymphoma is uncommon.

Objective: The objective of this study is to characterize the pathologic changes, distribution, and immunophenotype of 10 cases of primary NS lymphoma in cats.

Methods: Cases were retrospectively searched from two academic institutions. Pathology reports and archived histopathology glass slides were reviewed. Tissue sections were subjected to immunohistochemistry (IHC) for CD3, CD20, PAX5, and MUM1 for immunophenotyping.

Results: The mean age of affected cats was 9.2 years and no sex or breed predilection was observed. Neoplasms were characterized by well- or poorly-demarcated soft white masses that occurred in the brain (3 cases), spinal cord (3 cases), spinal nerves (1 case), or concomitantly in multiple areas within the NS (3 cases). Histologically, all neoplasms were composed of sheets of monomorphic round cells that infiltrated the epidural fat (2 cases), leptomeninges (2 cases), neuroparenchyma, (4 cases), epineurium, perineurium, and endoneurium (1 case), or a combination of these structures (1 case). CD3, CD20, and PAX5 IHC revealed that five neoplasms were B-cell lymphomas, three were T-cell-rich B-cell lymphomas, and two were T-cell lymphomas. No neoplasms were immunoreactive for MUM1.

Conclusions: In this study, 80% of the primary feline NS lymphomas were B-cell in origin and no relationship was detected between tumor type and overall distribution within the NS.

Clinical Pathology Focused Scientific Session I

November 5, 2017 | 1:30 PM – 4:30 PM
Session Chair: K. Zimmerman
Committee Members: E. Behling-Kelly, S. Clark, S. Connolly, M. Fry, N. Tripathi, C. Wagg

November 5, 2017
3:15 PM – 3:30 PM

COMPARISON OF METHODS FOR PREPARATION OF BIOFLUIDS IN DOGS FOR SMALL RNA SEQUENCING

Candice P. Chu, Mary B. Nabity

Background: The relative scarcity of microRNAs (miRNAs) in biofluids and consequently higher demand for input volume are the most challenging limiting factors for small RNA sequencing (RNA-seq) research.

Objective: To compare 6 commercial RNA isolation kits and 2 library preparation methods for serum and urine using small RNA-seq.

Methods: Serum and urine were collected from 7 carrier female dogs with X-linked hereditary nephropathy. Total RNA from 2 ml pooled serum and 10 ml pooled urine were isolated in triplicate using 3 methods each for urine (exoRNeasy, Norgen Urine Exosome, miRCURY Exosome) and serum (Direct-zol, mirVana PARIS, miRCURY

Biofluids). The 2 kits yielding the highest RNA concentration (based on fragment analyzer analysis) were selected for each sample type, and small RNA-seq was performed using 2 library preparations: TruSeq and NEXTflex. Data were analyzed using a newly updated computational platform, CPSS 2.0.

Results: Samples with total RNA concentration ≥0.8 ng/ul were satisfactory for library preparation. For urine, exoRNeasy combined with NEXTflex outperformed other combinations in total number of reads, miRNA percentage, and number of miRNAs identified. For serum, Direct-zol combined with NEXTflex was the only combination that enabled successful library preparation. In total, 198 and 115 miRNAs were identified in the serum and urine, respectively. Results from urine were distinct from serum, and library preparation method influenced urine results more than isolation method.

Conclusion: Different isolation and library preparation methods demonstrate detectable differences in miRNA results from urine and serum. Small RNA-seq provides a global assessment for comparing these methods for biofluids.

November 5, 2017
3:30 PM – 3:45 PM

INCREASED IGG AND PHOSPHATIDYLSERINE ON MARROW ERYTHROID PRECURSORS OF DOGS WITH PRECURSOR-TARGETED IMMUNE-MEDIATED ANEMIA AND ON BLOOD RBCS OF DOGS WITH IMMUNE-MEDIATED HEMOLYTIC ANEMIA

Cynthia A. Lucidi, John A. Gerlach, Leonard Ari Jutkowitz, Michael A. Scott

Background: Persistent nonregenerative anemia in dogs with ineffective erythropoiesis and evidence of selective erythroid precursor (nRBC) destruction is suspected or presumed to be immune mediated, whether accompanied by immune-mediated hemolytic anemia (IMHA) or not. Many dogs with this suspected precursor-targeted immune-mediated anemia (PIMA) appear to respond to immunosuppressive therapy, but the specific pathogenesis of the anemia is unknown. As in IMHA, cell destruction may be mediated by immunoglobulins or complement, and phosphatidylserine (PS) may be involved.

Objective: To test the hypotheses that IgG and phosphatidylserine (PS) are increased on nRBCs of dogs with PIMA, and that PS is increased on RBCs of dogs with IMHA.

Animals: Bone marrow aspirates and/or blood samples were collected from healthy colony dogs or client-owned dogs with PIMA, IMHA, or other conditions.

Methods: Flow cytometric assays were used to assess for IgG and PS on marrow nRBCs and blood RBCs.

Results: Five of 17 and 0 of 7 PIMA and non-PIMA dogs, respectively, had increased nRBC IgG compared to limits based on 10 healthy dogs, and nRBC PS was significantly increased in PIMA dogs compared to 5 healthy dogs. Based on 20 healthy dogs, RBC

IgG was increased in 9 of 11 and 0 of 9 IMHA and non-IMHA dogs, respectively, while 10 of 11 and 2 of 9 IMHA and non-IMHA dogs, respectively, had increased RBC PS.

Conclusions and Clinical Importance: Our data suggest a role for both IgG and PS in canine PIMA and IMHA. Additional studies are warranted.

November 5, 2017
3:45 PM – 4:00 PM

VIRTUAL MICROSCOPY IS MORE EFFECTIVE THAN CONVENTIONAL MICROSCOPY FOR TEACHING CYTOLOGY TO VETERINARY STUDENTS: A RANDOMIZED CONTROLLED TRIAL

Samantha J. McDonnel Evans, A. Russell Moore, Christine S. Olver, Paul R. Avery, Andrew B. West

Background: Virtual microscopy (VM) using scanned slides and imaging software is increasingly used in medical curricula alongside instruction in conventional microscopy (CM). Limited reports suggest that VM is feasible in the veterinary education setting, and generally well-tolerated by students. Whether students can apply knowledge gained through VM to practical use is unknown.

Objective: To determine whether instruction using VM, compared with CM, is a successful method of training veterinary students for the application of cytology in practice (i.e. using light microscopes).

Methods: Seventy-one veterinary students from Colorado State University who attended a voluntary 3-hour cytology workshop were randomized to receive the same instruction with either VM (n=35) or CM (n=36). These students were compared to a control group (n=22) of students who did not attend a workshop. All students took an assessment involving interpretation of 4 cases on glass slides with CM, designed to simulate the use of cytology in general practice. Students also took a survey consisting of 18 questions related to the effectiveness of the workshop, their opinions on cytology instruction in the curriculum, and their learning preference (VM or CM).

Results: The mean assessment score of the VM group (14.18 points) was significantly higher than the control group (11.33 points, p=0.003), whereas the mean of the CM group (12.77 points) was not significantly different from controls (p=0.17).

Conclusions: VM is not only an effective method of teaching cytology to veterinary students, which can translate to a real-world case scenario, but it outperformed CM instruction in this study.

November 5, 2017
4:00 PM – 4:15 PM

INTRANUCLEAR CYTOPLASMIC INVAGINATIONS IN SMALL ANIMALS: A RETROSPECTIVE STUDY OF 18 CASES

Charalampos Attipa, Balázs Szladovits

Background: In human medicine, intranuclear cytoplasmic invaginations (INCI) are a well-established cytologic feature of thyroid malignancy. In veterinary medicine, INCI have been sporadically reported in meningiomas, canine testicular tumors and neoplastic or aging hepatocytes of mice and rats.

Objective: Perform a retrospective study on small animal cytology cases where INCI were identified.

Methods: The digital database of the Diagnostic Laboratory at the Royal Veterinary College was searched using the terms "invagination" or "nuclear inclusion" or "pseudoinclusion". A total of 22 cytology reports were found. The cytological preparations from these cases were retrieved and reviewed for the presence of INCI to be included in the final analysis. In cases where cytological preparation were not available any more, the histology section from the same lesion were evaluated for the inclusions, and if present, were also included in the analysis.

Results: Eighteen cases were included in the final statistical analysis, with 89% (16/18) being neoplastic processes, 69% (11/16) being malignant and 44% (7/16) being neoplasms of epithelial origin. The two most commonly involved organs included the liver 17% (3/18) with two hepatocellular carcinomas and one cholangiohepatitis, and the brain 17% (3/18) with three meningomas. Other cases included three sarcomas, a histiocytic sarcoma, a plasmacytoma and a case of CSF with epithelial lining cells with INCI. Four cases were misdiagnosed, three in liver aspirates with vacuolar hepatopathy.

Conclusions: The INCI is an uncommon cytologic finding that occurs more commonly with neoplasia, especially malignant, but further larger studies are needed to truly characterize the significance.

November 5, 2017
4:15 PM – 4:30 PM

FOLLOW-UP MONITORING IN A CAT WITH LEISHMANIOSIS AND CO-INFECTIONS WITH HEPATOZOON FELIS AND "CANDIDATUS MYCOPLASMA HAEMOMINUTUM"

Charalampos Attipa, Kyriaki Neofytou, Christos Yiapanis, Pamela Martínez Orellana, Gad Baneth, Yaarit Nachum Biala, Harriet Brooks Brownlie, Laia Solano Gallego, Séverine Tasker

A 6-year-old, female neutered, domestic shorthair cat from Cyprus was presented with multiple ulcerated skin nodules. Cytology and histopathology of the lesions revealed granulomatous dermatitis with intracytoplasmic organisms, consistent with amastigotes of *Leishmania* species.

Biochemistry identified a mild hyperproteinaemia. Blood extraction and PCR detected *Leishmania* species, *Hepatozoon* species and "*Candidatus* Mycoplasma haemominutum" (CMhm) DNA. Subsequent sequencing identified *Hepatozoon felis*. Additionally, the rRNA internal transcribed spacer 1 (ITS1) locus of *Leishmania infantum* was partially sequenced and phylogeny showed it to cluster with species derived from dogs in Italy and Uzbekistan, and a human being in France.

Allopurinol treatment was administered for six months. Clinical signs resolved in the second month of treatment with no deterioration eight months post-treatment cessation.

Quantitative PCR and ELISA were used to monitor *L. infantum* blood DNA and antibody levels. The cat had high *L infantum* DNA levels pre-treatment that gradually declined during treatment but increased eight months post-treatment cessation. Similarly, ELISA revealed high levels of antibodies pre-treatment which gradually declined during treatment and increased slightly eight months post-treatment cessation. The cat remained PCR positive for CMhm and *Hepatozoon* species throughout the study. There was no clinical relapse twenty-four months post-treatment.

To our knowledge this is the first clinical report of feline leishmaniosis with *H. felis* and CMhm co-infections. The high *L. infantum* DNA levels post-treatment cessation might indicate that although lesions resolved, prolonged or alternative treatment could have been considered. In addition, the presence of two other pathogens might have had an effect on immune response.

Clinical Pathology Focused Scientific Session II

November 7, 2017 | 8:00 AM – 12:00 PM
Session Chair: K. Zimmerman
Committee Members: E. Behling-Kelly, S. Clark, S. Connolly, M. Fry, N. Tripathi, C. Wagg

November 7, 2017
8:00 AM – 8:15 AM

SERUM PROTEIN ELECTROPHORESIS IN HEALTHY AND INJURED WHITE RHINOCEROS (CERATOTHERIUM SIMUM)

Emma H. Hooijberg, Carolyn Cray, Michele Miller, Peter Buss, Gerhard Steenkamp, Amelia Goddard

Background: Due to the poaching crisis in southern Africa, an increasing number of wounded white rhinoceros are being presented for veterinary care. Inflammation and the acute phase response have not been investigated, although concentrations of total protein (TP) and globulins are higher in healthy white rhinoceros than other perissodactyls. Analysis of electrophoretic serum protein fractions is the first step towards evaluation of the proteome and identification of clinically useful inflammatory markers.

Objectives: To develop reference intervals (RIs) for agarose gel serum protein electrophoresis (SPE) for the white rhinoceros and to compare these results to those from animals with inflammation.

Methods: RIs for TP and agarose gel electrophoretic albumin and globulin fractions were generated using serum samples from 49 healthy wild adult white rhinoceros. Results were compared to those from 30 animals with various degrees and chronicity of soft tissue trauma/ inflammation.

Results: Six globulin fractions were identified and named provisionally as alpha-1a, alpha-1b, alpha-2, beta-1, beta-2 and gamma. RIs were generated for TP (76-111 g/L), albumin (10-27 g/L) and globulin fractions (alpha-1a: 1.6-3.2 g/L; alpha-1b: 1.7-3.6 g/L; alpha-2: 16.1-26.6 g/L; beta-1: 6.6-18.2 g/L; beta-2: 11.8-30.4 g/L; gamma: 10.4-23.1 g/L). Animals with trauma/ inflammation had lower concentrations of TP, albumin, total globulin, alpha and beta-1 globulins, lower percentages of alpha-2 and beta-1 globulins, and higher percentages of beta-2 and gamma globulins.

Conclusion: Healthy wild white rhinoceros have high concentrations of alpha-2, beta and gamma globulins. Animals with inflammatory lesions do not have alpha-globulin changes consistent with a classic acute phase response.

November 7, 2017
8:15 AM – 8:30 AM

A NANOPARTICLE VECTORED 35KD MEMBRANE PROTEIN FROM MYCOBACTERIUM PARATUBERCULOSIS ELICITS DEVELOPMENT OF CYTOTOXIC CD8 T CELLS WITH ABILITY TO KILL INTRACELLULAR BACTERIA.

Cleverson D. Souza, Gaber S. Abdellrazeq, Mahmoud M. Elnaggar, John P. Bannantine, William C. Davis

Background: Analysis of the immune response to a *Mycobacterium avium paratuberculosis* (*Map*) candidate *relA* deletion mutant vaccine in cattle revealed a 35 kD membrane protein (MMP) is a major component of the immune response to *Map.*

Objectives: Determine the functional activity of effector T cells that develop following stimulation with blood dendritic cells (bDC) and monocyte derived DC (MoDC) pulsed with MMP alone or incorporated into nanoparticles (NP) comprised of poly lactic-co-glycolic acid and monophosphoryl lipid A (PLGA/MPLA).

Methods: Effector T cells were generated ex vivo from monocyte depleted PBMC (mdPBMC) by 2 rounds of stimulation, first with bDC then with MoDC pulsed with MMP alone or incorporated into PLGA/MPLA NP. Primed mdPBMC were incubated with monocyte derived macrophages (MoMΦ) uninfected and infected with *Map,* then processed to determine the level of killing mediated by cytotoxic lymphocytes (CTL). Flow cytometry was used to phenotype primed PBMC and MoMΦ.

Results: Comparison of mdPBMC stimulated with MMP alone or incorporated in PLGA/MPLA NP showed the CTL proliferative response was enhanced when mdPBMC were primed with MPLA vectored MMP. Depletion studies demonstrated CD8 CTL activity only developed if CD4 and CD8 T cells were present at the time of Ag presentation. Intracellular killing was mediated through the perforin granzyme granulysin pathway.

Conclusions: Development of CD8 CTL is complex and involves coordinate stimulation of CD4 and CD8 T cells by antigen presenting cells. The ex vivo CTL response to MMP is enhanced when incorporated into a nanoparticle vector.

November 7, 2017
8:30 AM – 8:45 AM

MYCOPLASMA HAEMOLAMAE AND FECAL EGG COUNT RELATIONSHIPS WITH ERYTHROCYTE VALUES IN CLINICALLY HEALTHY CAMELIDS

Lisa C. Viesselmann, Ricardo Videla, John J. Schaefer, Bente Flatland, Deanna M.W. Schaefer

Background: *Mycoplasma haemolamae* and intestinal nematodes can cause anemia in camelids. Parasite control programs aim to suppress parasite loads to subclinical levels without excessive deworming that promotes resistance, but there are few evidence-based guidelines for acceptable parasite loads in camelids.

Objectives: In clinically healthy camelids: 1) Correlate *M. haemolamae* PCR status and RBC values. 2) Determine the fecal egg count (FEC) threshold above which RBC values are consistently below the median of the reference interval.

Methods: EDTA-anticoagulated blood was collected from 118 clinically healthy, non-anemic adult alpacas and llamas. PCV was measured by centrifugation, RBC and HGB were measured by ADVIA, and *Mycoplasma haemolamae* was detected by real-time PCR. The number of strongyle eggs per gram (epg) was determined in a freshly collected fecal sample using the Modified McMaster's test. Significant differences in RBC values between *M. haemolamae* positive and negative animals, and between various FEC thresholds, were assessed by independent T-test or Mann-Whitney rank sum test.

Results: PCV, HGB, and RBC were not significantly different between *M. haemolamae* positive (N=41) and negative animals (P>0.5). PCV, HGB, and RBC were significantly lower in animals with FEC>600epg (P<0.05). For 10/10 animals with FEC>600epg, PCV, HGB, and RBC were below the median of the respective reference interval.

Conclusions: Because positive *M. haemolamae* PCR is not associated with lower RBC values in healthy camelids, treatment for *M. haemolamae* may not be necessary in those animals. FEC>600epg has a negative effect on RBC values, so maintaining FEC below that level is recommended in camelids.

November 7, 2017
8:45 AM – 9:00 AM

CAMELID ERYTHROCYTE VALUES BY AUTOMATED VERSUS MANUAL METHODS

Deanna M. W. Schaefer, Lisa C. Viesselmann, tracy Stokol, Ricardo Videla, Bente Flatland

Background: The rigid, elliptical shape of camelid RBCs presents unique challenges for hematology analyzers. There are no published studies validating optimal ADVIA settings for camelids, although anecdotal reports indicate the equine setting is preferred.

Objective: Determine whether preprogramed ADVIA equine, bovine, or goat settings can be used to accurately determine erythrocyte values from camelid blood.

Methods: EDTA-anticoagulated blood was collected from 158 llamas and alpacas. Hematocrit, RBC count, MCV and MCHC were measured on the ADVIA 120 or 2120 using equine, bovine, and/or goat settings. Results were compared to manual methods, i.e. PCV, RBC count by impedance (Coulter Counter, 8fl threshold), calculated MCV, and calculated MCHC. Sample sizes for each comparison ranged from 13-131.

Results: Hematocrit was higher than PCV by an average of 3.1-13.2 percentage points depending on the ADVIA species setting. The ADVIA RBC count was lower than the manual count by an average of 0.71 $x10^6$/µl for the equine setting and 1.62 $x10^6$/µl for the goat setting. Depending on the species setting, automated MCV and MCHC were 6.0-16.7 fl higher and 3.2-11.8 g/dl lower than the calculated values, respectively. The only automated value for which the majority of results were within the recommended allowable total error range for the corresponding manual result was the RBC count using the ADVIA equine setting.

Conclusions: The RBC count using the ADVIA equine setting corresponds fairly well to the manual RBC count in camelids. Automated values for hematocrit, MCV, and MCHC should be replaced with PCV, and calculated MCV and MCHC.

November 7, 2017
9:00 AM – 9:15 AM

BIOLOGICAL VARIATION OF COMMON HEMATOLOGICAL MEASURANDS IN 9 HEALTHY HORSES

Luca Giori, Bente Flatland, Xiaocum Sun, Karen A. McCormick, Kellie Fecteau

Background: Biological variation is inherited random fluctuation of measurand concentration around a homeostatic set point unique to an individual (CV_I) and in a group of individuals (CV_G). Index of individuality (II) defines whether a reference change value (RCV) could be a more sensitive tool to assess clinically relevant deviation from "the individual normality" compared to population based reference intervals (PRIs).

Objectives: Goals of this prospective study were to characterize CV_I, CV_G, analytical variation (CV_A), II and RCV of hematological measurands in horses and assess usefulness of PRIs for interpretation of test results.

Methods: Nine clinically healthy horses were sampled for hematological analysis once weekly for 6 weeks. All samples were processed in triplicate within 30 minutes of collection. Standardized protocols were used to reduce preanalytical and analytical variation.

Results: 16 hematological measurand (WBC, RBC, HGB, HTC, MCV, MCH, MCHC, CHCM, CH, RDW, PLT, MPV, NEUT#, LYMP#, MONO# and EOS#) were evaluated. Coefficients of variation CV_I, CV_G, and CV_A were calculated using mixed model analysis, after removing outlier data. CV_I values ranged between 0.32 to 28.30%; CV_G from 1.05 to 26.72%; and CV_A from 0.62% to 11.35%. II was then calculated for each measurand using the formula $CV_G/(CV_A^2 + CV_I^2)^{1/2}$, and ranged from 0.79 to 9.03.

Conclusion: No II values were <0.7. II of CH, MCH, MCV and PLT (>1.7) suggests that RCV is better for patient data interpretation. Remaining measurands (n = 12) had intermediate individuality, so PRIs should be assessed in relation to subject-based reference intervals.

November 7, 2017
9:15 AM – 9:30 AM

COMPARISON OF TOTAL ALLOWABLE AND TOTAL OBSERVED ERROR FOR 2 HEMATOLOGY ANALYZERS

Yolandi Rautenbach, Emma H. Hooijberg

Background: The American Society for Veterinary Clinical Pathology (ASVCP) recently circulated draft guidelines for total allowable error (TEa) for hematology analyzers.

Objective: Compare the ASCVP TEa with total observed error (TEobs) and to generate sigma metrics (σ) for a point- of-care and bench-top hematology analyzer.

Materials and Methods: The bench-top analyzer was a Siemens ADVIA 2120 on which 3 levels of quality control material (QCM) were run once daily. The POC analyser was an Abaxis Vetscan HM5 on which 1 of 2 levels of QCM was run on alternate days. Internal quality control data over a 7-month period were reviewed. TEobs was calculated for each level of QCM and compared to ASVCP TEa and σ was determined. The quality goal index (QGI) was calculated when $\sigma<3.0$.

Results: For the ADVIA, TEobs <TEa for all measurands. σ was < 3.0 for hematocrit (Hct), platelet count (Plt), and mean cell hemoglobin (MCH). The QGI indicated the cause as imprecision for Plt and bias for Hct and MCH. For the HM5, TEobs < TEa for all measurands except MCH and mean cell hemoglobin concentration (MCHC) (both, TEa 10% versus TEobs 12%). The σ was < 3.0 for red blood cell count, Hct, MCHC, MCH, and Plt. Excess bias and imprecision were implicated for all measurands except Plt (imprecision).

Conclusions: ASVCP TEa goals were achievable for most measurands on both analyzers; the ADVIA performed better. For measurands with σ < 3.0, QCM preparation and stability as well as calibration procedures should be investigated.

November 7, 2017
9:30 AM – 9:45 AM

GROWTH OF ADENOCARCINOMA CELLS FROM CANINE PLEURAL FLUID ON AEROBIC MICROBIOLOGY CULTURE

Allison F. Dusick, Jonathan F. Bach, Pen-Ting Liao, Faye A. Hartmann

Background: A 9-year-old female spayed puggle was presented for anorexia, lethargy, and icterus. Thoracic imaging revealed pleural effusion and a pulmonary nodule. Cytologic examination of pleural fluid revealed an exudate (20,860 nucleated cells/uL, TP 4.1 g/dL) dominated by inflammatory cells with occasional clusters of cells suspicious for carcinoma. The fluid was submitted for aerobic bacterial culture; somatic cells grew on agar plates and in broth.

Objective: Our objective was to determine if the cells growing in microbiological culture media were consistent with suspect carcinoma cells seen on cytologic preparations of pleural fluid.

Methods: Cytologic smears were prepared from colonies growing on a blood agar plate. Cells were stained with Wright-Giemsa and for expression of cytokeratin. Growth in thioglycolate broth was concentrated, preserved in formalin, embedded in paraffin wax, and stained with hematoxylin and eosin and for expression of cytokeratin. Histologic sections of tissues obtained at necropsy were also examined.

Results: Cytomorphology of colonies from blood agar was consistent with suspicious epithelial cells seen on cytologic examination of pleural fluid. Cultured cells were mildly immunopositive for cytokeratin. Paraffin-embedded cells from broth showed scattered cells immunopositive for cytokeratin. The histologic diagnosis at necropsy was pancreatic adenocarcinoma with pulmonary, pleural, nodal, and gastric metastasis.

Conclusions: Cultured cells were cytologically consistent with those seen in pleural fluid. Though rare, malignant somatic cells can grow under conditions used for aerobic bacterial culture. Recognition of cultured colonies as somatic cells may aid the diagnosis of malignancy when cytologic preparations are not definitive.

November 7, 2017
11:15 AM – 11:30 AM

IDENTIFICATION OF A SINGLE BASE DELETION IN THE GLYCOPROTEIN IIB GENE CAUSING GLANZMANN THROMBASTHENIA IN A GOLDEN RETRIEVER

Pete William Christopherson, Amanda Hill, Marjory B. Brooks, Madeline Scofield, Kevin B. King, Mary K. Boudreaux

Glanzmann thrombasthenia (GT), an inherited intrinsic platelet disorder, results from a quantitative or qualitative defect of the platelet membrane glycoprotein complex IIb-IIIa (GPIIb-IIIa, integrin $\alpha_{IIb}\beta_3$, fibrinogen receptor). Over 500 causative mutations are known in people. In veterinary medicine, GT has been documented in Otterhounds, Great Pyrenees, two related Beagle-crosses, and several horse breeds. Mutations identified in animals have all been in the gene encoding GPIIb (*ITGA2B)*. This report involves an inbred, 2-month-old female Golden Retriever that developed petechiae, ecchymoses and oral bleeding. Platelet count, coagulation screening tests, and VWF:Ag were within normal limits. Flow cytometry showed the patient's platelet GPIIb-GPIIIa expression was severely reduced based on CD61 (GPIIIa) detection. Genomic DNA was subjected to PCR to amplify the coding regions of the genes encoding GPIIb and GPIIIa in the patient and focal coding areas in both parents, 3 siblings, and 2 of the dam's sisters. A single nucleotide deletion at position (1924delC) in *ITGA2B* was identified in the patient. This change results in a frameshift and premature termination codon 24 bases downstream. Parents were heterozygous for this mutation, as were 2 of 3 siblings tested and one of the dam's sisters. This is the first clinical description of GT and associated *ITGA2B* mutation in Golden Retrievers. Together with previously published characterizations of GT in 3 other breeds of dog, our findings suggest that GT should be included in the differentials for platelet-type bleeding in any dog breed. As with people, GT is likely to be mutationally heterogeneous in dogs.

November 7, 2017
11:30 AM – 11:45 AM

HIGH AGGREGATION ACTIVITY OF PLATELETS FROM AKITA DOGS

Sakurako Neo, Aoi Tokunaga, Eri Ogawa, Fumitoshi Asai

Background: Thrombocytopenia in healthy Akita dogs without any coagulation abnormalities has been found sporadically. However, there is little information about platelet function in Akita dogs.

Objectives: The aim of this study was to compare the platelet count, morphology and platelet aggregatory activity between Akita dogs and research Beagle dogs.

Methods: Blood samples were collected from 13 healthy Akita dogs and 13 laboratory Beagle dogs using EDTA or sodium citrate as the anticoagulant. CBC, platelet morphology on the peripheral blood smear, and platelet function were evaluated. In the aggregation assay, platelets in platelet-rich plasma were activated by adding ADP or collagen in the Born's optical aggregometer.

Results: Platelet numbers were significantly lower in Akita dogs compared to those in Beagle dogs (Akita dogs: 95.8 ± 11.2×103/µL, Beagle dogs: 331 ± 25.6×103/µL), whereas no evidence of hemorrhagic tendencies was observed in Akita dogs. No morphologic abnormalities such as macroplatelets or dysplastic platelets were seen in platelets from Akita dogs. Platelets from Akita dogs showed significantly higher maximum aggregation responses to all ADP concentrations tested compared with the control Beagle dogs. The lag phase of aggregation was significantly shorter in Akita dogs than in Beagle dogs when collagen at 5, 10 and 20 µg/mL was used as the agonist.

Conclusion: Akita dogs have lower platelet numbers and higher aggregation activities compared with research Beagle dogs. High aggregation activity of platelets may compensate for low mass in this breed.

November 7, 2017
11:45 AM – 12:00 PM

CHARACTERIZATION OF THE WAVEFORM DATA IN CANINE AND FELINE CONGENITAL HEMOSTATIC DISEASE

Angela D. Gwynn, Kim J. Little, Davis M. Seelig, Kevin D. Williams, Leslie C. Sharkey

Background: It is well known that prothrombin time (PT) and activated partial thromboplastin time (aPTT), save for extreme values, are poor predictors of bleeding and are neither sensitive nor specific for many hemostatic disorders. As hemostatic abnormalities can be a common finding with potentially significant clinical implications, categorizing deficiencies (e.g. primary versus secondary, acquired versus congenital) with mild to moderate changes, while important, can be difficult. However, advances in coagulation analyzers, specifically those using photo-optical density clot detection and providing clot waveform analysis (CWA), may facilitate the diagnosis of hemostatic disease in veterinary medicine.

Objective: To characterize the clot waveform parameters in five veterinary patients with confirmed congenital factor deficiencies using the turbidometric ACL-TOP CTS analyzer.

Methods: Coagulation data including PT, aPTT, fibrinogen, visual inspection of waveform curves, and waveform raw data (first and second derivative curves / deltas) was reviewed using internally generated reference intervals.

Results: In two cases of canine factor deficiency, unusual waveforms, markedly decreased second derivatives, and mildly elevated deltas were identified. Mild interference of the second derivative waveform and mild alterations of the second derivativeswere seen in two of three Factor XII deficientcats.

Conclusions: Inspection of waveform data may provide information which can assist with identifying congenital factor deficiencies. The waveform of canine hemophiliac A patients is similar to those reported in humans. Additional studies, including evaluation of the maximum velocity of the aPTT waveform, which may be used to predict severity of bleeding in human hemophiliac A patients, are warranted.

Clinical Pathology Focused Group Poster Session

C-01: AMASTIGOTES OF LEISHMANIA SPP. IN PERIPHERAL BLOOD OF A 5-MONTH-OLD DOG

George M. S. do Rego, Gabriel A. Sousa, Izabelle T. S. Carvalho, Sandy M. Honorato, Thais M. Ziober, Giane R. Paludo

Background: Visceral canine leishmaniasis (LVC) is an endemic zoonosis caused by an obligate intracellular protozoan of the genus *Leishmania* transmitted by the bite of sandflies (Lutzomyia sp.). A 5-month-old female mixed-breed dog presented to the UF Veterinary Hospital for weakness and peritoneal ascites.

Objective: The aim of this study was to report the occurrence of amastigote forms of *Leishmania* spp. in peripheral blood of a 5-month-old dog.

Methods: CBC was performed and a buffy coat preparation was also examined.

Results: Laboratory abnormalities included nonregenerative, normocytic, normochromic, anemia, (HCT: 27%, reference interval 37-55%), neutrophilia (12,963/µL, 3,000-11,500/µL), lymphopenia (298/µL, 1,000-4,800/µL), eosinopenia (0/µL, 100-1,250/µL), and monocytosis (1639/µL, 150-1350/µL). Serum biochemistry showed increased ALT (224U/L, 21-73U/L) and ALP (391U/L, 20-156U/L). Both blood and buffy coat smears showed the presence of amastigote forms of *Leishmania* spp. within the cytoplasm of monocytes and also as forms circulating free.

Conclusion: LVC is an emerging zoonosis that affects different ages of animals and its incidence is increasing in urban centers, threatening public health.

C-02: APPLYING DATA SCIENCE PRINCIPLES AND MACHINE LEARNING TO OPTIMIZE MANUAL SLIDE REVIEW OF AUTOMATED COMPLETE BLOOD COUNTS

Jennifer M. Hayes, Mitchell R. Hayes, Kristen R. Friedrichs, Heather A. Simmons

Background: There is a strong interest in the optimization of manual slide review (MSR) criteria for complete blood counts (CBCs) in human medicine. This interest has coincided with the emergence of modern data science and machine learning. However, these approaches are sparse in veterinary literature.

Objective: To investigate the application of machine learning and other data science approaches to the optimization of veterinary MSR criteria

Methods: 679 Sysmex XS-1000i CBCs with paired MSRs were performed at the Wisconsin National Primate Research Center (WNPRC) on samples from *Macaca*

mulatta and Macaca fasicularis. Criteria for positive slide classification were created in consultation with WNPRC veterinarians. The dataset was analyzed using a custom pipeline designed to perform supervised classification in order to establish criteria that would yield a sensitivity of >95% while minimizing MSR.

Results: A supervised classification algorithm was trained on a subset of the paired CBCs/MSRs. A subset of Sysmex flags was determined to have low sensitivity and specificity. However, a second subset of Sysmex flags was found to be moderately sensitive with low specificity.

Conclusions: Modern data science tools including machine learning can be useful for optimizing MSR criteria and validating/invalidating automated hematology analyzer flags. Benchmarks were not fully achieved; therefore, dataset expansion and further refinements are in progress to optimize sensitivity while maximizing specificity.

C-03: BONE MARROW CYTOLOGY PARAMETERS IN ADULT HEALTHY CATS

Naila Cristina Duda, Stella Valle, Nilson Nunes, Juliana Matheus, Ana Paula Borenstein, Felipe Okano

Background: Bone marrow assessment is an important tool for identifying many hematologic disorders. For cats, the current reference parameters were obtained using a low number of animals.

Objectives: To verify bone marrow cytology parameters from adult healthy cats.

Methods: 20 adult healthy cats were submitted to a clinical evaluation, CBC, clinical chemistry, bone marrow cytology (from humerus) and FIV/FeLV test. CBC samples were evaluated in a hematology analyzer, followed by differential count in light microscopy. Bone marrow smear samples were stained with Wright-Giemsa and submitted for microscopic evaluation included cellularity, iron stores, M:E ratio by 500 to 1000 cell count and cell morphology (erythroid, myeloid, megakaryocytic and other cells).

Results: All cats were negative for retroviruses and clinical chemistry findings were within reference limits. The most observed hematologic alteration was eosinophilia (25%; n=5). Comparing with current reference values of bone marrow evaluation, the mean of cellularity was 55.5% (25 to 80%), M:E ratio of 0.91 (0.31 – 2.05). Hemosiderosis (60%, n=12), mild erythroid hyperplasia (60%; n = 12) without evidence of anemia or reticulocytosis, erythrophagia (10%, n = 2), eosinophilic hyperplasia (25%, n = 5), mild lymphoid hyperplasia (5%, n = 1), granulocytic hypoplasia (20%, n = 4), megakaryocytic hyperplasia (15%, n=3) and megakaryocytic hypoplasia associated with mild hemodilution (40%, n = 8) were observed.

Conclusion: Most of the bone marrow parameters are different from reference values in previous studies. Hemosiderosis in healthy cats is uncommon and is associated with erythroid lineage alterations, not observed in the evaluated cats.

C-04: DETECTION OF BAND NEUTROPHILS AND TOXIC CHANGE- HOW CONSISTENT ARE WE?

Nicole J. Fernandez, Cornelia V. Gilroy, Catherine R. Wagg, Marie-France Roy

Detection of band neutrophils and toxic change is vitally important in the complete evaluation of veterinary patients. Traditionally, this is done via microscopic review of a blood smear by trained personnel. However, consistency of evaluation between raters is not known. In this study, 3 board-certified clinical pathologists each evaluated 108 blood smears from adult horses with acute disease. Pathologists determined if bands were present and if so, the percentage of bands. Pathologists also determined if toxic change was present and if so, the grade of toxic change (after Lambert 2016). Inter-rater agreement for detection of bands and high-grade toxic change was determined using Fleiss' kappa (FK). Agreement between pathologists for the detection of bands was moderate (FK= 0.60), but when pathologists agreed on the presence of bands there was excellent agreement on the percentage of bands and mature neutrophils. Agreement between pathologists for the detection of high grade, clinically relevant toxic change was minimal (FK=0.34). When pathologists agreed on the presence of high-grade toxic change, there was good agreement on cytoplasmic basophilia and fair agreement on Döhle bodies and cytoplasmic vacuolation. This study highlights the difficulty in consistently detecting bands and toxic change, even for highly trained personnel. In light of this, it may not be surprising that blood smear evaluation is frequently limited in private veterinary practice.

C-05: DISSEMINATED PHAEOHYPHOMYCOSIS CAUSED BY CURVULARIA LUNATA IN A DOG WITH MASTICATORY MUSCLE MYOSITIS

Francisco Conrado, Heather Madden, Richard C. Hill, Sarah S. K. Beatty

A 3-year-old, neutered male Doberman was presented to the University of Florida's Emergency and Critical Care Service for a one-month history of lesions on his right paws, progressive swelling of the right thoracic and pelvic limbs, and lameness. The dog was previously diagnosed with masticatory myositis and treated with immunosuppressive doses of prednisone and cyclosporin. Despite multiple courses of different antibiotic medications, the lesions failed to resolve and the dog developed a fever (103.5 °F), and began limping on his swollen, right paw. A fine needle aspirate of the enlarged right popliteal lymph node revealed moderate pyogranulomatous lymphadenitis with a low number of fungal elements admixed in dense areas. These structures were round or oval, measured roughly 8.0 to 14.0 µm in diameter, and had a pale basophilic rim with a pale greenish-blue cytoplasm that occasionally contained a few fine pinkish granules and/or had a wrinkled appearance. The lymphoid population appeared reactive. Colonies from a potato-flake agar culture were submitted for phenotypic characterization, PCR, and DNA sequencing, and the fungus *Curvularia lunata* was identified. *Curvularia* is an extremely rare pathogen in the dog despite being abundant in vegetative matter and soil, and most infections arise from traumatic implantation or wound contamination. *Curvularia* and other etiologic agents associated with phaeohyphomycosis are considered opportunists and, among the few case reports from the human literature, most have occurred in immunocompetent individuals. The dog in the case herein is undergoing antifungal treatment with terbinafine, amphotericin B, and itraconazole, with progressive clinical improvement.

C-06: DOES CALCULATION OF PLASMA PROTEIN: FIBRINOGEN RATIOS PROVIDE USEFUL INFORMATION IN EQUINE EMERGENCY ADMISSIONS?

Nicole J Fernandez, Marie-France Roy

Early recognition of systemic inflammation is vital to providing effective treatment. Fibrinogen is a moderate acute phase protein that is easy and inexpensive to measure in plasma by means of heat precipitation, and it is frequently evaluated in equine practice. The plasma protein: fibrinogen (PP:F) ratio was proposed to aid in the interpretation of hyperfibrinogenemia by removing the influence of dehydration, but is inconsistently used in practice. We compared the interpretation of fibrinogen concentrations and PP:F ratios in 81 adult equine emergency admissions. Minimum and maximum kappa values were calculated to account for the equivocal zone in the PP:F ratio. Agreement was non-existent to moderate (Kappa=-0.06-0.65). Most of the disagreement was in cases with a fibrinogen concentration of 4 g/L (upper limit of the reference interval), which, due to the methodology, may reflect a true fibrinogen concentration of 3-5 g/L. Evaluation of the clinical features of the discrepant cases was not useful in determining whether the PP:F ratio could discriminate inflammation vs dehydration. For this reason, and because of the near perfect agreement when fibrinogen concentration was >4 g/L, calculation of the PP:F ratio was deemed not to provide additional useful information in equine emergency admissions.

C-07: EPITHELIAL RICH THYMOMA, SPINDLOID VARIANT, WITH MULTIPLE THYMIC CYSTS IN A DOG

Charalampos Attipa, Hannah Wong, Henny Martineau, Balázs Szladovits

A 10-year-old, male neutered, cross breed dog was presented to the Royal Veterinary Collage with a history of acute lethargy and collapsing. Computerized tomography revealed a mediastinal mass.

Fine needle aspirate of the mass yielded admixed populations of cells including: 1) abundant partially necrotic fusiform cells, 2) aggregated large, plump, moderately pleomorphic spindle cells, 3) high numbers of uniform ciliated columnar epithelial cells 4) rare sheets of reactive mesothelial cell and 5) low numbers of variably degenerate neutrophils and macrophages. Cytological interpretation was challenging but neoplasia with necrosis and possible branchial cyst were suspected.

Histopathology of the excised mass revealed a densely cellular neoplasm composed of spindloid cells arranged in bundles and trabeculae with occasional rosettes. Multifocally within the neoplasm there were cysts lined by ciliated epithelial cells that contained eosinophilic to basophilic acellular material. Immunohistochemistry found that the neoplastic spindloid cells labelled positively for cytokeratin and were negative for vimentin, supporting the diagnosis of an epithelial rich thymoma of spindloid variant.

The multiple cysts lined by ciliated epithelium likely represent thymic cysts, however bronchogenic or secondary branchial cleft cysts cannot be ruled out since they can also be lined by ciliated epithelial cells. While there are occasional histopathological reports of canine thymomas with thymic cysts, such cytological reports are lacking. Veterinary cytopathologists should be aware of this possibility, and consider it as a differential for mediastinal masses in dogs.

C-08: EVALUATION OF AN IN-HOUSE AUTOMATED URINE SEDIMENT ANALYZER FOR QUALITATIVE EVALUATION OF ERYTHROCYTE AND LEUKOCYTE NUMBERS

Rick Alleman, Graham Bilbrough, Jeremy Hammond, Donald McCrann, Myrick Celine, Anusha Ramanujam, Jennifer Ogeer, Dennis B. DeNicola

Background: In 2016, the IDEXX SediVue Dx™ Urine Sediment Analyzer (SDx) was introduced to the in-house veterinary market. It performs automated gentle centrifugation and microscopic analysis of urine. Veterinary-specific image evaluation software automatically processes 70 high-resolution microscopic images.

Objective: To determine the sensitivity/specificity of SDx performance for determining the presence/absence of significant RBC/WBC numbers, using manual review of SDx images as the reference method.

Methods: 300 urine samples including both canine and feline samples processed by the SDx were selected to assure adequate representation of common urine formed elements. Two experienced microscopists, blinded to image selection and SDx performance (SW1.0.0.0 and SW1.0.1.3), characterized all 70 images for each sample. Samples with ≥6 cells/hpf were classified as positive for both RBC and WBC. Manual results were a consensus between the microscopists.

Results: There were 104/300 and 71/300 positive samples for RBC and WBC, respectively. Contrasting the performance of SW1.0.0.0 and SW1.0.1.3, sensitivity of the SDx for detection of RBC was 72.1%/76.0%, while specificity was 95.4%/95.9%, respectively. For WBC, sensitivity was 81.7%/85.9%, while specificity was 91.3%/93.0%, respectively. Decreases in false-positive (9 to 8) and false-negative (29 to 25) for RBC and decreases in false-positive (20 to 16) and false negative (13 to 10) for WBC were observed between SW1.0.0.0 and SW1.0.1.3.

Conclusions: Moderate to high sensitivity and specificity for the detection of urine RBC and WBC numbers are possible with the SDx. Mild improvement in performance was documented with the recently updated software (SW1.0.1.3).

C-09: FELINE MYELODYSPLASTIC SYNDROME WITH EXCESS BLASTS ORIGINALLY MASQUERADING AS ACUTE LEUKEMIA

Pierre L. Deshuillers, Andrea P. Santos, Ashley L. Leisering, Mara S. Varvil, Craig A. Thompson, John A. Christian, Paul Rossman, Christopher M. Fulkerson, Michael O. Childress, Andrea L. Vanderpool, Margaret Miller, Joanne B. Messick

A 6-year-old, castrated male Domestic Medium Hair cat was presented to the Purdue University Veterinary Teaching Hospital with a presumptive diagnosis, made 10 months previously, of acute leukemia. The initial L-asparaginase-CHOP (cyclophosphamide, doxorubicin, vincristine, and prednisone) chemotherapy regimen was changed to lomustine due to lack of clinical improvement. On multiple occasions, testing was negative for feline leukemia virus and positive for feline immunodeficiency virus. A complete blood count revealed a mild, macrocytic, nonregenerative anemia, as well as moderate neutropenia and lymphopenia, but platelet variables were within the reference intervals. Trilineage dysplasia, including megaloblastic erythrocytes, giant platelets, and hyposegmented granulocytes, was a prominent feature in the blood smear. The bone marrow was hypercellular with a moderately increased myeloid:erythroid ratio of 3, and blasts represented 10% of all nucleated cells. Trilineage dysplasia was also remarkable. Hypolobulated cells in both the myeloid and megakaryocytic lineages, plus megaloblastic erythroid precursors, were common (>10%). Together, these findings were most consistent with a myelodysplastic syndrome with excess blasts (MDS-EB). The cat received several rounds of an investigational treatment over the next 9 months; however, it developed an acute myeloid leukemia and was euthanized. Myelodysplastic syndromes are a group of rare clonal myeloid neoplasms characterized by ineffective hematopoiesis, morphologic dysplasia in at least 10% of hematopoietic cells, and peripheral cytopenias. MDS-EB has a poor prognosis and may transform into acute myeloid leukemia. There is currently no efficient therapeutic option for MDS in veterinary medicine and the role of the treatment administered to this cat needs further studying.

C-10: IMPROVING CLONALITY TESTING FOR FELINE T-CELL LYMPHOMA

Araya Radtanakatikanon, Stefan Keller, Peter F. Moore, Nikos Darzentas, William Vernau

Background: T-cell receptor (TCR) clonality assays are useful adjuncts for the diagnosis of T-cell lymphoma. In veterinary medicine, assays targeting the TCRγ (TRG) locus are currently used and their design is based on limited sequence knowledge, resulting in incomplete primer coverage and suboptimal sensitivity. In human medicine, assay sensitivity has been improved by evaluation of two additional T cell receptor loci, TCRδ (TRD) and TCRβ (TRB).

Objective: To improve the sensitivity of feline T-cell molecular clonality assays by using high-throughput sequencing (HTS) to inform primer design.

Methods: Thymic, splenic and mesenteric lymph node mRNA was isolated from normal cats and 5'RACE and multiplex PCR was used to amplify full-length TRG, TRB and

TRD rearrangements. The library was sequenced using an Illumina MiSeq v3 PE300. Germline sequences were identified using StochHMM, and V/J gene usage frequency and pairing were determined and visualized using the ARResT online tool and Circos software.

Results: Feline TRG, TRB and TRD loci rearranged 6V/6J, 19V/2D/10J and 6V/3D/2J genes, respectively. Six TRG, 27 TRB and 5 TRD primers were designed for multiplex PCR assays of respective loci. In a preliminary assessment of 143 intestinal biopsy samples, the new TRG assay and our existing assay were concordant in > 90% of instances. The new TRG assay produced a clonal result in 8/143 samples (5.6%) when our existing assay did not.

Conclusion: Newly developed feline T-cell molecular clonality assays will improve diagnostic testing for feline T-cell lymphoma. Further testing is required to determine optimal usage of these assays.

C-11: MICRORNA PROFILING OF LYMPH NODE ASPIRATES IN CANINE MULTICENTRIC LYMPHOMA

Karlee Craig, Geoffery A. Wood, Stefan M. Keller, Anthony J. Mutsaers, Robert Darren Wood

Background: MicroRNAs (miRNAs) are small non-coding RNAs that participate in post-transcriptional gene expression and regulate biological processes. Aberrant miRNA expression is a hallmark of cancer. The goal of cancer-specific miRNA profiling is to identify biomarkers that may be used for diagnosis, prognosis and identification of future therapeutic targets.

Objective: The aim of this study was to characterize miRNA profiles from lymph node aspirates of healthy dogs compared to dogs with high grade, multicentric B or T cell lymphoma.

Methods: A microarray was used to profile the expression of 277 canine miRNAs in lymph node aspirates from 10 dogs for each group. Relative quantification was calculated using the delta-delta Ct method.

Results: For B cell lymphoma, expression of 15 miRNAs was increased and 3 were decreased, while for T cell lymphoma 17 were increased and 6 were decreased by at least >8-fold compared to healthy dogs. Nine of these miRNAs were similarly altered in both groups. Notably, miR-155, which is highly overexpressed in human diffuse large B cell lymphoma, was >14-fold increased in the B cell lymphoma group.

Conclusions: Identification of these differentially regulated miRNAs allows for development of a customized PCR array for profiling larger sample numbers. Future goals include analysis of lymph node aspirates in parallel with plasma samples from the same lymphoma patients, and correlation of miRNA profiles with clinical outcome.

Development of miRNA profiles may be useful for prediction of response to therapy and for improving prognostication.

C-12: PRESUMPTIVE DIAGNOSIS OF HEPATOCELLULAR CARCINOMA WITH SPINDLOID FEATURES IN A DOG

Kelsey P. Legendre, Dorian Lara, Rudy W. Bauer, Stephen D. Gaunt

A 12 year-old female spayed Shih Tzu/Maltese mix dog was referred to Louisiana State University Veterinary Teaching Hospital for an abdominal mass. Microscopic examination of smears obtained via ultrasound-guided fine-needle aspiration revealed a population of individualized to frequently aggregated spindle cells that exhibited moderate anisocytosis and anisokaryosis, open chromatin, and 1-3 large, prominent nucleoli (and several macronucleoli). The cells contained a moderate amount of granular cytoplasm with 1-2 cytoplasmic tails that frequently contained small, globular, brightly eosinophilic pigment granules (interpreted as matrix). A cytologic diagnosis of sarcoma was made, with recommendation for histopathology for further classification. An exploratory surgery was performed and a mass on the left lateral liver lobe was identified and removed (no other abdominal masses were noted). Histopathology revealed a mass composed predominantly of polygonal neoplastic cells, resembling hepatocytes, arranged in sheets and rarely cords. The neoplastic cells had an eosinophilic, grainy to vacuolated cytoplasm, often with indistinct cell borders, especially in areas of more spindle-shaped cells. The histopathologic findings were consistent with hepatocellular carcinoma. Given the described spindle cells, a vimentin immunohistochemical stain was performed which was positive in the areas with spindle cells. To confirm hepatocellular origin of these cells, hepatocyte paraffin 1 (Hep Par 1) was applied to the spindle cells as well, but the cells were negative. Despite this finding, the presumptive diagnosis of hepatocellular carcinoma with spindloid features was made. Epithelial-to-mesenchymal transformation of the malignant cells with loss of hepatocellular markers was suspected.

C-13: PROSTATIC UROTHELIAL CARCINOMA WITH METASTASIS TO THE CARPUS IN A DOG

Ashley Leisering, Craig Thompson

A ten-year-old, neutered male Golden Retriever was presented to its primary care veterinarian due to chronic hematuria and chronic left forelimb lameness. On physical examination, a firm swelling over the left carpus was noted as well as lymphadenomegaly of the left prescapular lymph node. Radiographs of the left limb showed a soft tissue swelling with irregularity along the edges of the carpal bones. An abdominal ultrasound revealed the prostate to be enlarged with irregular, heterogenous regions and lymphadenomegaly of the sublumbar lymph nodes. Fine needle aspirates were obtained from the prostate, carpal mass, right sublumbar lymph node, and left prescapular lymph node. The cytologic interpretation from all sites was a carcinoma that was most consistent with a urothelial carcinoma due to the presence of intracytoplasmic Melamed-Wolinska bodies within low numbers of the neoplastic cells. Melamed-Wolinska bodies are a highly characteristic feature of urothelial carcinomas. On cytologic evaluation of the carpal mass, occasional osteoclasts and frequent wind-rowing of erythrocytes were noted indicating likely bone and joint involvement, respectively. Canine urothelial carcinomas are known to occasionally metastasize to bone, but there are very few reports of metastasis to a joint. To the authors' knowledge, this is the second published report of a canine urothelial carcinoma with metastasis to a joint diagnosed by cytology.

C-14: THE DETERMINATION OF RETICULATED PLATELETS AS A NOVEL PLATELET BIOMARKER TO DIFFERENTIATE THROMBOCYTOPENIAS OF VARIOUS PATHOLOGICAL ETIOLOGIES

Michael Winter, Francois Christen

Reticulated platelets, released from megacaryocytes in the bone marrow into circulationrepresenta fraction ofimmature platelets, which containresidual strands of mRNA or rRNA. Thematurationinto fully mature platelet takes less than 1 day and peripheral counts correlate therefore with megacaryocyte activity in bone marrow. The accurate determination of reticulated platelets was so far restricted to special flow cytometry methods, but can now also be automatically quantified via specific setup modifications in some modern hematology autoanalyzeras a novel non-routine platelet parameter.

Here we present reticulated platelet counts observed under two distinct pathological conditions in a rat model of thrombocytopenia, determined by the Sysmex XT hematology analyzer. Consumptive coagulopathy was induced by a single dose of LPS and platelet parameters were compared with results following repeat dosing over 1 week of a test compound, which induced marked maturation and differentiation arrests at various stages of megacaryopoiesis.

Marked thrombocytopenia was noted under both experimental conditions. Absolute reticulated platelet counts increased within 24h after LPS challenge already at peak thrombocytopenia phases, consistent with platelet consumption and regenerative bone marrow megacaryopoiesis. Conversely, thrombocytopenia induced by interference with megacaryocyte maturation and differentiation was associated with a marked depletion of both platelet and reticulated platelet counts. Under these conditions reticulated platelets only increased during the recovery phases when also platelet counts normalized.

The data demonstrated the high utility of reticulated platelet counts as a novel platelet marker for the differential diagnosis and monitoring of thrombocytopenias at various pathological conditions.

C-15: ZINC TOXICOSIS - ASSOCIATED HEMOLYTIC ANEMIA AND PANCREATIC DISEASE IN TWO DOGS

Kimberley D. Foote, Cornelia V. Gilroy, Shelley A. Burton, Barbara S. Horney

Zinc toxicosis in dogs is commonly associated with hemolytic anemia with or without Heinz body formation (~33%) or spherocytosis (~20%). Pancreatic disease has also been uncommonly reported. Two canine cases of zinc toxicosis involving pancreatic disease are presented.

The University of Prince Edward Island received blood samples from two dogs. The first was from a 4-month old intact female Australian Shepherd with a history of vomiting, anorexia, lethargy, pigmenturia and possible rodenticide exposure. The second was from a 1 year old intact female Lhasa Apso with a history of progressive weakness, pigmenturia and removal of a penny from the mouth the previous day. Both dogs had regenerative anemias with metarubricytosis but only the second dog had Heinz bodies, eccentrocytosis and spherocytosis. Both were azotemic, had elevated total bilirubin concentration and increased amylase and lipase activities. Prothrombin and partial thromboplastin times were prolonged only in the first case. The Australian Shepherd was euthanized following unsuccessful treatment for possible anticoagulant rodenticide poisoning and a metallic gastric foreign object was discovered on post-mortem examination. Pancreatic necrosis and renal nephrosis were diagnosed. Tissue zinc levels were very high. The Lhasa Apso had an American penny and Canadian quarter removed gastroscopically and pancreatic enzyme activities returned to normal 3 days later. Serum zinc concentration was appreciably elevated at the time of presentation but returned to a near normal level by 3 weeks later. These cases serve as a reminder that zinc toxicosis should be considered in dogs with concurrent hemolytic anemia and pancreatic disease.

Experimental Disease Focused Scientific Session I

November 5, 2017 | 1:30 PM – 5:00 PM
Session Chair: L. Baseler
Committee Members: O. Foreman, S. Gumber (Co-chair), L. Janke (Past Chair), S. Montgomery, Sara Santagostino, and Manu Sebastian

November 5, 2017
1:30 PM – 1:45 PM

USING NEXT GENERATION SEQUENCING TO IDENTIFY DRIVERS OF OSTEOSARCOMA METASTASIS THAT CAN BE TESTED IN A PATIENT-DERIVED XENOGRAFT MODEL SYSTEM

Amanda L. Koehne, Leanne C. Sayles, Alex Lee, Marcus R. Breese, Stanley G. Leung, Aviv Spillinger, Alejandro Sweet-Cordero

Background: Osteosarcoma is the most common primary bone tumor in humans and domestic dogs. Early metastatic spread to the lungs is a hallmark feature of the disease in both species. Complications from metastasis remain the most common cause of cancer-related death, yet the process remains poorly understood.

Objective: The objective of the study was to develop a systematic approach to identify genetic drivers of osteosarcoma metastasis and a model for testing and validation.

Methods: Both primary tumor site and metastatic site human OS samples were processed for whole genome sequencing, RNA sequencing, patient-derived xenograft (PDX) generation, and cell line generation. Potential drivers of metastasis were identified by comparing sequencing data from the metastatic site to primary site. These genes were validated with immunohistochemistry and tested *in vitro* and *in vivo* using cell lines and PDXs in various metastasis assays. Our study contained 52 patient samples from 35 patients, 21 PDXs, and 6 cell lines.

Results: Transcriptomic analysis of patient samples and PDXs defined a distinction between the primary and metastatic sites, and illustrated that the PDXs and cell lines resembled their parent patient sample. Several potential drivers of OS metastasis were identified including transcription factors, cytoskeletal proteins, and ECM regulators.

Conclusion: We demonstrated the utility of using multiple next generation sequencing approaches to identify drivers of OS metastasis. Additionally, our PDX and cell lines models recapitulated the genomic landscape of the parent tumor sample. Taken together, this approach provided both hypotheses generation and a PDX model to test the hypotheses.

November 5, 2017
2:30 PM – 2:45 PM

MULTIORGAN PATHOLOGIC CHARACTERIZATION OF MURINE XENOGENEIC GVHD

Sara Francesca Santagostino, Julie White, Alessandra Piersigilli, Carol Meschter, Melody Smith, Melissa Docampo, Eric Smith, Nai-Kong V. Cheung, Sayed Shahabuddin Hoseini, Elisa de Stanchina, Veron Browne, William Cohen, Sebastien Monette

Background: Highly immunodeficient mice support the engraftment of cells or tissues of human origin; however, this may lead to the development of xenogeneic GVHD (xGVHD), a recognized limitation of these models. Limited reports on the pathology of murine xGVHD are available, but a comprehensive characterization of the xGVHD lesional spectrum is missing.

Objectives: We aim to fully characterize the pathology of murine xGVHD associated with different strains of immunodeficient mice and types of xenografts, and compare those findings to lesions occurring with murine allogeneic GVHD (aGVHD).

Methods: Mice were enrolled in different experiments and transplanted with cells or tissues of murine or human origin. Necropsies were performed at scheduled time points or earlier when significant clinical signs occurred. Upon histologic evaluation, selected organs were stained with immunohistochemistry for human CD45, murine CD45, CD20, CD3, IBA1, Ki67, and cleaved caspase-3.

Results: Mice affected by xGVHD typically presented with multiorgan inflammatory lesions characterized by the presence of human T cells and apoptotic cells, with interface hepatitis and interface dermatitis being most prominent. Other organs significantly affected included thyroid, kidneys, lacrimal and salivary glands, and oral mucosa, but not intestine. Conversely, mice affected by aGVHD commonly exhibited inflammatory lesions in the intestine, with prominent crypt apoptosis.

Conclusions: While murine xGVHD and aGVHD overlap in their pathogenesis and lesions, the pathologic presentation of xGVHD has some distinct features. Given the use of humanized mouse models in preclinical discovery and safety assessment studies, this study provides a complete guide for the pathologic assessment of xGVHD.

November 5, 2017
2:45 PM – 3:00 PM

IDENTIFYING PLACENTAL PHENOTYPES OF PREECLAMPSIA FOLLOWING VASOPRESSIN INFUSION IN PREGNANT MICE

Katherine N. Gibson-Corley, Jeremy A. Sandgren, Katherine J. Perschbacher, Guorui Deng, Donna A. Santillan, Mark K. Santillan, Justin L. Grobe

Human preeclampsia (PE) is associated with elevated secretion of arginine vasopressin (AVP), and chronic infusion of AVP into pregnant mice is sufficient to model PE by causing hypertension, renal glomerular endotheliosis, proteinuria, fetal placental hypoxia, and growth restriction. Early stages of PE are associated with defective trophoblast invasion of maternal spiral arteries, leading to decreased artery diameter and placental oxygenation. AVP infusion (24 ng/hr, sc) caused placental hypoxia (chromatin-bound HIF1α on gestational day (GD) 17.5) and reduced placental growth factor mRNA. Therefore, we performed histopathologic analyses of placentas collected from mice at GD12.5 infused with saline or AVP, with the hypothesis that AVP leads to early placental PE phenotypes. This preliminary cohort demonstrated an increased urine protein content and reduced plasma osmolality, trends toward elevated mid-gestational systolic blood pressure, reduced fetal and placental masses, and similar changes in heart rate. GD12.5 placentas were then routinely processed and stained with HE or cytokeratin-8 to examine morphological changes induced by AVP. AVP infusion had no significant effects on labyrinth, junctional zone, or decidua layer thicknesses. AVP caused a reduction in average maximum spiral artery diameter within the decidua and a trend toward reduced total spiral artery number, but no difference in the maximum invasion depth of CK8-positive trophoblasts. We conclude that AVP infusion is sufficient to induce mid-gestational features of PE in pregnant mice, including reduced spiral artery diameter. Such morphological changes may be associated with the placental hypoxia and reduced placental growth factor expression in this model.

November 5, 2017
4:30 PM – 4:45 PM

A TISSUE EXPLANT MODEL USING LUNG AND LYMPH NODE TO STUDY THE PATHOGENESIS OF CANINE DISTEMPER VIRUS

Omar Antonio Gonzales-Viera, Kevin Douglas Woolard, Kevin Keel

Background: Canine distemper virus (CDV) causes severe lymphoid and respiratory infection in dogs. Due to the unknown steps in CDV transmission and the important relationship between viral and host receptor (SLAM) distribution, development of an accurate model to simulate natural host-virus interaction is important.

Objective: To develop a lung and lymph node canine explant model to study the CDV transmission, replication and virus-receptor interaction.

Methods: We collected lung and lymph nodes from four recently euthanized dogs. Tissues were infected with CDV and cultured for five days. Immunohistochemistry and immunofluorescence (IF) was performed to demonstrate the viral antigen distribution, double IF to observed the co-localization of infected cells and SLAM receptor, and real-time RT-PCR to quantify the viral replication rate.

Results: CDV was able to replicate in lungs and lymph nodes since the first day post infection (dpi) and the replication was increasing during the time frame. In the lungs, alveolar macrophages were the only cells positive for CDV, it is similar to that observed in other morbilliviruses, which are carried by macrophages from lung to regional lymph nodes. In the lymph nodes, only the histiocytic cells were positive for CDV with a preferential localization in the subcapsular sinus and randomly in the medulla and within afferent lymphatic vessels. CDV-positive syncytial cells were observed in the subcapsular sinus since the fourth dpi. The co-localization of CDV and SLAM receptor was observed since the first dpi.

Conclusion: Lung and lymph node explants are accurate models to study features about the CDV pathogenesis.

November 5, 2017
4:45 PM – 5:00 PM

A SYNGENEIC IMMUNOCOMPETENT DOUBLE-LABELED B16F10 MURINE MELANOMA MODEL FOR EFFICACY AND SAFETY TESTING OF ONCOLYTIC VIROTHERAPY USING THE HSV-1 VC2 LIVE-ATTENUATED VACCINE STRAIN

Natalie Wall Fowlkes, Dominique Townsend, Brent Stanfield, Paul Rider, Vladimir Chouljenko, Rafiq Nabi, Ramesh Subramanian, Michael Mathis, Konstantin Kousoulas

A mutated Herpes Simplex Virus-1has been FDA approved and currently used for the treatment of human melanoma (Imlygic; T-VEC; Amgen). We have shown that the HSV-1 VC2 strain developed in our laboratory induces robust humoral and cellular immune responses in mice and protects against lethal intravaginal and ocular challenge by virulent HSV-1 and HSV-2 strains. VC2 contains mutations in gK and UL20 that prevent virus entry into neuronal axons preventing establishment of latent disease, while replicating efficiently in epithelial and fibroblastic cells. To test efficacy of VC2 and VC2-based viruses as an oncolytic and immunotherapy for the treatment of melanoma, we have developed an immunocompetent, syngeneic mouse model system that allows for the dynamic monitoring of melanoma growth and metastasis. The system consists of double-labeled B16F10 murine melanoma cells, which express luciferase and eGFP, allowing *in vivo* imaging and fluorescent microscopy of the tumors. B16F10 cells are known to be non-permissible to HSV-1 and do not express nectin-1 receptors. To enhance virus entry and spread, B16F10 cells have been stably transfected with human nectin-1 tagged with mCherry. VC2 is being tested for the ability to induce anti-melanoma immune responses in mice. Preliminary studies suggest a systemic immune response to virotherapy with VC2, and therapy has been well-tolerated in mice thus far.

Experimental Disease Focused Scientific Session II

November 7, 2017 | 1:30 PM – 5:00 PM
Session Chair: L. Baseler
Committee Members: O. Foreman, S. Gumber (Co-chair), L. Janke (Past Chair), S. Montgomery, Sara Santagostino, and Manu Sebastian

November 7, 2017
1:30 PM – 1:45 PM

PATHOLOGY OF AEROSOLIZED SUDAN EBOLAVIRUS INFECTION IN THREE SPECIES OF NONHUMAN PRIMATES

Donald K. Nichols, Elizabeth E. Zumbrun, Aysegul Nalca

Sudan ebolavirus (SEBOV) and *Zaire ebolavirus* are responsible for the majority of ebolavirus infections of humans. Our objectivefor this study was to determine the effects of infection by aerosolized SEBOV. We exposed rhesus macaques (*Macaca mulatta*; rhesus), cynomolgus macaques (*Macaca fascicularis*; cynos), and African green monkeys (*Chlorocebus sabaeus*; AGMs) to either a high dose or a low dose of aerosolized virus. Thirty-five of the 36 animals died due to SEBOV infection. One rhesus in the low dose group survived to the end of the study. On average, cynos had the shortest survival time and rhesus had the longest. AGMs had the most severe liver lesions and there was more viral antigen in their livers, as demonstrated by immunohistochemistry, than the macaques. Rhesus had relatively mild liver lesions and much less viral antigen in their livers and spleens. AGMs developed mild broncho-interstitial pneumonia whereas most of the macaques had moderate to marked fibrinosuppurative pleuropneumonia. Lung lesions tended to be more severe in rhesus than cynos, and much more viral antigen was present in the lungs of both macaque species compared to AGMs. Within a primate species, the only remarkable differences between high and low viral dose groups occurred with survival time in cynos and the presence of extensive necrosis in mediastinal lymph nodes of rhesus. Although all of the monkeys in this study developed systemic SEBOV infections, there were notable differences between the species; which species represents the best model for aerosolized SEBOV in humans has yet to be determined.

November 7, 2017
2:30 PM – 2:45 PM

OPTIMIZED AND VALIDATED COLLECTION OF GANGLIA FOR TRANSLATIONAL STUDIES IN PIGS

David K. Meyerholz, Leah R. Reznikov

Genetically engineered pig models are increasingly used in translational research to study a variety of human diseases including cystic fibrosis, cancer, neurofibromatosis, neurodegenerative and cardiovascular diseases. The nervous system has been implicated as playing a role in several of these diseases and thus sensory/autonomic ganglia are studied for markers of disease and candidate targets for pharmacotherapeutics. Because the trigeminal, vagal and dorsal root ganglia have important roles in sensory and autonomic function, we targeted these ganglia to optimize and validate collection techniques. Neonatal pigs (3 male and 3 female) were euthanized with institutional animal care and use committee approval. Vagal (nodose) ganglia were collected using a ventral midline incision and subsequently identified cranial and lateral to the larynx. Dorsal root ganglia were collected following isolation of a targeted region of the spine and laminectomy of the dorsal spinal canal. Two variations of the technique were identified: one permitted optimal ganglion morphology and the other maximal tissue harvest for molecular studies. For trigeminal ganglia, the calvarium and brain were excised followed by removal of the temporal bone, the globe and optic nerve. After minor soft tissue dissection, the trigeminal nerve and ganglia were readily observed. For all ganglia (100%) histological examination validated successful collection of the ganglia and each had a timely extraction at necropsy. These techniques provide quick and reproducible approaches to collect ganglia for research studies.

November 7, 2017
2:45 PM – 3:00 PM

A FERRET MODEL OF CYSTIC FIBROSIS (CF)-INDUCED PANCREATIC PATHOLOGY AND ENDOCRINE ISLET REMODELING

Pavana G. Rotti, John F. Engelhardt, Katherine N. Gibson-Corley

Loss of cystic fibrosis transmembrane conductance regulator (CFTR) function leads to progressive destruction of the exocrine pancreas, exocrine pancreatic insufficiency and, in some cases, cystic fibrosis-related diabetes (CFRD). CFRD affects nearly 50% of adult cystic fibrosis (CF) patients, however, the onset and advancement of this disease is not well characterized. We have developed a CF ferret model that has four distinct phases of glycemic tolerance in which animals develop diabetic level spontaneous hyperglycemia in Phase II which rapidly normalized around two months of age and remains near normal during Phase III. In order to relate the changes in glycemic tolerance to the histopathologic changes in the pancreas, we analyzed markers of tissue remodeling. Similar to that in CF patients, Phase III CF ferrets displayed almost complete replacement of exocrine tissue by fat. Additionally, a 10-fold increase in ductal area was observed in Phase III CF ferrets compared to control. A sudden increase in percent endocrine cells in Phase III correlated to the previously reported transient recovery in hyperglycemia. The typical distribution of islets in the exocrine mass was lost in CF ferrets, with islets found clustered within the remaining pancreas amidst fibrotic tissue. Expression of tissue remodeling markers like MMP7 increased in islets in the presence of inflammation in Phase II. A congruent increase in α-SMA and desmin expressing myofibroblasts was observed in CF ferrets, consistent with stellate cell activation. These dynamic remodeling events likely contribute to variations in glycemic tolerance of CF ferrets during exocrine pancreas decline.

November 7, 2017
4:30 PM – 4:45 PM
CHARACTERIZATION OF AUSTRALIAN LABRADOODLE DYSTROPHINOPATHY
Stephanie M. Shrader, Seungwoo Jung, Bruce F. Smith

Duchenne muscular dystrophy (DMD) is an X-linked recessive disorder that is caused by mutations in the dystrophin gene.Dystrophinopathieshave been reported in various dog breeds; however, as is typical of most animal models for human disease, none of the models completely recapitulate all aspects of DMD. We have identified a novel dystrophin mutation in exon 21 in a line of Australian Labradoodles that develop rapidly progressive clinical signs and a natural lifespan of roughly six months. To characterize this model, we utilized seven dystrophin-deficient and five control (normal) male littermates from multiple related litters. Affected dogs had poor weight gains and eventual weight loss, a plantigrade stance with gait abnormalities, exercise intolerance, skeletal muscle atrophy,macroglossa,ptyalism, dysphagia, and kyphosis. Echocardiographic findings included hyperechoic foci in the left ventricular papillary muscles, septal hypokinesis, and statistically significant decreases in the normalized left ventricular end systolic and end diastolic volumes. Second degree atrioventricular (AV) block type II was also identified in one of the affected dogs. Consistent necropsy findings included peripheral diaphragmatic hypertrophy with central atrophy/fibrosis,glossalhypertrophy, and atrophy of the head, body, and limb musculature. Histopathologic skeletal muscle changes consisted of fibrofatty infiltration, myocyte degeneration and necrosis, regeneration,lymphohistiocyticinflammation, and foci of mineralization. Microscopic cardiac changes were limited to a focal area of mineralization adjacent to the sinoatrial node in the dog with the AV block. Based on these findings, the dystrophin-deficient Australian Labradoodle may be a useful model for the study of both DMD-related myopathy and cardiomyopathy.

November 7, 2017
4:45 PM – 5:00 PM
PROTEOMIC ANALYSIS OF LOCAL DISEASE-SPARING RESPONSES TO BOVINE RESPIRATORY SYNCYTIAL VIRUS IN INTRANASALLY-VACCINATED AND CHALLENGED CALVES
John A. Ellis, Darren Gray, Sheryl P. Gow, Mark H. Mooney

Bovine and human respiratory syncytial viruses (BRSV, HRSV) are primary causes of pneumonia in calves and children, respectively. Parenteral and intranasal (IN) vaccines confer protection against BRSV infection, which is associated with systemic and local antibody and cellular immune responses. To better understand the response engendered by IN vaccination, 3-8 day old calves received a single component BRSV vaccine, a "3-way" vaccine (BRSV, BHV-1, BPIV-3), or placebo, IN, and were challenged via aerosolization of BRSV 42 days later. Both groups of BRSV-vaccinated calves had significantly less pulmonary lesions and significantly higher arterial PO_2 concentrations compared to controls. 2D-DIGE (24cm, pH 3-10NL) proteomic analysis of pharyngeal tonsil lysates (n=8/gp) indicated a differential proteome response among the groups. Principal component analysis (PCA) of 864 detected protein spots revealed clustering of treatment groups (20.41% R2X variation – PC1 and PC2). Three placebo and 2 vaccinated calves that required euthanasia on days 6 and 7 post challenge were separated from other calves within the PCA scores plot. Vaccinated calves that overlaped on the PCA plot with the placebo group had lower serum IgG post-challenge, suggesting that alterations to the pharyngeal tonsil proteome indicate protective immunity. Seventy-six protein spots were significantly different (ANOVA, $p < 0.05$) between vaccinated and placebo groups, and 105 between animals euthanized early versus on day 8. Of these, 67 protein spots were selected based on FDR testing ($q < 0.2$) for MALDI identification. These data indicate differential responses in the pharyngeal tonsil proteome which correlates with vaccination and disease-sparing.

Experimental Disease Focused Group Poster Session

E-01: ONTOGENY OF TOLL-LIKE 3 AND 4 RECEPTOR EXPRESSION AND WHITE MATTER DEVELOPMENT IN THE FERRET BRAIN

Jessica M. Snyder, Pratik Parikh, Thomas Wood, Kylie Corry, Megan Larmore, Brian Johnson, Daniel Moralejo, Sandra Juul

Background: Inflammation caused by perinatal infection with superimposed hypoxia and/or hyperoxia appears to be important in the pathogenesis of preterm neonatal encephalopathy. The inflammatory response is likely mediated through Toll-like receptor (TLR)-dependent mechanisms. White matter is particularly vulnerable during the third trimester.

Objective: To determine TLR3 and TLR4 expression and accumulation in the brain and characterize white matter development in the neonatal ferret.

Methods: Ferret kits (n=3-4 per time point) were sacrificed at postnatal days (P) 1, 5, 10, 15, 21, and 40 for immunohistochemistry and gene expression studies. Behavioral tests were performed 3x/week from P21-P43 to generate a composite behavioral score. Statistical analyses included one way ANOVA with Tukey's post-test.

Results: TLR3 and TLR4 mRNA expression increased significantly from P1 to P40. Little TLR4 positive staining was present, although percent positive staining increased from P5 to P40. TLR3 positive staining was also low at all ages. Myelin basic protein (MBP) immunostaining and mRNA expression markedly increased from P15 to P40. There was little change in NG2 mRNA expression over time and PDGFRα mRNA expression peaked at P10 and then declined. Performance on behavior tests improved in the 4th week of life.

Conclusions: TLR3 and TLR4 expression is present at low levels in the neonatal ferret brain, which suggests that there should be a response to TLR3 and TLR4 agonists such as polyinosinic:polycytidylic acid and lipopolysaccharide. Rapid myelination occurs between the 3rd and 6th weeks after birth, which corresponds to the development of reflexes and basic motor skills.

E-02: PATHOLOGICAL LESIONS IN THE CENTRAL NERVOUS SYSTEM AND PERIPHERAL TISSUES OF MICE WITH STREET RABIES VIRUS (1088 STRAIN)

Kazunori Kimitsuki, Kentaro Yamada, Nozomi Shiwa, Satoshi Inoue, Akira NISHIZONO, Chun-Ho PARK

Background: Most studies on rabies virus pathogenesis in animal models have employed fixed rabies viruses.

Objective: To clarify the pathogenesis of street rabies virus in mice.

Methods: The street virus (1088 strain, 10^6 FFU) was inoculated into the right hindlimb of six-week-old *ddY* mice. Five mice were sacrificed per day at 3, 5, 8 and 11 days post-inoculation (DPI). Transverse sections of CNS were subjected to histopathology, immunohistochemistry and TUNEL. For electron microscopic observation, lumbosacral spinal cord including ganglion cells embedded in epoxy resin and sections were stained with uranyl acetate and lead citrate.

Results: At 5 DPI, ganglion cells in the right lumbosacral spinal dorsal root ganglia showed chromatolysis. Axonal degeneration and inflammatory cells increased with infection progress in the spinal dorsal horn and dorsal root ganglia. Right hindlimb paralysis was observed from 7 DPI, which progressed to quadriparalysis. No pathological changes were observed in the ventral horn and root fibers of the spinal cord. Viral antigen was first detected in the right hindlimb muscle at 3 DPI, followed by the right lumbosacral dorsal root ganglia, dorsal horn of spinal cord, left red nuclei, medulla oblongata and cerebral cortex at 5 DPI.

Conclusions: These results suggested that the 1088 virus ascended the lumbosacral spinal cord via mainly afferent fibers at early stage of infection and moved to cerebral cortex using descending spinal tract. It was concluded that significant pathological changes occur in the sensory tract of the spinal cord; this selective susceptibility results in clinical features of the disease.

E-03: LATENT TOXOPLASMA GONDII INFECTION PROMOTES BRAIN ATROPHY IN WILD-TYPE AND HUNTINGTON'S DISEASE MICE

David Donley, Teal Jenkins, Marley J. Realing, Jason Gigley, Jonathan Fox

Human latent toxoplasmosis has a prevalence of ~25 % in the USA. While the latent infection has historically been considered harmless, there is recent evidence that it is associated with a number of brain disorders. Huntington disease (HD) is a chronic progressive neurodegenerative disorder caused only by a CAG repeat expansion in the huntingtin gene that manifests with motor and cognitive symptoms. The purpose of this study was to evaluate the effect of chronic latent *Toxoplasma gondii* infection in wild-type mice and also in mice transgenic for the HD mutation. We have previously found that mice on the FVB background develop latent toxoplasmosis with cysts in brain and striated muscle, without mortality during the acute infection. We therefore studied the YAC128 HD mouse model that is maintained on this genetic background, together with wild-type litter-mate mice. Wild-type and HD mice were orally infected at 2 months of age with ME49 strain *T. gondii* or vehicle. Behavior was monitored longitudinally and mice were sacrificed at 12-months of age. Brain weights were significantly decreased in infected wild-type and HD mice at 12-months of age. There was a significant interaction between genotype and infection status indicating that the loss of brain weight was significantly greater in HD as compared to wild-type mice. The results show that latent *T. gondii* can promote brain atrophy in FVB mice and that this effect is potentiated in the presence of the HD mutation. Ongoing studies are using unbiased stereology to characterize the basis for the brain atrophy.

E-04: EARLY EVENTS OF BLUETONGUE VIRUS INFECTION IN THE SKIN OF EXPERIMENTALLY-INFECTED SHEEP AND GOATS

Virginia Gamino, Eleonora Melzi, Joshua Leach, Marco Caporale, Pamela Johnston, Massimo Palmarini

Background: Bluetongue is a major disease of ruminants caused by an arbovirus known as bluetongue virus (BTV). The clinical outcome of BTV infection substantially differs between different ruminant species and ranges from asymptomatic to severe hemorrhagic disease. While sheep are highly susceptible to bluetongue disease, the infection in goats is commonly asymptomatic. This variability has been attributed to viral and host factors but numerous questions still remain unanswered. Early events during BTV infection have been shown to be critical in BTV pathogenesis.

Objective: To investigate differences in viral replication and immune response in the skin of sheep and goats inoculated intradermally with BTV-8.

Methods: Skin samples were collected between 8 and 168 hours post-infection (pi). Immunohistochemistry was used to evaluate viral replication, infiltration of T lymphocytes and macrophages, and expression of MX1. Expression of IFN-β was determined using RNAscope®. Immunofluorescence, confocal microscopy and RNAscope® aided the identification of infected cells. SlidePath Tissue IA and HALO were used for quantitative analysis.

Results: BTV-8 infected similar cell types in both species, which included endothelial cells and macrophages; however, sheep showed more virus infected-cells (especially endothelial cells) at different times pi. Infiltration of macrophages was more prominent in goats at most times pi. Infiltration of T lymphocytes and levels of IFN-β and MX1 expression were also higher in goats, especially at later times pi.

Conclusions: Variation in the dermal immune response between sheep and goats, especially macrophage infiltration and IFN-β expression, may impact the level of local replication and persistence of BTV-8.

E-05: DIVERSITY OUTBRED MICE IDENTIFY NOVEL GENETIC LOCI ASSOCIATED WITH TUBERCULOSIS

Gillian L. Beamer, Daniel M. Gatti, Claudia Abeijon, Igor Kramnik

Eight different inbred strains of mice (5 laboratory and 3 wild-derived) selected for maximum genetic differences were bred to establish the Diversity Outbred (DO) mouse population. The population is commercially available through the Jackson Laboratory (Bar Harbor, ME). When infected with a low dose of *Mycobacterium tuberculosis* (*M.tb*), the genetic diversity in the DO population contributes to a wide spectrum of phenotypes including features of human tuberculosis (TB) that are not well-modeled by standard inbred laboratory strains. Here, we infected DO mice with ~20 Colony Forming Units (CFUs) of *M.tb* by aerosol. Approximately 20% of the DO mice developed rapid morbidity by 6-7 weeks post-infection associated with high *M.tb* load, granuloma necrosis, and neutrophil influx into lung alveolar spaces. As expected from the light microscopy results, gene expression profiles of lung tissue identified strong upregulation of granulocyte activation/adhesion pathways in the highly susceptible DO mice as compared to other DO mice (including mice euthanized due to morbidity at much later time points of infection). When quantified in the lungs, neutrophil-related proteins such as the chemokine CXCL1, S100A8, and matrix metalloproteinase 8 (neutrophil collagenase) were strong correlates of disease. Furthermore, preliminary QTL mapping using these and other traits in the DO population identified multiple new genetic loci associated with host responses to *M.tb.* Ongoing studies are being performed to extend and validate these preliminary findings.

E-06: DIVERSE CAVITY TYPES AND EVIDENCE THAT NECROSIS-INDUCED TEARS ARE THE MECHANISTIC DRIVER OF CAVITATION DURING TUBERCULOSIS

Elizabeth A. Ihms, Michael E. Urbanowski, William R. Bishai

Background: Effacement of normal lung parenchyma by air-filled cavities is an important sequela of pulmonary tuberculosis. Despite their clinical significance, the pathogenesis of tuberculous cavitation is poorly understood, with controversy as to whether the fundamental mechanism involves matrix depletion, lipid pneumonia, or mechanical factors.

Methods: In this study, a novel repetitive aerosol infection model using *Mycobacterium tuberculosis* was used to generate cavities in 20 New Zealand White rabbits. Serial computed tomography was performed to monitor cavity progression over 14 weeks. 3D reconstructions were compiled for each time point, allowing comprehensive four-dimensional cavity mapping. Terminally, cavities were processed for histopathology, and quantification of fibrosis was performed by Trichrome staining.

Results: We observed that cavities progress rapidly from areas of consolidation, and often show a pattern of explosive growth followed by gradual contraction. Cavities formed preferentially in the caudodorsal lung fields, and were frequently subpleural. Cavitation was invariably associated with necrosis, and a subset of cavities displayed histologic evidence of direct airway communication. Histomorphology revealed four distinct cavity types – smooth, rough, mixed and fibrous – providing insight on early and transitional lesions.

Conclusions: Our study, with serial 3D radiographic monitoring of tuberculous cavity development, reveals that cavitation is a highly dynamic process with preferential formation at sites of high mechanical stress. These findings suggest a new paradigm for the pathogenesis of tuberculous cavitation, in which mechanical stress acts on the necrotic granuloma to produce acute tears in structurally weakened tissue, with subsequent air trapping.

E-07: BOVINE HERPESVIRUS-4 AS A NOVEL DELIVERY PLATFORM FOR THEILERIA PARVA ANTIGENS IN CATTLE.

Laura B.A. Williams, Lindsay M. Fry, Giulia Tebaldi, Gaetano Donofrio

East Coast Fever (ECF), caused by the tick-borne protozoan parasite *Theileria parva*, is a severe lymphoproliferative disease that results in significant bovine morbidity, mortality, and production losses in sub-Saharan Africa. The development of a sustainable next-generation subunit vaccine is critical to the long term control of ECF. BoHV-4 can accommodate a large amount of foreign genetic material and establish a persistent infection to facilitate the long-term cyclical expression of *T. parva* antigens. The aim of this study is to test the feasibility of using BoHV-4 as a viral platform for delivering *T. parva* antigens to incite an antigen-specific immune response in cattle.

Recombinant BoHV-4 expressing the *T. parva* antigens Tp2 and Tp9 were constructed from a nonvirulent strain of BoHV-4 genome cloned as a bacterial artificial chromosome. The *in vitro* stability of the viral constructs was determined by nested PCR across the inserted genes, and cattle were immunized with the recombinant viruses. Nested PCR was used for detection of BoHV-4 and Tp2 and Tp9 DNA in peripheral blood of cattle. The humoral immune response to BoHV-4 was evaluated with a commercial indirect ELISA. Antibody to the Tp2/Tp9 antigen was detected by immunoblot.

The recombinant viruses were developed and the constructs maintained the recombinant insertion *in vitro* and *in vivo*. Cattle developed sustained serologic responses to BoHV-4. Antibody to Tp2/Tp9 was detected by immunoblot analysis.

Preliminary data suggest that BoHV-4 can be used as a novel viral vector for use in cattle.

E-08: HISTOPATHOLOGIC FEATURES OF AN INDWELLING CARDIAC CATHETERIZATION PROCEDURE IN YORKSHIRE CROSS BARROWS

David S. Rotstein, Oscar Chiesa, Raoul Gonzales, Anne J. Lewandowski, Andrea Kouneski, Michael J. Myers

Background: One hundred Yorkshire cross barrows were used for an E. coli lipopolysaccharide (LPS)-challenge study to mimic an acute inflammatory response. To enable frequent blood draws, LPS administration, and end-of-study euthanasia, a novel indwelling catheterization was utilized.

Objective: To determine the host response to a novel catheterization procedure.

Methods: Quarantined and socialized pigs were segregated into groups of 10 prior to catheterization and LPS challenge. Clinical chemistry and serum protein analysis, and body temperature measurement was done prior to LPS administration. Catheterization involved minimally invasive ultrasound guided placement in the jugular vein and tunneling into the subcutaneous tissues to the dorsum. Pigs were sacrificed 24 to 28 hours after LPS administration and necropsy occurred within 10 minutes. Target organs, vena cava, and three sites of catheterization (exterior entry site, mid catheter site, and distal near the jugular vein) were sampled.

Results: Prior to LPS administration, no infection or inflammation was detected and body temperatures were normal. Tract site Inflammation was observed and involved fibrin with enmeshed neutrophils on the lumen edge and fibroblasts, and macrophages around the lumen. There was myofiber degeneration, regeneration, and dystrophic mineralization. Suture granulomas were also observed. Vena cava endothelial hyperplasia, subendothelial fibrosis, and phlebitis were observed in 20% of animals. The first study group had abscess formation at the distal site (9/10). Local antibiotic administration led to fewer cases (3/90) in subsequent study groups.

Conclusions: The indwelling catheter procedure enhanced study activities, minimized stress, and had a localized host inflammatory and reparative response.

E-09: PHENOL-RICH PEQUI FRUIT EXTRACT REDUCES OXIDATIVE INJURY IN RATS FED HYPERCALORIC DIET

Fernanda Figueiredo Mendes, Vanessa Souza Cruz, Leandro Lopes Nepomuceno, Karla Marcia da Silva Braga, Eugenio Goncalves de Araujo

Background: Pequi (*Caryocar brasiliense*) is an edible fruit from Cerrado, a savanna-like biome in Central Brazil. Pequi mesocarp retains a high concentration of phenols capable to neutralize in vitro reactive oxygen species linked to cardiovascular ischemic diseases, a worldwide leading cause of death We investigated the effect of pequi mesocarp ethanolic extract on global ischemia/reperfusion induced brain damage in rats fed hypercaloric diet.

Methods: Rats (48) were fed hypercaloric (24) or commercial diet (24) for 60 days and sorted into two subgroups of 12, treated or untreated with pequi extract. After a 30-day extract supplementation period, rats were subjected to global cerebral ischemia followed by reperfusion. We performed brain lesions microscopic evaluation (H&E), Nissl staining for counting viable and non-viable neurons in both cerebral cortex and hippocampus, and quantification of cerebral cortex neurons marked either by anti-p-ERK1/2 or p-AMPKα antibodies.

Results: Hypertriglyceridemia and significant increase in visceral fat were observed in the hypercaloric diet group. Ischemia and reperfusion brain lesions were less intense ($p<0.05$) in the hypercaloric diet, extract-treated animals. Although extract supplementation did not change the amount of viable and nonviable cells in the cerebral cortex and hippocampus, p-ERK1/2 and p-AMPKα cell labelling was significantly reduced ($p<0.05$) in the hypercaloric group.

Conclusions: Ethanolic extract of pequi mesocarp reduces induced oxidative brain lesions in rats fed a hypercaloric diet and has modulatory effect on ERK1/2 and AMPKα activation in the cerebral cortex.

E-10: STREPTROCOCCUS PNEUMONIAE INFECTION IN RESPIRATORY SYNCYTIAL VIRUS INFECTED NEONATAL LAMBS

Sarhad SA Alnajjar, Panchan Sitthicharoenchai, Jack Gallup, Mark Ackermann, David Verhoeven

Background: Respiratory syncytial virus (RSV) is the main cause of viral bronchiolitis resulting in hospitalization and the most frequent cause of secondary respiratory bacterial infection especially by *Streptococcus pneumoniae (Sp)* in infants. While murine studies have demonstrated enhanced morbidity during a viral/bacterial co-infection, human meta-studies have been mixed. Moreover, less is known about pathogenesis of *Sp* serotype 22 and especially the co-pathologies between RSV and *Sp* dual infections.

Objective: Here, we sought to examine mechanisms contributing to co-pathogen-induced morbidity using a large neonatal lamb animal model naturally permissive to infection by both pathogens.

Methods: Colostrum deprived lambs (aged 3-5 days) were randomly divided into four groups. Two of the groups were nebulized with RSV M37 (1.27×10^7 IFFU/mL), and the other two group nebulized with cell–conditioned mock media. At day 3 post-infection, one RSV group (RSV/*Sp)* and one mock-nebulized group (*Sp only*) were infected with ($2x10^6$ cfu *of Sp*) intratracheally. At day 6 post-infection all lambs were humanely euthanized and bacterial/viral pathogeneses were assessed by culture, focus forming unit, qPCR, IHC, and histopathology.

Results: Lambs dually infected with RSV and *Sp* had higher RSV titers by qPCR but lower *Sp* than the other comparable groups. Additionally, lung lesions were more intense in the RSV/*Sp* group as characterized by increased interalveolar wall thickness with neutrophils and lymphocyte infiltration.

Conclusions: Despite lower *Sp* in lungs, lambs co-infected with RSV exhibited greater morbidity and tissue histopathology. Thus, enhanced disease severity may be due more to elevated immunopathogenesis than elevated bacterial pathogenesis.

E-11: BOVINE RESPIRATORY EXPLANTS AS A TOOL TO STUDY CONTAGIOUS BOVINE PLEUROPNEUMONIA

Giovanni Di Teodoro, Giuseppe Marruchella, Anna Rita D'Angelo, Andrea Di Provvido, Gianluca Orsini, Gaetano Federico Ronchi, Massimo Scacchia

Contagious bovine pleuropneumonia (CBPP) is a severe disease caused by Mycoplasma mycoides subsp. mycoides (Mmm). The knowledge of the pathogenesis of CBPP is poor and hampered by the limited availability of laboratory animal and in vitro models of investigation.

The purpose of the present study is to assess respiratory explants as useful tool to study the early stages of CBPP.

Explants were obtained from trachea, bronchi and lungs of slaughtered cattle, which proved to be negative for Mycoplasma spp. and for the major bacterial and viral respiratory pathogens. The interaction of Mmm with specific cells was studied by immunohistochemistry (IHC), double-labelling indirect immunofluorescence (DLIIF) and laser scanning confocal microscopy (LSCM). Mmm capability to survive and proliferate within the explants was evaluated by standard microbiological procedures. Finally, the putative cellular internalization of Mmm was investigated by the gentamicin invasion assay.

IHC and DLIIF indicate that Mmm can colonize explants, showing a marked tropism toward the lower airways. In particular, Mmm was detected on/inside the bronchiolar and alveolar epithelial cells, the alveolar macrophages and the endothelial cells. The interaction between Mmm and explants was abolished by the pre-incubation with anti-Mmm immune sera. Mmm was able to survive and proliferate in all respiratory explants, during the entire time course of the experiments (120 hours). LSCM and gentamicin invasion assay both confirmed that Mmm can enter the host's cells.

Our data supports bovine respiratory explants as suitable tool to investigate CBPP, complementary and hopefully alternative to experimental infection of cattle.

E-12: PULMONARY FUNCTION IN COTTON RATS AFTER RSV INFECTION

Margaret E. Martinez, Olivia Harder, Lucia Rosas, Ian Davis, Stefan Niewiesk

Background: Human respiratory syncytial virus (RSV) is a leading cause of bronchiolitis and viral pneumonia in infants and young children. RSV is associated with increased airway resistance, decreased compliance, airway hyperresponsiveness, and hypoxemia.

Objective: Our goal was to characterize pulmonary function in normal adult cotton rats, as well as at 2, 4, 6 and 8 days post-RSV infection (d.p.i.).

Methods: The gold standard for measuring lung mechanics is through the forced oscillation technique, which had not previously been applied to the small animal model for RSV infection, the cotton rat (*Sigmodon hispidus*). Pulmonary edema was assessed by measuring the lung wet:dry ratio, and mucus production using PAS/Alcian blue special staining and Aperio software (color deconvolution algorithm). Carotid arterial oxygenation was measured by the i-Stat pediatric blood gas analyzer.

Results: Female cotton rats at 4 d.p.i. had a significant increase in baseline airway resistance, decreased baseline static compliance, and airway hyperresponsiveness after maximum methacholine challenge, which corresponds to the time point of peak viral replication. There wasn't a significant difference in lung wet:dry weight ratios, arterial oxygen saturation, or mucus production between normal and RSV infected cotton rats.

Conclusions: Our findings suggest that perturbations in pulmonary function that occur 4 d.p.i. may not be severe enough to cause hypoxemia: this may explain why, unlike RSV-infected infants with bronchiolitis, cotton rats do not display clinical signs after RSV infection. Therefore, pulmonary function measurements via forced oscillation technique may be more sensitive in detecting pulmonary abnormalities in cotton rats after RSV infection.

E-13: EQUINE PLATELET LYSATE AFFECTS THE IMMUNOMODULATORY CAPACITY OF MESENCHYMAL STEM CELLS

Maria Naskou, Scarlett Sumner, Ian Copland, Jacques Galipeau, John Peroni

Background: The mesenchymal stem cell (MSCs) secrotome plays an important role to regulate influx of endogenous progenitor cells, mediate apoptosis, fibrosis and tissue revascularization and, most notably, to decrease of inflammation and promote immune-regulation. Thus, MSCs have been proposed as a therapeutic for use in systemic and local inflammatory injuries. MSCs produced for clinical use rely on the use of the xenogeneic culture media fetal bovine serum (FBS) which has the potential to cause immune reactions. Platelet lysate (PL) is an attractive homologous alternative to FBS which is showing promise as a cell culture supplement.

Objective: To determine the effect of equine PL and conditioned media on the ability of MSCs to regulate the activation of LPS-stimulated monocytes.

Methods: Equine MSCs were cultured in FBS or equine PL and PL or FBS conditioned media were generated. LPS-stimulated equine monocytes were exposed to the following conditions: (a) MSCs cultured with FBS or PL, (b) FBS and PL conditioned media (no cells) (c) PL alone. Cell culture supernatants were collected at certain time pointsand assayed for the production of the pro-inflammatory cytokine tumor necrosis factors-alpha (TNF-α) through ELISA.

Results: LPS-stimulated monocytes exposed to MSCs cultured in PL or FBS and PL conditioned media produced significantly less TNF-α than LPS-stimulated monocytes alone. Exposure to PL achieved greater suppression of TNF-α release.

Conclusions: We show that media alone without MSCs may suppress monocyte activation better than MSCs, therefore, biological products such as PL may be considered for the regulation of cell-mediated immune responses.

E-14: REACTIVE OXYGEN SPECIES IN THE PATHOGENESIS OF CONTAGIOUS BOVINE PLEUROPNEUMONIA

Giovanni Di Teodoro, Francesco Mosca, Giuseppe Marruchella, Andrea Di Provvido, Pietro Giorgio Tiscar, Massimo Scacchia

Contagious bovine pleuropneumonia (CBPP) is a severe disease caused by Mycoplasma mycoides subsp. mycoides (Mmm). The pathogenesis of CBPP is poorly understood. Some data suggests that reactive oxygen species (ROS), produced by Mmm via the metabolism of glycerol and/or by phagocytic cells, could contribute to the tissue damage.

On the basis of what above, the present study aims to evaluate the generation of ROS by Mmm and/or bovine neutrophils in vitro.

Blood samples were collected from slaughtered cattle. Neutrophils were purified by standard procedures and adjusted to 5 x 10^6/ml. Three Mmm strains were tested: (1) "Caprivi", a highly virulent African strain; (2) 57/13, isolated in Italy in 1990s'; (3) T1/44, an attenuated African strain used as vaccine. The final Mmm concentration was adjusted to 10^8 colony forming units/ml. A luminol-based chemiluminescence assay was carried out to quantify ROS generated by Mmm or by neutrophils incubated with Mmm.

The incubation with all Mmm strains significantly increased the production of ROS by neutrophils, and that was notably enhanced by the Caprivi strain. The generation of ROS by Mmm was negligible in absence of glycerol, while it significantly increased after adding glycerol and reached the highest level with the Caprivi strain. However, in absolute terms, Mmm produced very low amounts of ROS when compared with neutrophils.

Assuming that ROS are relevant for CBPP pathogenesis, our results argue in favour of a major role for neutrophils as the main factory of such toxic metabolites.

E-15: CHRONIC TRICHLOROETHYLENE AND ARSENIC EXPOSED MDR-NULL MICE DEVELOP RENAL DAMAGE BUT NO TUMORS

Amie L. Perry, David W. Threadgill

Trichloroethylene (TCE) and inorganic arsenic (iAs) are environmental contaminants and carcinogens that target the kidney. Chronic exposure to TCE is associated with increased incidence of renal cell carcinoma. While co-exposure to TCE and iAs is likely in certain populations, such as those near Superfund sites, a review of the literature did not reveal any studies on co-exposure or co-carcinogenesis of these toxicants. Our objective was to determine whether multidrug resistance gene-null mice chronically exposed to TCE and iAs could be used to model the development of renal cell carcinomas in similarly exposed human populations. An F3 mouse population derived from the FVB/N-*Abcb1a*^{tm1Bor}, *Abcb1b*^{tm1Bor} and CAST/EiJ strains was exposed to trichloroethylene and/or arsenic for one year and kidneys from these mice were processed for H&E staining before being analyzed for neoplastic and pre-neoplastic changes. Despite use of TCE and iAs in combination and at environmentally relevant concentrations, primary renal cell tumors failed to develop. A significant increase in histologic evidence of renal disease was observed overall with any level of iAs exposure, but not with TCE exposure. We expected that inclusion of multiple toxicants in environmentally relevant concentrations as well as knock-out of the multidrug resistance gene would increase susceptibility to renal cell tumors and more accurately model human disease. While this model more accurately reflects human exposure conditions, the development of primary renal tumors observed in humans following chronic trichloroethylene exposure was not reproduced after inclusion of a second, potentially co-carcinogenic agent in this model.

E-16: RENAL INVOLVEMENT IN CONTAGIOUS BOVINE PLEUROPNEUMONIA: PATHOLOGICAL AND IMMUNOHISTOCHEMICAL FINDINGS

Giovanni Di Teodoro, Giuseppe Marruchella, Anna Rita D'Angelo, Andrea Di Provvido, Geoffrey Muuka, Massimo Scacchia

Contagious bovine pleuropneumonia (CBPP) is caused by Mycoplasma mycoides subsp. mycoides (Mmm) and typically affects the lung and the pleural surfaces. However, the involvement of kidneys is also frequently reported.We describe herein the pathological findings observed in kidneys of CBPP-affected animals.

A total of 43 cattle were studied: 41 with experimental CBPP (induced by endotracheal intubation) and 2 with natural CBPP. All animals were subjected to detailed post-mortem investigations. In addition, tissue samples were routinely processed for histopathological and immunohistochemical studies. Pleural and/or pulmonary lesions, acute-to-subacute (n=10) or chronic (n=33), were consistently detected. Grossly, renal lesions were seen in 13 cases, consisting of infarcts (n=8) or tubulointerstitial nephritis (n=5).

Histologically, different lesions were observed in the kidneys of 38 cattle: lympho-histiocytic tubulointerstitial nephritis (n=36), tubular mineralization (n=24), hyaline casts (n=10), glomerulonephritis (n=9), necrosis (n=6), fibrosis (n=3) and ectatic tubules (n=3).

Immunohistochemistry demonstrated Mmm antigens in all animals under study, with different, often combined distribution patterns: tubular (n=24), glomerular (n=8), mixed glomerular/tubular (n= 11), vascular (n=14). The glomerular pattern was usually observed in cattle showing acute pleuropneumonia, while the tubular pattern was always associated with chronic lung lesions. In 4 cases, Mmm-specific immunoreactivity was seen in glomeruli affected by glomerulonephritis.

Taken together, this data confirms the common involvement of kidneys in CBPP. Considering that immune-complex deposition could contribute to CBPP, glomerulonephritis should be carefully regarded to better understand the pathogenesis of the disease.

E-17: CYP26A1 AND CYP26B1 DELETION IN MICE CAUSES SEVERE MULTISYTEMIC PATHOLOGY

Jessica M. Snyder, Cathryn Hogarth, Nina Isoherranen

Retinol (vitamin A) has important roles in development, vision, reproduction, and immune system function. Retinol is metabolized to retinoic acid, which subsequently is degraded by two cytochrome P450 enzymes encoded by *Cyp26a1* and *Cyp26b1*. To further investigate the roles of these enzymes, 21 day old mice generated using a Cre-Lox breeding scheme received 5 consecutive once daily intraperitoneal injections of 80 mg/kg tamoxifen to generate *Cyp26a1*, *Cyp26b1* or *Cyp26a1* and *Cyp26b1* dual knockout mice. Compared to tamoxifen and Cre expression controls, experimental double knockout animals had dramatic systemic pathology including severe proliferative and ulcerative dermatitis of the skin and ear pinna, blepharitis, hyperkeratosis of the external ear canal with mild otitis media, marked lymphadenomegaly and splenomegaly, fat atrophy, moderate to severe hyperkeratosis of the nonglandular stomach, and chronic-active gastritis of the glandular stomach. In male mice, there was variably severe testicular degeneration. Increased eosinophilic crystals involving the nasal and gastric epithelium were observed in experimental mice. Mice lacking *Cyp26b1* alone had similar, although less severe, changes. Mice lacking *Cyp26a1* alone appeared essentially normal. *Cyp26b1* is involved in skin homeostasis, and the lesions affecting squamous epithelium are consistent with previously described retinoid dermatitis. Additional roles in hematopoietic stem cell homeostasis and adipose metabolism are suspected. Studies in these mice will help inform future studies investigating the role of *Cyp26a1* and *Cyp26b1* in hematopoietic stem cell and adipose metabolism as well as potential systemic effects in mice treated with Cyp26 inhibitors for other therapeutic indications.

E-18: IN VITRO VALIDATION OF THE HIPPO PATHWAY AS A PHARMACOLOGICAL TARGET FOR CANINE MAMMARY GLAND TUMORS

Marilène Paquet, Samantha Guillemette, Charlène Rico, Philippe Godin, Derek Boerboom

Canine mammary tumors (CMTs) are the most common neoplasias in intact female dogs. The clinical and molecular similarities between CMTs and breast cancer make them a valuable model for the study of the human disease. As misregulated Hippo signaling is thought to play an important role in breast cancer development and also occurs in CMT, we sought to determine if Hippo represents a valid pharmacological target for the treatment of CMT. Six CMT cells lines were assessed for their expression of the Hippo pathway effectors YAP and TAZ and for their sensitivity to verteporfin, an inhibitor of YAP-mediated transcriptional coactivation. Four cell lines that expressed YAP (CMT-9, -12, -28, -47) were found to be very sensitive to verteporfin treatment, which killed the cells through induction of apoptosis with ED50 values of 14-79 nM. Conversely, two cell lines that did not express YAP (CF-35, CMT-25) were an order of magnitude more resistant to verteporfin. Verteporfin suppressed the expression of YAP/TAZ target genes, particularly *CYR61* and *CTGF*, which play important roles in breast cancer development. Verteporfin was also able to inhibit cell migration and anchorage-independent growth. Likewise, verteporfin efficiently suppressed tumor cell invasiveness in the CMT-28 and -47 lines, but not in CF-35 cells. Together, our findings provide proof of principle that pharmacological targeting of the Hippo pathway compromises the viability and attenuates the malignant behavior of CMT cells. These results will serve as the basis for the development of novel chemotherapeutic approaches for CMT that could translate to human medicine.

E-19: WIDESPREAD SEVERE MYODEGENERATION IN A HOMOZYGOUS FEMALE DOG WITH DYSTROPHIN DEFICIENCY

Jessica S. Fortin, Chady H. Hakim, Scott Korte, Gayle C. Johnson, Dongsheng Duan

Dystrophin mutations are the most common cause of human Duschene muscular dystrophy (DMD). Skeletal and cardiac muscle pathology results from absence of functional dystrophin. The dystrophin mutation of Golden Retrievers has been outcrossed onto a Labrador Retriever background, which initially resulted in a milder phenotype. Additional outcrossing to other breeds has produced relatively outbred colony to study this disease. Necropsy was performed on a 10-month-old female Labrador-mixed breed canine derived from a research colony and found unexpectedly death. Macroscopic findings were limited to reduced skeletal and laryngeal muscle volume and mild dilatation of the esophagus. Microscopic findings consisted of extensive degeneration and regeneration of the axial skeletal, tongue, esophagus and laryngeal muscles that were characterized by considerable central nucleation, individual fiber mineralization and interstitial fibrosis. The muscle heart findings consisted of adipose cells interstitial infiltration. Genotype examination demonstrated a splice-site mutation in the dystrophin gene (chrX:27926946-27926946) that resulted in aberrant RNA processing. In the other allele, a long interspersed repetitive element-1 (LINE-1) insertion in intron 13 (in-frame stop codon) was present. The splice-site mutation and

LINE-1 insertion are similar to that found in Golden Retriever and Pembroke Welsh corgi with DMD, respectively. Both mutations were introduced through artificial insemination mating of a carrier female and an affected male. Although the cause of death in this animal was undetermined, female Golden retriever carriers can succumb to cardiomyopathy as could have happened herein. This case demonstrates that severe myodegeneration can occur in female Golden retriever dogs devoid of ameliorating normal X chromosome.

E-20: CHARACTERIZATION OF AUSTRALIAN LABRADOODLE DYSTROPHINOPATHY

Stephanie M. Shrader, Seungwoo Jung, Bruce F. Smith

Duchenne muscular dystrophy (DMD) is an X-linked recessive disorder that is caused by mutations in the dystrophin gene.Dystrophinopathieshave been reported in various dog breeds; however, as is typical of most animal models for human disease, none of the models completely recapitulate all aspects of DMD. We have identified a novel dystrophin mutation in exon 21 in a line of Australian Labradoodles that develop rapidly progressive clinical signs and a natural lifespan of roughly six months. To characterize this model, we utilized seven dystrophin-deficient and five control (normal) male littermates from multiple related litters. Affected dogs had poor weight gains and eventual weight loss, a plantigrade stance with gait abnormalities, exercise intolerance, skeletal muscle atrophy,macroglossa,ptyalism, dysphagia, and kyphosis. Echocardiographic findings included hyperechoic foci in the left ventricular papillary muscles, septal hypokinesis, and statistically significant decreases in the normalized left ventricular end systolic and end diastolic volumes. Second degree atrioventricular (AV) block type II was also identified in one of the affected dogs. Consistent necropsy findings included peripheral diaphragmatic hypertrophy with central atrophy/fibrosis,glossalhypertrophy, and atrophy of the head, body, and limb musculature. Histopathologic skeletal muscle changes consisted of fibrofatty infiltration, myocyte degeneration and necrosis, regeneration,lymphohistiocyticinflammation, and foci of mineralization. Microscopic cardiac changes were limited to a focal area of mineralization adjacent to the sinoatrial node in the dog with the AV block. Based on these findings, the dystrophin-deficient Australian Labradoodle may be a useful model for the study of both DMD-related myopathy and cardiomyopathy.

E-21: EVALUATION OF EFFICACY OF DUAL PI3K/AKT/MTOR PATHWAY INHIBITION IN CANINE OSTEOSARCOMA CELLS IN VITRO AND XENOGRAFT MOUSE MODEL.

Travis K. Meuten, Douglas H. Thamm

Canine appendicular osteosarcoma is a highly aggressive malignancy carrying a poor prognosis for which novel therapies are needed. The PI3K/Akt/mTOR pathway is a prototypic survival pathway associated with resistance to chemotherapy, more rapid tumor growth, and a higher incidence of metastases. Multiple downstream targets of pAkt and signaling cascade feedback loops necessitate the investigation of novel multinodal pathway inhibitors for a robust blockade of the pathway.

We hypothesized that constitutive pathway activation in canine osteosarcomas could be effectively blocked with a dual-acting PI3K/mTOR inhibitor, VDC-597, *in vitro* and *in vivo* (xenograft mouse model).

In canine osteosarcoma cell lines, western immunoblots and immunohistochemistry were used to assess indicators of pathway activation, including phosphorylation and localization of Akt and 4EBP1, and localization of FoxO1a. Cells were treated with VDC-597 alone and in combination with other chemotherapeutic agents. Cytotoxicity, growth inhibition, and anti-migratory activity were evaluated using fluorescent live-cell imaging. A xenograft mouse model was employed to examine in vivo efficacy of VDC-597 and cytotoxic chemotherapy in pathway inhibition, as well as suppression of tumor growth.

In canine osteosarcoma cells with high PI3K/Akt/mTOR pathway activation, VDC-597 dose- and time-dependently reduced phosphorylation of Akt and 4EBP1, cytosolic pAkt immunoreactivity, cellular proliferation, and migration, while enhancing nuclear FoxO1a localization, apoptosis and chemosensitivity. Reduction of tumor growth and Akt/4EBP1 phosphorylation was observed in xenografts.

This evidence is currently serving as the basis for further intraosseous xenograft modeling for in vivo osteosarcoma chemotherapy, and indicates potential useful strategies for treatment of dogs with osteosarcoma.

E-22: ECTONUCLEOTIDE PYROPHOSPHATASE/PHOSPHODIESTERASE 1 IS A DRIVER OF METASTASIS IN A SUBSET OF HUMAN OSTEOSARCOMA

Amanda L. Koehne, Leanne C. Sayles, Alex Lee, Stanley G. Leung, Aviv Spillinger, Alejandro Sweet-Cordero

Background: Ectonucleotide Pyrophosphatase/Phosphodiesterase 1 (ENPP1) is a transmembrane protein on osteoblasts that inhibits mineral deposition in the ECM. We have seen increased ENPP1 expression in a subset of human OS metastasis samples and patient-derived xenografts (PDX).

Objective: The objective of the study was to characterize the contribution and mechanism of action of ENPP1 on OS metastatic spread.

Methods: ENPP1 expression was knocked down (shRNA) and/or knocked out (CRISPR) in 3 human OS cell lines and 1 PDX. The metastatic propensity of the cells was interrogated using *in vitro* migration and invasion assays as well as *in vivo* lung colonization assays. RNA sequencing of the shENPP1 cell lines was performed to identify ENPP1 associated pathways that may point to a mechanism.

Results: Knockdown and knockout of ENPP1 decreased OS colony formation and migration *in vitro* and decreased metastatic colonization of the lungs *in vivo*. RNA sequencing of the shENPP1 cell lines identified several known cancer and metastasis genes, including L1CAM and TMEFF2, which might be involved in ENPP1 signaling in

OS. Pathway analysis demonstrated enrichment in genes associated with cellular movement, cell morphology, and cellular assembly and organization.

Conclusions: ENPP1 is a driver of metastasis in a subset of OS and is necessary for tumor cell colonization of the lung.

E-23: A SYNGENEIC IMMUNOCOMPETENT DOUBLE-LABELED B16F10 MURINE MELANOMA MODEL FOR EFFICACY AND SAFETY TESTING OF ONCOLYTIC VIROTHERAPY USING THE HSV-1 VC2 LIVE-ATTENUATED VACCINE STRAIN

Natalie Wall Fowlkes, Dominique Townsend, Brent Stanfield, Paul Rider, Vladimir Chouljenko, Rafiq Nabi, Ramesh Subramanian, Michael Mathis, Konstantin Kousoulas

A mutated Herpes Simplex Virus-1has been FDA approved and currently used for the treatment of human melanoma (Imlygic; T-VEC; Amgen). We have shown that the HSV-1 VC2 strain developed in our laboratory induces robust humoral and cellular immune responses in mice and protects against lethal intravaginal and ocular challenge by virulent HSV-1 and HSV-2 strains. VC2 contains mutations in gK and UL20 that prevent virus entry into neuronal axons preventing establishment of latent disease, while replicating efficiently in epithelial and fibroblastic cells. To test efficacy of VC2 and VC2-based viruses as an oncolytic and immunotherapy for the treatment of melanoma, we have developed an immunocompetent, syngeneic mouse model system that allows for the dynamic monitoring of melanoma growth and metastasis. The system consists of double-labeled B16F10 murine melanoma cells, which express luciferase and eGFP, allowing *in vivo* imaging and fluorescent microscopy of the tumors. B16F10 cells are known to be non-permissible to HSV-1 and do not express nectin-1 receptors. To enhance virus entry and spread, B16F10 cells have been stably transfected with human nectin-1 tagged with mCherry. VC2 is being tested for the ability to induce anti-melanoma immune responses in mice. Preliminary studies suggest a systemic immune response to virotherapy with VC2, and therapy has been well-tolerated in mice thus far.

Diagnostic Pathology Focused Scientific Session I

November 5, 2017 | 1:30 PM – 5:00 PM
Session Chair: L. Kennedy
Committee Members: E. Burrough (Co-Chair), R. Derscheid, F. Oliveria, A. Rodrigues Hoffman, D. Rotstein (Past Chair), H. Snyman

November 5, 2017
1:35 PM – 1:40 PM

A SURVEY OF GROSS AND HISTOLOGIC LESIONS OBSERVED IN UNIVERSITY OF WASHINGTON'S AXENIC MICE (MUS MUSCULUS)

Isaac M. Barber-Axthelm, Adeline M. Hajjar, Stacey M. Meeker, Charlie C. Hsu, Piper M. Treuting

Background: Following the recent exponential growth of microbiome research, the use of axenic mice in biomedical research has also experienced a rapid growth. One important consideration when conducting research with axenic mice is understanding the gross and histopathologic changes that can be present in axenic mice in the absence of experimental manipulation.

Objective: The aim of this study was to survey the gross and histologic lesions observed in axenic mice from the University of Washington Gnotobiotic Animal Core.

Methods: Records from the University of Washington Veterinary Diagnostic Laboratory were searched for submissions from the Gnotobiotic Animal Core (2014-Present), which included both cull animals which were clinically healthy as well as clinically ill animals. Archived slides from axenic mice were reviewed to evaluate histologic lesions.

Results: As expected for anexic animals, cecal distention was the most common gross lesion identified at necropsy, with roughly one-half of the cases with documented cecal distention having a concurrent cecal volvulus. The most common histologic changes observed were consistent with previously reported changes. Notably there were systemic lymphocytic aggregates, which were observed in the mesentery, salivary gland, pancreas, and thyroid gland. Additional histologic changes included prominent gut-associated lymphoid tissue, and induced bronchiolar associated lymphoid tissue.

Conclusion: A variety of gross and histologic lesions can be present in axenic mice without additional experimental manipulations. These lesions should be considered when performing gross or histopathologic evaluations of axenic mice, as these changes may reflect individual variation in axenic mice, and are not a result of experimental manipulation.

November 5, 2017
1:40 PM – 1:45 PM

METALLOTHIONEIN EXPRESSION IN DOGS WITH CHRONIC HEPATITIS AND ITS CORRELATION WITH HEPATIC FIBROSIS, INFLAMMATION, AND KI-67 EXPRESSION

Santhi Sridharan, Andrew Allen, Beverly Kidney, Ahmad Al-Dissi

Background: Chronic primary hepatitis occurs commonly in dogs, and the etiology is rarely found. Metallothionein (MT) is a heavy metal-binding protein found in many organs, including the liver. MT was recently shown to enhance liver regeneration and decrease hepatic fibrosis in human beings.

Objective: To examine the expression of MT in 24 chronic hepatitis sections and to understand its relationship to hepatic inflammation, fibrosis, bile duct proliferation, and regeneration.

Methods: Immunohistochemistry was used to examine MT and Ki-67 expression which were also scored. Fibrosis, inflammation and bile duct proliferation were also scored.

Results: Regression analysis revealed a significant positive correlation between MT labeling intensity and Ki-67 (r^2 = 0.29, $P < 0.05$). The percentage of MT-positive cells and the overall MT expression were both positively correlated with growth fraction (r^2= 0.25 and 0.26, respectively; P <0 .05). A negative correlation was found between the overall MT labeling and fibrosis (r^2 = 0.18, P <0 .05). A similar trend of negative correlation was also found between the percentage of MT-positive cells and fibrosis, but the P value was not statistically significant (r^2 = 0.14, P = 0.0684).

Conclusions: These findings suggest a protective role of MT in dogs affected by chronic hepatitis, similar to its role in human beings. These dogs may respond to treatment modules focusing on enhancing the expression of MT.

November 5, 2017
1:45 PM – 1:50 PM

CONGENITAL PULMONARY AIRWAY MALFORMATION-LIKE ANOMALY IN A DOG

Jolie A. Demchur, Carol A. Margolis, Adam S. Yoskowitz, Charles W. Bradley

Congenital pulmonary airway malformation (CPAM) is a rare developmental disorder of the lower respiratory tract, most commonly diagnosed in prenatal or neonatal children. In veterinary medicine, there is a single case report of a young boxer dog with similar pulmonary lesions and concurrent pneumothorax. Multiple congenital pulmonary malformations in a 10-month-old male intact, merle Australian Shepherd mixed breed dog with bilateral microphthalmia are described. Thoracic imaging (radiographs and CT) identified pneumonia and markedly abnormal lung parenchyma composed of numerous cysts, bronchiectasis, regional atelectasis and hyperinflation. The dog was euthanized due to pneumonia refractory to treatment. On gross examination, multiple lung lobes consisted of solid to spongy tissue with multiloculated cysts and regions composed of thick-walled cysts up to 5.5 cm diameter with sparse pulmonary parenchyma. In lobes with more normal parenchyma, the lungs were firm and consolidated with apical emphysema. The pulmonary vasculature and bronchi had a normal distribution. Secondary and tertiary bronchi were ectatic with slightly flaccid, collapsible walls. Histologically, the cysts comprised a central cavity lined by a columnar to cuboidal ciliated epithelium supported by a fibrovascular wall containing bundles of smooth muscle. Bronchi and bronchioles often had absent, hypoplastic, or disorganized cartilage plates, irregular bands of smooth muscle, and columnar to cuboidal epithelium. A severe bronchopneumonia affected the abnormal, cystic lung lobes and more normal parenchyma. The imaging, gross and histologic lesions are consistent with CPAM in humans. This is a rare developmental abnormality to be considered when diagnosing congenital pulmonary anomalies in the dog.

November 5, 2017
1:50 PM – 1:55 PM

PANFUNGAL NEXT-GENERATION SEQUENCING OF FORMALIN-FIXED, PARAFFIN-EMBEDDED ANIMAL TISSUES ENABLES IDENTIFICATION OF MIXED FUNGAL INFECTIONS

Courtney Meason-Smith, Aline Rodrigues Hoffmann

Background: Molecular methods have proven valuable for identification of fungi observed on histopathology due to subtle nuances in some fungal morphologies and the fastidious nature of fungal cultivation. A recent panfungal PCR assay on formalin-fixed, paraffin-embedded (FFPE) animal tissues using conventional sequencing was unable to identify 30% of fungi for which PCR amplification was successful.

Objectives: We hypothesized the inability to sequence PCR product was due to heterogenetiy of fungal sequences. The objective of this study was to employ panfungal next-generation sequencing (NGS) on DNA previously extracted from 50 FFPE blocks for which conventional sequencing was unsuccessful.

Methods: The FFPE blocks contained a variety of tissues from a variety of host animals with varying amounts of fungi observed on histopathology. Panfungal NGS was performed on an Illumina Mi-Seq instrument using primers targeting the Internal Transcribed Spacer (ITS) region. Sequences were processed using open-source bioinformatics software, QIIME, and a fungal ITS sequence database.

Results: Results thus far identified heterogenous populations of fungal DNA in three block extracts, and the etiologic agents have been consistent with medical histories and histologic findings. Forty-seven block extracts have been successfully amplified at the sequencing center using increased cycle numbers, and sequencing results are pending.

Conclusions: This is the first known application of fungal NGS on FFPE tissues to identify mixed fungal infections. NGS has improved the sensitivity of the conventional assay and will aid in identification of emerging fungal pathogens as well as prompting future investigations into the mechanism and implications of mixed fungal infections.

November 5, 2017
1:55 PM – 2:00 PM

CANINE RETROPHARYNGEAL LYMPHADENITIS

Charles-Antoine Assenmacher, Stephen D. Cole, Curtis Schelling, Charles W. Bradley

Enlargement of the medial retropharyngeal lymph nodes (MRPLN) is associated with neoplasia, reactive and inflammatory processes. Diagnosis and treatment of retropharyngeal lymphadenitis poses a clinical challenge due to a paucity of information as to cause or outcome. The aim of this study is to identify common clinical correlates and causes for retropharyngeal lymphadenitis in dogs. A biopsy database search for canine MRPLN was performed (January 1, 2004 to March 1, 2017). Seventy-six cases were included, and medical records were reviewed as available. Twenty-eight cases were diagnosed as a drainage reaction, 22 as neoplastic and one as infarcted. Twenty-five cases of lymphadenitis were diagnosed, the majority of which were pyogranulomatous (16) and the remaining (9) were necrosuppurative. The most common abnormalities at time of biopsy included pyrexia (21/22), neutrophilia (19/22), mild anemia (5/20), thrombocytopenia (3/20), and mild alkaline phosphatase elevations (4/17). Three cases (3/20) progressed to fatal regional cellulitis with respiratory complications.

Standard diagnostic tests including aerobic (16/20), anaerobic (14/20) and fungal (3/20) cultures of the MRPLN and histochemical stains for infectious agents (25/25) at biopsy are generally unrewarding, with *Staphylococcus pseudintermedius* cultured in one case, and intralesional gram negative bacilli identified in another. Serology for *Bartonella spp.* was positive in 2 of 3 cases tested. Polymerase chain reaction for genus level detection of *Bartonella* 16s rRNA gene was performed on twenty cases of formalin fixed paraffin embedded tissues and was negative. A cause for medial retropharyngeal lymphadenitis in dogs is rarely identified and supports a noninfectious etiology in some cases.

November 5, 2017
2:05 PM – 2:10 PM

CAROLI SYNDROME IN A 6-YEAR-OLD ROTTWEILER

Nathan D. Helgert, Mee-Ja M. Sula

A 6-year-old intact female Rottweiler was submitted for necropsy after several days of lethargy, ascites, and dyspnea following a lifelong history of liver disease. Grossly, the liver was firm, orange-tan with an enhanced reticular pattern and a bosselated surface, and was smaller than expected at 1.9% of body weight. On cut section, the bile ducts were markedly ectatic and contained viscous light green fluid. The gall bladder was distended, diffusely thickened, and white. There was also a single renal cortical cyst. Within the lungs, there were multiple regions of atelectasis with raised light tan, firm circular nodules.

Histologically, the liver had abundant bridging fibrosis that extended through the limiting plate replacing hepatocytes. Within these areas, there were abundant biliary profiles and biliary epithelial cells that did not form lumens and occasionally, were in direct contact with the surrounding hepatocytes. Some bile ducts were markedly ectatic and lined with hyperplastic epithelium. Scattered throughout the kidneys, there was mild interstitial nephritis with fibrosis, multiple ectatic renal tubules, and a single large cyst.

Gross and histologic findings paired with the early onset of liver disease is consistent with Caroli syndrome, a ductal plate malformation caused by persistence of embryonic hepatic and biliary epithelial progenitors. In humans, ductal plate malformations are frequently associated with adult polycystic kidney disease and pancreatic cysts. In veterinary species, ductal plate malformations are frequently, but not necessarily, associated with renal cysts. A renal cyst was present in the current case.

November 5, 2017
2:10 PM – 2:15 PM

MYELOID SARCOMA (MS) PRESENTING AS A PARA-OVARIAN MASS IN A JUVENILE DOG

Jessica J. Bailey, Jey W. Koehler

A 15-month-old intact female Dachshund dog with a previous 1-week history of dyspnea, lethargy, and anorexia presented with radiographic evidence of pleural effusion, a diffuse interstitial lung pattern, a mottled spleen, enlarged lymph nodes, and an enlarged right ovary. At necropsy, there was a right para-ovarian mass that surrounded the proximal ureter and adrenal gland, and infiltrated the regional hypaxial muscles. Multiple 1-2-mm flat, round, subpleural foci were scattered throughout the lungs, and there were extensive pleural adhesions. Histologically, the para-ovarian mass was composed of neoplastic round cells, and there was multi-organ infiltration and intrasinusoidal presence of neoplastic cells in the liver, supportive of leukemia. The morphology and immunohistochemical profile (Iba1, myeloperoxidase, and elastase) of the neoplastic cells are most consistent with myelomonocytic differentiation. The final diagnosis was acute myelomonocytic leukemia (AML-M4) with widespread tissue invasion and right para-ovarian tumor formation (myeloid sarcoma). Myeloid sarcoma, also known as granulocytic sarcoma (GS) or chloroma, is described in humans as a focal mass of extramedullary proliferation of myeloid precursors seen either preceding or following a diagnosis of acute myeloid leukemia. However, in domestic animals, myeloid sarcomas are largely incidental findings and not typically associated with the subsequent onset of AML. This case is of interest because there are very few reports of myeloid sarcoma in domestic animals, especially in combination with leukemia, and none describing this anatomic location. A review of the current literature for myeloid sarcomas in conjunction with AML in humans and domestic animals will be discussed.

November 5, 2017
2:15 PM – 2:20 PM

IMMUNOHISTOCHEMICAL EXPRESSION OF MULTIPLE MYELOMA ONCOGENE 1 IN CANINE CUTANEOUS HISTIOCYTOMA

Justin M. Stilwell, Dan R. Rissi

Background: Multiple myeloma oncogene-1 (MUM-1) immunohistochemistry (IHC) is mainly used for diagnostic confirmation of plasma cell tumors (PCTs) in dogs. Previous reports indicate MUM-1 is not expressed in histiocytic tumors.

Objective: The objective of this study was to demonstrate MUM-1 expression in cases of canine cutaneous histiocytoma (CH) and compare the staining pattern with canine PCTs submitted to the same protocol.

Methods: Twenty cases of CH and ten cases of PCTs were submitted to MUM-1 and ionized calcium-binding adapter molecule 1 (Iba1) IHC. Distribution of MUM-1 immunostaining in CHs and PCTs was scored 3 (>60% of positive cells), 2 (30-59%), 1 (10-29%), or 0 (<10%) and assessed with student's t-test.

Results: All CHs had patchy to diffuse (average score of 2.65), weak to strong nuclear and weak to moderate cytoplasmic staining for MUM-1 and strong cytoplasmic staining for Iba1. All PCTs had diffuse (average score of 3.00), strong nuclear and moderate cytoplasmic staining with MUM-1 and no staining with Iba1. Although cytoplasmic staining differed, both tumor sets had prominent nuclear staining, which is considered diagnostic for MUM-1 IHC. While nuclear staining score was significantly different ($p=0.03$), 15 of 20 CHs scored 3, making it an unreliable differentiating criterion.

Conclusions: This study confirms MUM-1 expression in canine CH. Although not typically a diagnostic challenge, MUM-1 expression should be used with caution or in conjunction with additional immunomarkers to distinguish between poorly differentiated round cell tumors, especially when a histiocytic or plasma cell origin is suspected.

November 5, 2017
2:20 PM – 2:25 PM

PITUITARY NULL-CELL ADENOMA WITH SUSPECTED STALK SYNDROME IN A FEMALE RHESUS MACAQUE (MACACA MULATTA)

Stephanie Myers, Logan France, Caroline Garrett, Sarah Beck

A 13-year old, female Rhesus macaque (*Macaca mulatta*) presented for progressive weight loss, vision loss, and wide circling to the right over the course of one year. She also exhibited galactorrhea, despite not having given birth in over two years. Significant bloodwork results included hyperprolactinemia (147.6 ng/ml; reference 20-50 ng/ml) and normal thiamine levels. Ophthalmologic exam revealed inability to fix or follow an object, searching behavior, and subjective optic nerve pallor OU on fundic exam. Subsequent MRI revealed a large (3.2x2.9x2.5 cm) contrast-enhancing hypophyseal mass protruding from the pituitary fossa with associated marked mass effect. Due to poor prognosis, humane euthanasia was elected. Gross necropsy revealed a 2.4x2.0x1.4 cm, tan, firm mass dorsal to the sella turcica with compression of surrounding neuropil and optic chiasm. On cut section, the mass extended as far rostrally as the basal ganglia and as far caudally as the pons. The mass also asymmetrically compressed the lateral ventricles, causing mild to moderate hydrocephalus. Histopathologically, the mass was most consistent with a pituitary adenoma. Immunohistochemical stains were strongly positive for synaptophysin and negative for prolactin, adrenocorticotropic hormone, growth hormone, follicle-stimulating hormone, and melanocytic-stimulating hormone. Transmission electron microscopy revealed poorly-demarcated neoplastic cells that contained numerous granules measuring 100-200 nanometers, confirming a pituitary null-cell adenoma. Pituitary tumors occur with varying frequency in many veterinary species, and clinical signs often depend on the size and functionality of the tumor. In this case, the hyperprolactinemia and galactorrhea is likely secondary to pituitary stalk compression syndrome.

November 5, 2017
2:25 PM – 2:30 PM

SPONTANEOUS ATYPICAL FIBROMAS/FIBROSARCOMAS IN DJUNGARIAN HAMSTERS (PHODOPUS SUNGORUS)

Hui-Wen Chang, Victor Fei Pang, Chian-Ren Jeng, Fun-In Wang, Chen-Hsuan Liu

Specimens of ventro-abdominal skin tumors derived from five adult, (18 to 24 month), male Djungarian hamsters (cases 1 to 5) were submitted to the Graduate Institute of Molecular and Comparative Pathobiology of Veterinary Medicine National Taiwan University. Histologically, tissues from cases 1 and 3 were well demarcated dermal growths that showed continuity with dermal nests of cutaneous ganglion cell-like cells (GC cells). Neoplastic GC cells were moderately pleomorphic containing an eccentrically located ovoid nucleus with conspicuous nucleoli and abundant foamy basophilic cytoplasm. A high degree of pleomorphism with frequent presence of multinucleated cells was present in case 4. Cases 2 and 5 were poorly demarcated and highly infiltrative that exhibited muscular invasion. Collagenous matrix was demonstrated by Masson's trichrome staining in all cases. Immunohistochemistry revealed positive reactivity to vimentin and androgen receptor. The morphologic and immunohistochemical features were consistent with atypical fibromas (case 1 and 3) or their malignant counterpart, fibrosarcomas (case 2, 4 and 5). Spontaneous atypical fibromas or fibrosarcomas are derived from the cutaneous GC cells of the thoracoabdominal skin. Previous studies have revealed their androgen dependency during puberty in Djungarian hamsters. The origin of GC cells remains uncertain. The most acceptable hypothesis suggested that these cells might be undifferentiated mesenchymal cells in the skin. The prevalence of atypical fibromas/fibrosarcomas might have been underestimated in the past; thus, they should be included in the list of differential diagnoses when hamster skin tumors are encountered.

November 5, 2017
2:30 PM – 2:35 PM

ARTICULAR HISTOPLASMOSIS IN CATS

Susan E. Fielder, Theresa E. Rizzi, James H. Meinkoth, Andrew S. Hanzlicek, Ruth M. Hallman

Background: Clinical signs of disseminated histoplasmosis are often non-specific. Lameness and joint effusion are rarely reported as a manifestation of this disease. However, clinical experience in endemic areas suggests that articular disease is common in cats with histoplasmosis.

Objectives: Describe clinical and diagnostic findings of articular histoplasmosis in cats.

Methods: Medical records between 2001 and 2017 were reviewed. Inclusion criteria required (1) diagnosis of histoplasmosis by cytology, histology, urine/serum antigen, or culture and (2) lameness or joint effusion as a presenting complaint or physical examination finding.

Results: Twenty five cases met inclusion criteria. Four had incomplete records, but available data was included when applicable. Lameness was a presenting complaint in 17/21 cats and the only complaint in 9/21 cats. Initial diagnosis was made by cytology in 22/25 cats and by culture, urine antigen and necropsy in one case each. Diagnostic cytology samples included: synovial fluid (13), lymph node (6), lung (1), bone (1) and skin (2). Two additional cases had synovial fluid examined but no organisms present. Inflammation was present in all synovial fluid samples examined. Biopsy was obtained in two cats and histologic diagnoses included osteomyelitis with no infectious organisms identified (including GMS) and severe lymphoplasmacytic synovitis consistent with feline periosteal proliferative polyarthritis. Histoplasma urine antigen test was positive in 7/12 cats.

Conclusion: Inflammatory arthritis is common in cats with histoplasmosis, with lameness a common presenting complaint. Organisms are found in synovial fluid cytology in most cases. If not, appropriate additional diagnostics must be pursued.

November 5, 2017
2:35 PM – 2:40 PM

ULCERATIVE & GRANULOMATOUS GLOSSITIS, PHARYNGITIS & ENTERITIS IN A HEIFER CAUSED BY RHODOCOCCUS EQUI

Ahmad A. Saied, Laura K. Bryan, David C. Bolin

Rhodococcus equi is a Gram-positive aerobic coccobacillus of the *Corynebacteriales* that is associated with clinical disease in horses and less frequently, other animal species including humans. *R. equi* infection in horses is common and is characterized by pyogranulomatous pneumonia and ulcerative enterocolitis. *R. equi* clinical disease in cattle, however, is rare and typically manifests as granulomatous lymphadenitis discovered in the abattoir. Here, we report an unusual pathological presentation of *R. equi* infection in a 19-month-old female Santa Gertrudis. The heifer had a history of intermittent inappetence, and weight loss for a 3-month-period before euthanasia. Gross and histological examination revealed severe, chronic, ulcerative and granulomatous inflammation in the tongue, pharynx, and small intestine. Also, the heifer had severe, granulomatous pharyngeal and mesenteric lymphadenitis. Bacterial cultures from the ileum, tongue, and liver yielded numerous to medium numbers of *Rhodococcus equi.* Polymerase chain reaction analysis of the isolate detected the linear virulence plasmid *vapN*, which is often identified in bovine isolates (*traA* and *vapN*-positive). The bacteria also lack the circular plasmids; *vapA*, and *vapB* that are associated with virulence in horses, and swine, respectively. This is the first report of ulcerative and granulomatous enteritis in a bovine due to *Rhodococcus equi* infection.

November 5, 2017
2:40 PM – 2:45 PM

PRIMARY GASTROINTESTINAL FOLLICULAR LYMPHOMA IN THREE DOGS

Michael A. Richardson, Tuddow Thaiwong, Matti Kiupel

In humans, primary gastrointestinal lymphomas are most commonly B-cell lymphomas with diffuse large B-cell lymphoma (DLBCL) and mucosa-associated lymphoid tissue lymphoma (MALTOMA) representing the majority of cases. Gastrointestinal follicular lymphomas (FL) are rare primary gastrointestinal neoplasms with an indolent behavior and good response to surgical resection or monoclonal antibody therapy. In dogs, enteropathy-associated T cell lymphomas (EATL) large cell type (type 1) is the most common gastrointestinal form of canine lymphoma; however, DLBCL also occurs as primary gastric and, to a lesser degree, intestinal lymphoma. Primary gastrointestinal FL has not been reported in dogs. We identified three cases of primary rectal FL in dogs with histomorphologic and immunohistochemical features similar to the human entity. The affected dogs were all castrated males, between 9 months to 2 years old and of varying breeds. Expanding the rectal submucosa in all 3 caseswere large, uniform, densely packed follicles that had no polarity, differentiating them from hyperplastic lesions. There were uniform proportions of centrocytes and centroblasts through all follicles with an absence of a dark zone of mantle cuff, and no tingible body macrophages. Immunophenotypically, centrocytes and centroblasts expressed typical B-cell antigens (CD20, CD79a, and PAX5), anti-apoptotic protein BCL2, and lacked expression of the T-cell antigen CD3. PCR for rearrangements of the immunoglobulin heavy chains confirmed a clonal expansion of B-cells in all cases. Surgical excision or chemotherapy appeared curative. Recognition and accurate diagnosis of this distinct manifestation of gastrointestinal lymphoma is important for appropriate therapeutic selection and prognosis.

November 5, 2017
3:30 PM – 3:35 PM

PRACTICAL APPLICATION OF MINION NEXT-GENERATION SEQUENCING OF PORCINE REPRODUCTIVE AND RESPIRATORY SYNDROME VIRUS FOR ROUTINE IDENTIFICATION AND STRAIN TYPING

James B. Stanton, Kevin K. Lahmers, Stephanie Michelle Todd, Tanya LeRoith

Detection and characterization of pathogens by their nucleic acid is a universally accepted approach. This is typically accomplished by polymerase chain reactions; however, this requires knowledge of suspected pathogens and is limited in pathogen genotyping. Next-generation sequencing allows for an unbiased identification of organisms and provides information that can be used for genotyping. The nanopore-based MinION was used to detect and correctly genotype multiple Porcine Reproductive and Respiratory Syndrome virus (PRRSV) infections in porcine fluids using a custom library for the, open-source software, Centrifuge, which includes prokaryotic and viral RefSeq genomes as well as additional sequences from targeted porcine viruses (e.g., 700+ PRRSV genomes). This commercially available sequencer provides long reads, which allows for spanning of repetitive regions and coverage of full viral genomes in a single read. Furthermore, it is powered by the laptop USB port, has a low capital investment and provides real-time sequence information, which can speed bioinformatics and allow for optimization of individual sequencing runs. In addition to identifying PRRS virus, co-infections with other viruses and bacteria were identified, demonstrating the unbiased nature of this assay. Analysis using Centrifuge can be completed on a laptop within 90 seconds of full sequencing runs. In summary, nanopore-based sequencing is a financially and technically achievable diagnostic tool, which allows for rapid and effective identification and genotyping of PRRSV with the added benefit of identifying co-infections.

November 5, 2017
3:35 PM – 3:40 PM

NOVEL LYSOSOMAL STORAGE DISEASE IN A 7-MONTH-OLD GOLDEN RETRIEVER DOG

Tyler A. Harm, Dana N. LeVine, Scott Christopher, Shannon J. Hostetter, Elizabeth M. Snella, N. Matthew Ellinwood, Gary S. Johnson, Jodi D. Smith

Lysosomal storage diseases (LSD) are a group of genetic and, infrequently, acquired disorders characterized by accumulation of macromolecules within lysosomes due to defective catabolism. Heritable LSDs have been reported in a number of veterinary species, which, given the relative rarity of these diseases in children, have become critical models for the study of LSD pathogenesis and potential therapies. To date, two LSDs have been reported in Golden Retriever dogs: Sandhoff disease (a GM2 gangliosidosis) and neuronal ceroid lipofuscinosis associated with a *CLN5* mutation. We report a novel LSD in a 7-month-old male Golden Retriever that presented with a history of progressive pelvic limb weakness, ataxia, difficulty prehending food, and diarrhea. On physical examination, lateral strabismus, nystagmus, head tremor, generalized muscle atrophy, and weakness were noted. Creatinine kinase was minimally elevated, and suspect lysosomal inclusions were noted within lymphocytes on a blood smear. One week later the dog was humanely euthanized and submitted for necropsy. Significant pathological findings included diffuse cytoplasmic vacuolation of neurons in the brain, spinal cord, and mesenteric ganglion, and in retinal ganglion cells. Ultrastructurally, the vacuoles contained lamellated membranous inclusions arranged in either a concentric or stacked pattern, consistent with gangliosides. Beta-hexosaminidase A activity was elevated (in contrast to decreased activity expected in Sandhoff disease), and the dog tested negative for the previously identified *CLN5* mutation. Findings are consistent with a novel Golden Retriever LSD, which may serve as a new model for LSD study. Whole genome sequencing is in progress to identify the causal mutation.

November 5, 2017
3:45 PM – 3:50 PM

NOVEL REAL-TIME SCOPE-BASED DIGITAL SCANNER TECHNOLOGY NOW A VIABLE ALTERNATIVE TO STATIC IMAGE 'STITCHING' OR BENCHTOP SLIDE SCANNING OF OVERSIZED TISSUES

James R. L. Stanley, Serge Rouselle, Armando Tellez

Background: Despite leaps in scope-based digital imaging technology, the imaging of large/oversized histologic tissues (e.g., bone sections, medical devices, whole mount embryos) requires either the 'stitching' together of numerous static images, a tedious to impossible process, or the use of benchtop slide scanners, which are cost prohibitive to many pathologists.

Objective: To investigate the efficiency and quality of imaging oversized/large histologic tissues using a dynamic microscope-mounted digital scanning technology compared a standard static digital imaging system.

Methods: Using the 10x objective, an oversized histologic bone section measuring 38 x 13mm was digitally imaged in its entirety using both static images captured with standard image capturing software (i.e., manual 'stitching') and a scope-based digital pathology scanning software.

Results: The digital imaging of an oversized histologic section of bone at 10x by the manual 'stitching' of static images required the capture of 75 images and 56 minutes compared to the use of the scope-based digital scanning software which required no repeated image capture and 6.5 minutes.

Conclusion: The emergence of the novel Pantoptiq scope-based slide scanning software by ViewsIQ has revolutionized the ability and productivity of pathologists who routinely work with oversized tissues (e.g., bone sections, medical devices, whole mount embryos), the imaging of which has heretofore been limited to either the tedious and time-consuming 'stitching' of static images or access to expensive slide scanners.

November 5, 2017
3:55 PM – 4:00 PM

FEMALE REPRODUCTIVE PATHOLOGY IN GERIATRIC FALLOW DEER (DAMA DAMA)

Colleen F. Monahan, Wynona Shellabarger, Kimberly A. Thompson, Kirk Suedmeyer, Dalen W. Agnew

Spontaneous reproductive neoplasms in cervids, and ruminants in general, are rare. This study included 15 female fallow deer (*Dama dama*) (age 11 to 17 years) from a single inbred and reproductively senescent population-submitted to the Michigan State University Veterinary Diagnostic Laboratory or the Reproductive Health Surveillance Program (Association of Zoos and Aquariums). These cases represented 72% of the total mortality among females in this population, with complete necropsies performed, from 2007-2017. An additional case was from another zoo. These deer were housed with sterilized males for many years. The objective of this study was to describe the reproductive pathology of a population of aged fallow deer which had been barren for many years. Reproductive tract examination yielded 44 lesions, including 18 neoplastic and 26 non-neoplastic. The most common neoplasm was endometrial carcinoma (9/18), 8 of which had extensive metastasis to multiple organs. Additional neoplasms included leiomyoma (3), leiomyosarcoma (2), endometrial stromal tumor (1), endometrial adenoma (1), uterine squamous cell carcinoma (1), and vaginal adenocarcinoma (1). One leiomyosarcoma had extensive metastasis to multiple organs. The most common non-neoplastic lesion was cystic endometrial hyperplasia (9). Other non-neoplastic reproductive lesions included ovarian subsurface epithelial cysts (3), ovarian follicular atrophy (3), paraovarian cysts (2), ovarian follicular cystic hyperplasia (2), serosal inclusion cysts (2), endometrial polyps (2), adenomyosis (1), endometrial atrophy (1), and vaginal cystic epithelial hyperplasia (1). This case series and a few case reports in the literature suggest that fallow deer are predisposed to reproductive neoplasia and/or that pregnancy may be protective.

November 5, 2017
4:00 PM – 4:05 PM

INTRAOCULAR MALIGNANT TUMOR WITH SECONDARY FUSARIUM INFECTION IN A MARINE FISH DIAGRAMMA PICTUM

Wen-Ta Li, Chieh Lo

Eye disorders are not uncommon in displayed fish and may be caused by trauma, microorganism infection, and/or neoplastic diseases. However, intraocular neoplasiawith secondary fungal infection is rarely reported in marine fishes. An adult *Diagramma pictum* was presented with unilateral buphthalmos of the right eye, which subsequently ruptured and was enucleated. Grossly, the cornea was edematous and irregularly thickened/proliferative. The intraocular architecture was effaced and replaced by grey to dark red tissue. Microscopically, the intraocular grey to dark red tissue was composed of neoplastic pleomorphic cells arranged in trabeculae and islands with the presences of partial whorls and storiform patterns. The neoplastic cells had a moderate amount of indistinctly bordered eosinophilic cytoplasm, and contained a round to oval nucleus with clumped chromatin and prominent nucleolus. Mitosis was infrequently observed. Aggregates of the neoplastic cells werefound around the optic nerve, indicative of invasive biological behavior. Considering the cell morphology and the immunohistochemical results (CK and S100 positive), melanoma was the most likely diagnosis. Furthermore, multifocal foci composed of mixed inflammatory cells with cell debris and fungal hyphae were noted within the melanoma. The fungi were molecularly identified as *Fusarium solani*. In the present case, the intraocular melanoma can increaseintraocular pressure by disrupting the aqueous humour circulation and thereby cause buphthalmos, increasing the risk of corneal abrasion and fungal infection. It is speculated that the fungal infection is mainly located within the melanoma because of the rich blood supply in tumor growth.

November 5, 2017
4:05 PM – 4:10 PM

CO-OCCURRENCE OF NASAL POLYPS AND PRIMARY NASAL CARCINOMAS IN DOGS

James C. Tarrant, David E. Holt, Amy C. Durham

Background: Nasal polyps in dogs occur as space occupying soft tissue masses, often with turbinate destruction and extension through nasal septum. Clinical signs include sneezing, nasal discharge, epistaxis, and stertor. Primary nasal carcinomas present with similar clinical signs and advanced imaging findings. Biopsy samples of nasal masses often have features of both a nasal polyp and carcinoma. Occasionally, the first biopsy sample is diagnostic of a nasal polyp; a repeat deeper tissue biopsy following irresolution of clinical signs has revealed carcinoma.

Objective: This study's aim is to examine the co-occurrence of nasal polyps and nasal carcinomas.

Methods: Case selection from the Penn Vet Diagnostic Laboratory included 220 dogs with a biopsy diagnosis of nasal carcinoma (72 adenocarcinomas, 21 transitional, 12 poorly differentiated, 9 squamous cell, 1 acinic cell, 1 adenosquamous, and 103 carcinomas not otherwise specified), and/or nasal polyp from 2004-2017. All slides with a diagnosis of nasal carcinoma were reviewed for key histological features of a polyp; namely normal ciliated epithelium, markedly edematous submucosal stroma, mucin accumulation, lymphangiectasia, and mild mixed inflammation.

Results: Of the 220 cases of nasal carcinoma, 47 dogs (21%) had large regions of tissue diagnostic of nasal polyp. A primary diagnosis of nasal polyp was made in 55 dogs, and of these, 6 dogs (11%) underwent a repeat biopsy that revealed a carcinoma.

Conclusions: Nasal polyps may occur in conjunction with nasal carcinoma. In the case of continued clinical signs, repeat biopsy to reveal possible neoplasia following a diagnosis of nasal polyp may be warranted.

November 5, 2017
4:10 PM – 4:15 PM

HISTOPATHOLOGICAL FINDINGS IN LIVER, GILLS AND KIDNEY OF ACARÁS (GEOPHAGUS BRASILIENSIS) FROM WATER RESERVOIR BILLINGS IN SÃO PAULO STATE, BRAZIL

Sandy L. Pulecio Santos, José Henrique Hildebrand Grisi Filho, Ivy Tasso Gomes, Ana Carolina Camachos Lopez, Vera Lisa Generosa Paiva Silva, Barbara Held, Marta Condé Lamparelli, Gilson Alves Quináglia, Lilian Rose Marques de Sá

Background: Billings, the largest water reservoir in São Paulo state and the largest source of water supply in the São Paulo Metropolitan Region. Histopathological evaluation of different aquatic species has been used as a biomarker in different contexts, since it may reflect the quality of the environment. Acará (*Geophagus brasiliensis*) is widely distributed in Brazil and used for human consumption.

Objective: To describe histopathological findings of gills, liver and kidney of acarás from Billings reservoir captured in 2014 and 2015 and their possible implications as a bioindicator in this context.

Methods: Tissues of 210 acarás captured at different control points of the reservoir were evaluated. Sampling was carried out in March, April and November of 2014 and 2015. Necropsy and tissue processing were performed for histopathological evaluation.

Results: In the gills (n=204) the most frequent lesions were pillar cell rupture in 110 fishes(53%), epithelial hypertrophy in 109 (53.43%), edema in 81 (39.71%), epithelial hyperplasia in 54(26.47%) and parasites in 49 (24.02%). In liver (n=208) hydropic vacuolar degeneration in 167 individuals (80.29%), congestion in 123 (59.13%), intracellular pigments deposition in 123 (59.13%), steatosis in 104 (50%), bile periductal fibrosis in 60 (28.85%). In the kidney (n=86), congestion in 41 fishes (47.67%), lymphocytic interstitial infiltrate in 39 (45.35%), hemorrhage in 11 (12.79%), tubular degeneration in 10 (11.63%) and membranous glomerulonephritisin 4 (4.65%).

Conclusion: Histopathological findings described mainly degenerative, nonspecific and reversible changes in liver and kidneys, but irreversible changes in gills indicating that water condition adverse affected acaras.Financial support:CNPq-Brasil, FEHIDRO–AT 603.

November 5, 2017
4:15 PM – 4:20 PM

NASAL ADENOCARCINOMA WITH INTESTINAL DIFFERENTIATION IN A BEAGLE

Chi-Fei Kao, Fei Victor Pang, Hui-Wen Chang, Chian-Ren Jeng, Fun-in Wang, Chen-Hsuan Liu

Background: Non-intestinal glandular malignancy with intestinal differentiation has been well documented in humans but not in animals. Aside from microscopic similarity, the neoplastic cells exhibit unique immunohistochemical and molecular characteristics indicating intestinal differentiation. Here we report the first case of a canine nasal adenocarcinoma histologically and immunohistochemically resembling colorectal adenocarcinomas.

Case Description: A 9-year-old, male intact beagle with 1-year unilateral nasal discharge presented with left facial swelling and a maxillary mass. Diagnostic imaging revealed osteolysis of the left maxilla and destruction of the nasal conchae by a soft tissue opacity involving the leftorbital cavity, frontal sinus and Eustachian tube. Thoracic and abdominal cavities appeared unremarkable. Several biopsies of the mass were preserved and processed routinely for histological evaluation.

Results: Microscopically, the mass comprised mucinous pools and tubulopapillary growths reminiscent of neoplasms arising from the intestinal tract with varying numbers of goblet cells. The neoplastic cells were cuboid to columnar, having eosinophilic cytoplasm and mostly basally placed nuclei. They were positive for pan-CK, CDX-2, Villin and CK20 but negative for CK7. Fine needle aspiration of the submandibular lymph node showed clusters of epithelial cells with mild pleomorphism, indicative of metastasis.

Conclusion: Although without necropsy, we would not be able to completely rule out a primary intestinal tumor, the absence of related history, failure to identify other primary neoplasm, and the extreme rarity of nasal metastasis from colorectal adenocarcinomas, made this unlikely. To our knowledge, this should be the first case demonstrating intestinal differentiation in nasal adenocarcinoma in the veterinary literature.

November 5, 2017
4:20 PM – 4:25 PM

ATHEROSCLEROSIS AND STENOSIS OF THE ASCENDING AORTA WITH MYOCARDIAL FIBROSIS AND LEFT VENTRICULAR DILATATION IN A YOUNG MEERKAT (SURICATA SURICATTA)

Chien-Hao Chen, Yen-Hsueh Lai, Hui-Wen Chang, Cho-Hua Wan, Chian-Ren Jeng, Fei Pang

Background: Atherosclerosis is primarily of interest to the comparative pathobiologist. Its development is a progressive process and usually becomes clinically manifest in middle age or later in humans. Herein we report a case of young meerkat with artherosclerosis in the ascending aorta that leadto aortic stenosis, dilated cardiomyopathy, and myocardial fibrosis.

Case Presentation: A one-year-old, pregnant meerkat that had a history of hypercholesterolemia with its littermates showed weakness and died within two days. Upon necropsy, dilated left ventricle and multiple white linear to nodular protruded plaques in the ascending aorta leading to severe luminal stenosis were observed. Congestion and hemorrhage were noted in multiple organs.

Results: Linear atherosclerotic plaques, characterized by a fibrous cap and central deposits of cholesterol clefts with scattered foam cells, mineralization, and collagenous fibrosis, involved the intima and upper media of the ascending aorta. Focal rupture of the fibrous cap with mild fibrin deposition was revealed. There was diffuse mild subendocardial and severe zonal myocardial fibrosis with scattered mineralization. The liver showed diffuse moderate centrilobular congestion and necrosis. Areas of congestion, fibrin thrombi, and hemorrhage were seen in the lungs, kidney, and uterus. Heart failure cells were widely distributed in the lungs.

Conclusion: Cholesterol granulomas with/without hypercholesterolemia and aortic atherosclerosis was reported in three 4-to-10-year-old meerkats. In both human and veterinary medicine, atherosclerosis in young individuals is a rare phenomenon. The mechanism for hypercholesterolemia in the meerkat is not known and the potential role of genetics in meerkats remains to be evaluated.

November 5, 2017
4:25 PM – 4:30 PM

METASTATIC LEIOMYOSARCOMA OF THIGH MUSCLE ORIGIN IN A BUDGERIGAR (MELOPSITTACUS UNDULATES)

Sonika Patial, Christine Higbie, Thomas Tully Jr., Nobuko Wakamatsu

An adult male budgerigar presented with a history of intermittent left leg lameness. Radiographs revealed severe soft tissue swelling surrounding the left femur. The coxofemoral joint was unidentifiable and the femur was present in multiple, thin, fragments. Due to poor prognosis, the budgerigar was euthanized and necropsy was performed. Grossly, the proximal portion of the left limb was markedly enlarged with a 3x1x1cm, intramuscular, multilobulated firm nodule which extensively effaced and replaced the femur; on cut surface, the nodule was yellow with caseous to firm material. The caudal half of the left kidney was moderately enlarged and contained a focal, raised, white nodule that on cut surface contained white to yellow caseous material. Histopathologically, infiltrating, expanding and largely replacing the left hind limb skeletal muscle and the bone was an un-encapsulated, abundantly cellular neoplasm composed of spindle cells arranged in interlacing streams and bundles, and separated by interconnecting rays of fibrovascular stroma. Neoplastic cells had indistinct cellular borders, moderate amounts of eosinophilic and finely fibrillar cytoplasm, irregularly oval to elongate centrally located nuclei, finely stippled to vesiculate chromatin, and indistinct nucleoli. Mitotic index was 50 per 10 high-power fields. Anisocytosis and anisokaryosis was moderate. Metastases were present within the left kidney, liver, spleen, and lungs. Based on the histopathological and immunohistochemical feature of strong immunoreactivity to alpha-smooth muscle actin, the neoplasm was consistent with a leiomyosarcoma. Leiomyosarcomas are extremely rare in budgerigars. To the best of our knowledge, this is the first case report of a metastatic leiomyosarcoma in budgerigar.

November 5, 2017
4:30 PM – 4:35 PM

ASPERGILLUS NIGER-ASSOCIATED CALCIUM OXALATE CRYSTALS IN AN EURASIAN EAGLE OWL (BUBO BUBO)

Patti Kiser, Danielle Meritet, Rob Bildfell

An adult male Eurasian eagle owl (*Bubo bubo*) housed at a wildlife rehabilitation facility in southern Oregon died after progressive ill-thrift. Radiographs taken prior to death demonstrated abnormal radiopaque material in the coelom and the owl was submitted for post-mortem examination. Black pigmented fungus was noted grossly, particularly in the respiratory tissues, with abundant oxalate crystal deposition associated with fungal infestation subsequently observed histologically. The fungus was later confirmed by culture to be *Aspergillus niger*. Oxalate crystals in the absence of hyphae were also identified in some tissues. There have been few reports of aspergillosis caused by

Aspergillus niger in avian species and the wide tissue distribution of oxalates is highly unusual.

November 5, 2017
4:35 PM – 4:40 PM

CO-INFECTION OF MYCOBACTERIUM MARINUM AND M. FORTUITUM IN A CAPTIVE ADULT DIAMONDBACK WATER SNAKE (NERODIA RHOMBIFER) CAUSING DISSEMINATED MYCOBACTERIOSIS WITH AN ACUTE CUTANEOUS MANIFESTATION

Tatiane Terumi Negrao Watanabe, Emi Sasaki, Gordon J. Pirie, Nobuko Wakamatsu

Background: Mycobacteriosis is a sporadic disease in reptiles and potentially zoonotic. *Mycobacterium marinum*, the most common isolate in reptiles, is a slow growing species commonly found in both fresh and salt water and causes cutaneous nodules and papules known as "aquarium/fish tank granuloma" in humans. *Mycobacterium fortuitum* is a ubiquitous, rapid growing species, and the infection has been reported in pet snakes and the owner suffering from lymphadenitis.

Case Description: An adult male captive diamondback water snake (*Nerodia rhombifer*) was found dead after presenting one day history of lethargy with multifocal, acute ulcerative cutaneous lesions throughout the body. The snake ate two sunfish (*Mola* spp) five days prior to death. In addition to the cutaneous lesions, gross examination revealed multifocal to coalescing, white to pale tan nodules in the lung and liver and segmental impactions of the small and large intestines with digesta containing fish's bones. Histopathologic examination confirmed severe granulomatous inflammation with numerous intrahistiocytic acid-fast bacteria in the skin and underlying muscle, lung, liver, and intestines. *M. marinum* and *M. fortuitum* were identified by culture of the hepatic granuloma, followed by PCR and rpoB gene sequencing.

Conclusions: This is a case report of mycobacteriosis in a water snake co-infected with slow and rapid growing mycobacterial species. Ingestion of infected fish/water and direct contact with contaminated water are considered most likely routes of the infection. In addition to forming granulomatous nodules in multiple organs, the mycobacterial infection caused acute ulcerative cutaneous lesions and intestinal impactions due to intestinal mural granulomas.

November 5, 2017
4:40 PM – 4:45 PM

EVALUATION AND OPTIMIZATION OF MOLECULAR CLONALITY ASSAYS FOR CANINE B CELL LYMPHOMA

Mei-Hua Hwang, Nikos Darzentas, Dorothee Bienzle, Peter F Moore, Stefan M Keller

Background: Molecular clonality assays can be a powerful diagnostic test for identifying canine B cell lymphoma when conventional means yield equivocal results. However, current assays suffer from a high false negative rate. We hypothesized that this is because the incomplete annotation ofthe canine immunoglobulin heavy chain

(IGH) locus and the limited characterization of somatic hypermutation haveresulted in only partial primer coverage of rearranged genes.

Objectives: (1) To identify potential causes for false negative results of canine B cell clonality assays based on deep sequencing data of the rearranged IGH repertoire and, (2) to improve existing assays by re-designing existing primer sets.

Methods: Immune repertoire sequencing was performed on cDNA from 3 lymphoid organs of 3 healthy dogs each on the Illumina MiSeq v3 PE300. Sequencing reads were analyzed using the ARResT/Interrogate platform. Primer sequences of existing assays were compared to sequencing data *in silico*. Novel primers were designed to minimize mismatches.

Results: Mismatches between primer sequences and the sequenced repertoire were due to incomplete coverage of germline sequences as well as somatic hypermutation. For most forward primers, covering an additional IGHV gene of family 3 reduced mismatches up to 20%. Substantial mismatches were also observed resulting from frequently rearranged members of family 1. Mismatches due to somatic hypermutation were less prominent.

Conclusion: The novel primer sets are anticipated to provide increased sensitivity due to improved germline coverage and reduced mismatches; however, further testing is required to determine exact sensitivity against existing assays.

November 5, 2017
4:45 PM – 4:50 PM
EQUID HERPESVIRUS 5-ASSOCIATED DERMATITIS IN TWO HORSES IN OREGON, USA
Christiane V. Löhr, Patti Kiser, Robert Bildfell, Jens Peter Teifke

Background: Equid herpesvirus 5 (EHV5)-associated dermatitis resembling herpes-associated erythema multiforme has been reported in an adult horse in Germany (Herder 2012). Viral etiology had been confirmed by molecular analyses including PCR, sequencing and in situ hybridization.

Objective: We report two cases of dermatitis associated with Equine Herpesvirus 5 (EHV5) in adult horses from separate premises in Oregon, USA.

Methods and Results: A male 19-year-old Thoroughbred gelding and an adult American Paint mare presented with crusty lesions over the nares. Punch biopsies from affected skin of each horse were examined and showed similar histopathology. The epidermis had marked epidermal hyperplasia and hyperkeratosis with scattered apoptotic keratinocytes and numerous intranuclear inclusion bodies in keratinocytes bordering areas of erosion. The dermis showed moderate lymphoplasmacytic interface dermatitis with pigmentary incontinence. The skin of the mare also had vascular sclerosis and marked elastosis consistent with damage due to chronic ultraviolet light exposure. In situ hybridization specific for EHV5 was performed as previously published and intensely labeled many keratinocytes in affected areas.

Discussion: Histopathology of the dermatitis in horses in Oregon was very similar to that described for a horse with EHV5-associated dermatitis in Germany, and EHV5 DNA was identified in affected keratinocytes. To the best of our knowledge this is the first report of EHV5-associated dermatitis in the North American horse population.

November 5, 2017
4:50 PM – 4:55 PM

SYSTEMIC AMYLOIDOSIS IN A CAPTIVE POPULATION OF PRONGHORN ANTELOPE (ANTILOCAPRA AMERICANA)

Margaret E. Martinez, Katie Seeley, Dawn Zimmerman, Priya Bapodra, Rachel Cianciolo

Background: Fourteen pronghorn antelope (*Antilocapra americana*) from a single captive herd underwent complete or partial autopsies between 1997 and 2016.

Objective: Our goal was to characterize the histologic and ultrastructural changes, as well as the underlying cause of the amyloid.

Methods: Histologic examination of all H&E and congo red stained microscopic slides, as well as transmission electron microscopy and mass spectrometry was performed in two cases. Four banked serum samples from affected pronghorns had serum amyloid A, haptoglobin, beta and gamma globulin levels measured. Pedigree analysis and retrospective investigation into the clinical histories was performed.

Results: Ten animals had histologic evidence of amyloidosis resulting in a prevalence of 77% of autopsied animals. The majority (90%) of animals had histologic amyloid in the kidneys often causing global to segmental expansion of the glomerular mesangium and capillary wall. Transmission electron microscopy demonstrated glomerular deposits of haphazardly arranged non-branching fibrils consistent with amyloid, as well as a coarser fibril. Mass spectrometry revealed the presence of serum amyloid A and fibronectin. The banked serum amyloid A, beta and gamma globulin levels from affected pronghorns were within normal ranges for healthy domestic cattle. Clinical commonalities between most of the cases included elevated fecal strongyle counts (*Haemonchus* spp.), anemia, hypoproteinemia and azotemia. At post-mortem examination, several (40%) animals had a diagnosis of pneumonia. There was no significant difference between the mean degree of relatedness and presence of amyloidosis.

Conclusions: Therefore, the systemic reactive amyloidosis was likely secondary to underlying chronic inflammation caused by haemonchosis and/or pneumonia.

November 5, 2017
4:50 PM – 5:00 PM

FROM THE FDA CASE FILES: PENTOBARBITAL TOXICOSIS AND EXOGENOUS THYROTOXICOSIS IN CANNED DOG FOOD

David S. Rotstein, Jennifer Jones, April Hodges, Mark Glover, Eric Nelson, Amber McCoig, Olgica Ceric, Sarah Nemser, Lee Anne Palmer, Lauren Carey, Jackie Queen, Jennifer Erickson, William Burkholder, Elizabeth Edwards, Ruth Yowell, John Buchweitz, Renate Reimschuessel

FDA Center for Veterinary Medicine (CVM) investigated two separate adulteration incidents that involved canine illness and death after exposure to canned pet food that resulted in subsequent recalls. The investigations involved owner interviews, medical record review, sample collection and analysis. The pathology component included necropsy or microscopic evaluation of the pet food. The first incident involved five dogs from a household that developed acute neurologic clinical signs following consumption of food contaminated with pentobarbital. One dog was euthanized secondary to complications and the remaining dogs recovered. Pentobarbital was detected in the gastric contents and open canned food by a contract veterinary diagnostic laboratory. No significant findings were found grossly and microscopically. Regulatory analysis using gas chromatography-mass spectrometry and liquid chromatography-mass spectrometry (LC-MS) confirmed the presence of pentobarbital in several lots at varying concentrations. In the second incident, a cluster exogenous thyrotoxicosis cases were reported. Affected dogs were exposed to two brands of dog food with a common manufacturer. All dogs had an elevated triiodothyronine (T3) level. Using liquid chromatography-inductively coupled plasma mass spectrometry and LC-MS, the canned product was found to contain the iodine species; T3, monoiodotyrosine, and diiodotyrosine. An island of thyroid follicles was present microscopically in the same brand, but different lot of canned food. Cytokeratin 19, thyroglobulin, and calcitonin immunohistochemistry were negative. The source of the contamination was gullet in the product. As a result of these two investigations, firms were inspected, the products were recalled, and FDA CVM issued web updates to provide public information.

Diagnostic Pathology Focused Scientific Session II

November 7, 2017 | 1:30 PM – 5:00 PM
Session Chair: L. Kennedy
Committee Members: E. Burrough (Co-Chair), R. Derscheid, F. Oliveria, A. Rodrigues Hoffman, D. Rotstein (Past Chair), H. Snyman.

November 7, 2017
1:30 PM – 1:40 PM

CHARACTERIZATION OF NATURAL CANINE DISTEMPER VIRUS INFECTION IN FIVE ADULT LINNEAEUS'S TWO-TOED SLOTHS (CHOLOEPUS DIDACTYLUS)

Allison M. Watson, Andrew C. Cushing, Julie D. Sheldon, Eman Anis, Rebecca P. Wilkes, Edward J. Dubovi, Linden E. Craig

Canine distemper virus infections have been documented in numerous captive zoo animal species. An outbreak of canine distemper virus in a private zoo in Eastern Tennessee in July 2016 led to fatal clinical disease in five adult, wild-caught Linneaeus's two-toed sloths (*Choloepus didactylus*). Clinical signs included hyporexia, lethargy, mucopurulent nasal discharge, and oral and facial ulcerations. At necropsy, affected animals had crusting and ulcerative lesions of the lips, nose, tongue, and oral cavity. Microscopically, the most severely affected organ was the liver; widespread, random, hepatic necrosis was present in all sloths. Epithelial cells and histiocytes within the oral mucosa, lips, nares, lung, gastrointestinal tract, trachea, spleen, liver, and urinary bladder contained discrete, eosinophilic intranuclear and intracytoplasmic inclusion bodies and occasional syncytial cells. Canine distemper virus was confirmed with immunohistochemistry and virus isolation. Sequencing of the virus confirmed the novel American 4 strain prevalent in Eastern Tennessee wildlife. All five sloths lacked gross and histopathologic lesions within the central nervous system, which is characteristic of canine distemper virus in commonly affected domestic and wild animal species. This is the first report of canine distemper virus infection in Superorder Xenarthra, and characterizes the significant morbidity and mortality in Linneaeus's two-toed sloths.

November 7, 2017
1:40 PM – 1:50 PM

PATHOLOGIC LESIONS ASSOCIATED WITH NOVEL RAT POLYOMAVIRUS 2 IN A COLONY OF X-SCID RATS

Lora H. Rigatti, Tuna Toptan, Joseph T. Newsome, Patrick S. Moore, Yuan Chang

Polyomaviruses (PyVs) are known to infect a wide range of vertebrates and invertebrates and are associated with a broad spectrum of diseases, including cancers, particularly in immune-suppressed hosts. In 2016, our group identified a novel polyomavirus, designated rat polyomavirus 2 (RatPyV2), from a breeding colony of rats having X-linked severe combined immunodeficiency. The virus was initially identified using a human panpolyomavirus immunohistochemistry test (P-PIT), and the genomic sequence was subsequently amplified and identified as a novel 5.1-kb polyomavirus closely related to human Washington University (WU) and Karolinska Institute (KI) and vole polyomaviruses but notably divergent from Rattus norvegicus PyV1 (RnorPyV1; also designated RatPyV1).

Complete postmortem examinations were performed on 12 animals. Gross lesions were limited to the nasal turbinates and lungs while microscopic lesions were present in a variety of glandular tissues including salivary, harderian, uterine, prostate, and mammary glands in addition to respiratory tissues. Lesions included glandular necrosis, atrophy, fibrosis, inflammatory infiltrates, and prominent intranuclear epithelial inclusion bodies.

RatPyV2 is a newly discovered polyomavirus that causes significant disease in immunosuppressed rats. Although immune competent rats do not develop obvious clinical or pathologic disease, they do seroconvert after exposure, and the subclinical effects that RatPyV2may have on research studies are currently unknown. It is important for veterinary pathologists to be able to identify the lesions associated with this novel virus.

November 7, 2017
2:00 PM – 2:10 PM

A NOVEL NEORICKETTSIAL DISEASE IN THREE DOGS IN THE PACIFIC NORTHWEST

Gabrielle Pastenkos, Dan Bradway, Kevin Snekvik, Ilaria Cerchiaro, Susan Mehain, Denise Krytenberg, Chrissy Eckstrand

Background: The genus *Neorickettsia* encompasses species of obligate, intracellular bacteria responsible for significant disease in mammals, including Potomac Horse Fever and Salmon Poisoning Disease (SPD). The *Stellanchasmus falcatu* (SF) agent is a member of this genus previously associated with only mild clinical disease in dogs. Between 2013 and 2016, three dogs in Washington State presented with severe disease suggestive of SPD, but were negative for *N. risticii* by polymerase chain reaction (PCR).

Objective: To describe the clinicopathologic findings in three cases of SPD-like disease in dogs attributed to the SF agent.

Methods: Diagnosis and further characterization of disease associated with the SF agent was achieved by antemortem fine needle aspirates of mesenteric and submandibular lymph nodes, gross necropsy and histopathology, PCR using genus-specific primers and amplicon sequencing.

Results: Predominant clinical signs included depression, anorexia, regurgitation, and diarrhea. Hematologic findings include azotemia, phosphatemia, hyponatremia, hypokalemia, hypoproteinemia, leukocytosis, and lymphopenia.Cytology demonstrated granulomatous lymphadenitis with basophilic organisms suggestive of *Neorickettsia* species. Necropsy findings included lymphadenomegaly, hepatomegaly, and hepatic mottling. Histopathology revealed severe histiocytic and lymphoplasmacytic splenitis, hepatitis, lymphadenitis, enteritis, colitis, and tonsillitis with extensive necrosis. Bacteria were not identified in replicate, Giemsa-stained sections. *Neorickettsia* DNA was detected in all dogs using genus-specific primers; direct sequencing showed 100% sequence identity to the SF agent in two dogs.

Conclusions: This is the first clinicopathologic description of severe disease in multiple dogs attributed to the SF agent. These findings may suggest the emergence of a new Neorickettsial disease in the Pacific Northwest.

November 7, 2017
2:10 PM – 2:20 PM

INVESTIGATING THE CAUSE OF A NODULAR AND ULCERATIVE DERMATITIS IN DONKEYS IN MOROCCO

Caroline Louise Millins, Gigi Kay, Noursaid Tligui, Houssein Tligui, Derek Knottenbelt, Pawel Herzyk

Background: A nodular and ulcerative dermatitis of unknown aetiology has been observed in donkeys from Morocco with a characteristic distribution over the ears, with less severe lesions on the head and elsewhere on the body.

Objective: Skin biopsies were submitted for investigation of the cause of this dermatitis.

Methods: Skin biopsies from four donkeys were submitted between 2014 and 2017. Tissues were processed routinely for histological examination and stained with haematoxylin and eosin, Periodic acid Schiff(PAS) reaction and Grocott methylamine silver (GMS). Skin biopsies were cultured on Saborauds media and DNA extracted for molecular investigation of aetiological agents. Samples were PCR tested with general fungal primers and with primers to detect *Prototheca* spp.

Results: Histological changes were similar in all skin biopsies. Extending from the dermo-epidermal junction, the superficial and deep dermis were effaced by myriad round structures surrounded by pyogranulomatous inflammation, organisms were present within the cytoplasm of macrophages and free within the dermis. Organisms were approximately 15-20um in diameter with a 10-15um diametercentral, round, amphophilic nucleus surrounded by a 1-2um clear, PAS and GMS positive capsule. Division of organisms by an internal septum into two structures was present multifocally, occasional endosporulation was seen. There was no growth on Saborauds media and PCR testing with general fungal and *Prototheca* spp. primers was negative.

Conclusion: The results of histopathology, molecular testing and culture suggest that these donkeys are infected with the same etiologic agent. Utilization of next generation shotgun sequencing could provide genetic characterization of the infectious agent.

November 7, 2017
2:20 PM – 2:30 PM

SMALL B CELL LYMPHOMA CASE SERIES: CLINICAL, HISTOPATHOLOGY, AND FLOW CYTOMETRIC FEATURES

Kelly L. Hughes, E.J. Ehrhart, Lauren J. Harris, E.D. Rout, Janna A. Yoshimoto, Alana Kuzmik, Paul R. Avery, Anne C. Avery

Background: The most common type of lymphoma in the dog is diffuse large B-cell lymphoma (DLBCL). Other less common B-cell lymphomas, marginal zone lymphoma and follicular lymphoma, have unique features which are readily identified by histopathology. Uncommon B-cell lymphomas include mantle cell lymphoma, small-cell lymphocytic lymphoma, and B-cell lymphoblastic lymphoma and are poorly described in the dog and difficult to distinguish from one another.

Objective: Describe clinical, histopathology, and flow cytometric features of small B-cell lymphomas in order to better subclassify these diseases

Methods: Sixteen cases of small B-cell lymphoma that were not classified as marginal zone or follicular lymphoma, and where immunophenotyping by flow cytometry was also available, were characterized.

Results: The median age of cases was 10 years old (range 7-14). Lymphocytosis was identified in 36% (4/11) cases, splenomegaly was identified in 55% (5/9) cases, and hepatomegaly was noted in 50% (5/10) cases. A variety of dog breeds were represented. Lymph node biopsies consisted of diffuse and occasionally nodular infiltrates of neoplastic lymphocytes. Cells were small to intermediate (nucleus 1.25-2 X RBC), with indented nuclei, coarse chromatin, indistinct nucleoli and scant cytoplasm. Flow cytometric features were heterogeneous, but the cells were consistently smaller than histologically confirmed DLBCL. These characteristics prevented ready classification according to human WHO criteria.

Conclusion: Small B-cell lymphoma subtypes are difficult to further sub-classify by histopathology and flow cytometry. A small B-cell lymphoma subtype is described that may represent a single entity which does not fit readily into established WHO categories.

November 7, 2017
2:30 PM – 2:40 PM

CONGENITAL CATARACTS AND MICROPHAKIA WITH RETINAL DYSPLASIA, AND OPTIC NERVE HYPOPLASIA IN A CALF

Christopher L. Siepker, Karen Paige Carmichael, Jennifer L. Zimmer, Kathleen M. Bedard, Kelsey A. Hart

Background: A two-month-old, female, Aberdeen-Angus calf was referred to the University of Georgia (UGA) ophthalmology service for evaluation of congenital cataracts and blindness. The dam has a history of visual defects and previously produced affected calves.Physical examination revealed opaque lenses, attenuation of the iridic granules, persistent pupillary membranes, and a left corneal ulcer with possible abscessation. Additional findings in the calf included; poor body condition, prognathism, dome-shaped head, excessive nasal drainage, mild contracture of the limbs, and fever. Despite attempts at medical therapy, euthanasia was elected and both eyes were submitted to the UGA Diagnostic Ophthalmic Pathology Service.

Results: Gross findings revealed bilateral wrinkling of the ventral retina, a small optic nerve, and a small, soft lens. Histologically, lesions in both eyes are similar.There is severe and diffuse lenticular degeneration that is more severe OD than OS. The posterior retina is dysplastic, characterized by numerous, variably sized folds that multifocally create rosette-like structures. Retinal blood vessels are multifocally surrounded by a mild inflammatory infiltrate. Multifocally, replacing and apparently continuous with the retinal pigmented epithelium, is a band of connective tissue composed of densely packed spindloid cells of variable thickness. Abundant collagenous stroma replaces nerve bundles in the small optic nerve head. Immunoreactivity for BVDV was primarily observed within the ciliary body and iris with scattered labeling also observed within the retina and choroid.

Conclusions: Gross and histologic ocular findings are consistent with viral induced congenital ocular defects in a calf.

November 7, 2017
2:40 PM – 2:50 PM

CLINICAL AND HISTOPATHOLOGICAL CLASSIFICATION OF FELINE INTRAOCULAR LYMPHOMA

Ayla R. Musciano, Matthew R. Lanza, Richard D. Dubielzig, Amy C. Durham

Background: Cats with intraocular lymphoma may present without any signs of systemic involvement at the time of enucleation (presumed solitary ocular lymphoma; PSOL). The proportion and outcomes of cats with PSOL versus ocular involvement of multicentric lymphoma are not fully understood.

Objective: This retrospective study's aims are to characterize and correlate clinical signs, histopathologic features, WHO lymphoma classification, and outcome in feline intraocular lymphoma.

Methods: 172 cases were diagnosed at Penn Vet Diagnostic Laboratory and Comparative Ocular Pathology Laboratory of Wisconsin. Slides were reviewed for neoplastic cell characteristics and concurrent ocular disease; 163 cases had complete subtyping via immunohistochemistry (CD79a, CD20, Pax5, CD3). Based on submission forms and follow-up data, cases were categorized as PSOL if only ocular lesions were identified. Cats with signs of systemic disease (e.g. lymphadenopathy, organomegaly, lethargy, anorexia) were categorized as suspected systemic involvement (SSI).

Results: The majority exhibited concurrent uveitis (75%) and glaucoma (58%). Sixty-six cases (38%) provided clinical data at enucleation: 48 were PSOL (73%); 18 were SSI (27%). Diffuse large B-cell lymphoma was most common (n=69;42%), followed by peripheral T-cell lymphoma (n=25;15%). Other subtypes included anaplastic large B (n=7;4%) and T-cell (n=5;5%) lymphomas. 16 cases (10%) were negative for all immunohistochemical markers. Survival data was available for 60/172(35%). The overall median survival time was 73 days; PSOL cases survived longer compared to SSI cases (101 vs 61 days, respectively). Lymphoma subtype and survival were not associated.

Conclusions: Cats with PSOL likely represent a greater proportion of cases and have better clinical outcomes.

November 7, 2017
2:50 PM – 3:00 PM

HISTOLOGIC AND IMMUNOHISTOCHEMICAL CHARACTERISTICS OF 19 CASES OF EQUINE CUTANEOUS PSEUDOLYMPHOMA AND THE ASSOCIATION WITH BORRELIA BURGDORFERI

Joseph M. Malatos, Kathleen Kelly, Jeanine Peters-Kennedy

Pseudolymphoma is a non-neoplastic, lymphoproliferative process difficult to distinguish from cutaneous lymphoma. These proliferations are poorly characterized in the veterinary literature, especially in horses. Histopathologic features and causes have been established in humans with pseudolymphoma. The cause in humans is thought to be due to various antigenic stimuli; drugs, infections (including *Borrelia burgdorferi*), arthropod bites and contactants. *Borrelia burgdorferi*-associated cutaneous pseudolymphoma has been reported in one horse. To characterize the histologic features of cutaneous pseudolymphoma in horses and evaluate the association of *B.burgdorferi*, a retrospective study was performed. Polymerase chain reaction (PCR) for *B.burgdorferi* and immunohistochemistry for anti-CD3 antibody (T cells) and anti-CD20 antibody (B cells) was performed on 19 archived formalin-fixed paraffin embedded (FFPE) equine skin samples with the diagnosis of pseudolymphoma (2007-2016). Fifteen of 19 cases were cutaneous to subcutaneous nodules composed of discrete follicles that were predominantly composed of CD20+ B cells, partially to completely rimmed by small CD3+ T cells with parafollicular areas composed of a mix of T and B lymphocytes with rare eosinophils. In 4/19 cases, follicle formation was less prominent with a diffuse cutaneous T and B lymphocyte mixed infiltration. In one case there were multifocal eosinophilic granulomas within the pseudolymphoma. PCR was performed on DNA extracted from FFPE samples using primers targeting the *B. burgdorferi OspA gene*. *B. burgdorferi* DNA was amplified from one case (1/19). These results provide morphologic and immunophenotypic characteristics of equine pseudolymphoma and demonstrate that *B. burgdorferi* is a potential cause of pseudolymphoma in the horse.

November 7, 2017
3:30 PM – 3:40 PM

THYMOMA-ASSOCIATED EXFOLIATIVE DERMATITIS IN A GOAT

Alex D. Byas, Amy Kunkel, Tanya Applegate, Chad B. Frank

Case Report: A nine-year-old Rock Alpine doe presented to the Colorado State University Veterinary Teaching Hospital for progressive scaling and ulceration over the withers, coronary bands and dew claws. Radiographs revealed a large cranial mediastinal mass. Necropsy revealed marked exfoliative to ulcerative lesions which affected the dorsum, ventrum, pinna, neck, teats, coronary bands and dewclaws and a large multifocally cystic, soft, white, encapsulated mass. Histopathology of the cranial mediastinal mass showed a cytokeratin-positive epithelial neoplasm with marked infiltrates of non-neoplastic CD3-positive lymphocytes, consistent with a lymphoepithelial (mixed) thymoma. Histopathology of the skin lesions demonstrated a hyperkeratotic interface dermatitis and folliculitis with apoptosis of the stratum corneum, spinosum and basale as well as the follicular epithelium.

Discussion: Thymoma-associated exfoliative dermatitis, a paraneoplastic syndrome, has been previously recognized in cats and a rabbit. The classic feline cutaneous lesion has been previously described as a cell –poor interface dermatitis. In this case, the dermatologic lesion was characterized by a cell-rich interface dermatitis with transepidermal and follicular apoptosis. Recognition of this entity in goats is diagnostically important as thymomas are prevalent in goats but frequently lack associated clinical signs.

Conclusion: This is the first report of thymoma-associated exfoliative dermatitis in a goat. This syndrome should be considered as a clinical differential in goats presenting for dermatologic lesions.

November 7, 2017
3:40 PM – 3:50 PM

CANINE OCULAR MELANOSIS: MORPHOLOGIC AND IMMUNOHISTOCHEMICAL EVALUATION OF PIGMENTED SCLERAL PLAQUES

Erica L. Noland, Ethan M. Dawson-Baglien, Simon M. Petersen-Jones, Matti Kiupel, Dodd G. Sledge

Ocular melanosis (OM) is a progressive disease most common in Cairn Terriers, in which large, densely pigmented round cells accumulate in the uvea. While clinical consequences of disease are often related to accumulation of free pigment and cells in the trabecular meshwork and impingement on aqueous drainage, extension of pigmented cells into the sclera and development of plaques suggests that the pathogenesis of lesions involves invasion. While the expansile pigmented round cell population in the anterior uvea has been immunophenotyped as melanocytes, the seemingly invasive pigmented cell population in scleral plaques has not been specifically examined. Goals of this study were to immunophenotype pigmented cells within scleral plaques of eyes from Cairn Terriers with OM and to compare results to the immunophenotype of cell populations in the anterior uvea. One enucleated globe with clinically and histologically confirmed OM and associated scleral plaques from six Cairn Terriers were immunolabeled for HMB-45, Melan-A, PNL2, CD18, CD204, and Iba-1. In pigmented plaques and the anterior uvea, the majority of large round pigmented cells was positive for HMB-45 and CD18, and negative for Melan-A, PNL2, CD204, or Iba-1. The immunolabeling for HMB-45, a specific melanocyte marker, confirms a melanocytic origin of the pigmented cells in both sites. Immunolabeling for CD18 is of interest as this surface marker may be important for melanocyte migration in OM. Thus, pigmented cells in scleral plaques are not immunophenotypically distinct from those in the uvea despite their invasiveness, but expression of CD18 by melanocytes of OM warrants further investigation.

November 7, 2017
3:50 PM – 4:00 PM
BACILLUS CEREUS GANGRENOUS MASTITIS IN GOATS
Rahul B. Dange, Santiago S. Diab, Kristin A. Clothier, Leslie W. Woods

Bacillus cereus has been clinically identified as an occasional cause of mastitis in cattle but details on the pathology of the condition in small ruminants is lacking. Two goats from different herds were diagnosed with severe gangrenous mastitis caused by *B. cereus* and clinicopathological findings are described. Goat A, a 4-year-old female Anglo-Nubian died following labored breathing, recumbency and bloody milk discharge. Goat B, an adult female Saanen that was euthanized following acute history of swollen and painful right mammary gland. Grossly, both goats had unilaterally swollen, firm mammary gland with a well demarcated reddish/bluish discoloration of the skin. Goat A had red pink mammary gland with multifocal pale grey foci. Goat B had multifocally dilated, pus-filled acini, marked subcutaneous edema that extended along the ventral midline to the thorax, and a firm, edematous, swollen red teat. Goat A had petechial and ecchymotic hemorrhages suggestive of septicemia or toxemia in various serosal surfaces. Histologically, both goats had diffuse severe necrotizing and suppurative mastitis with abundant intralesional Gram positive bacilli. Goat B had acute coagulative necrosis of the epithelium of the teat cistern with severe congestion, hemorrhage and edema and Gram positive bacilli lining the teat cistern. *B. cereus* was isolated from mammary swabs in pure culture. *B. cereus* is a sporadic cause of gangrenous mastitis in cattle and small ruminants. The condition has been reported to respond favorably to treatment if detected early, but delay may cause progression to severe gangrenous mastitis and fatal sepsis or endotoxemia.

November 7, 2017
4:00 PM – 4:10 PM
PATHOLOGY ASSOCIATED WITH A HIGHLY VIRULENT AEROMONAS HYDROPHILA STRAIN IN FARM-RAISED CATFISH
Wes A. Baumgartner

Recently in Mississippi a highly virulent strain of *Aeromonas hydrophila* has become an emerging disease in farmed catfish. This particular strain may have come from an Asian source, and is unusually virulent to catfish. At the aquatic diagnostic lab at Mississippi State University (Starkville, MS), this variety of motile aeromonad septicemia (MAS) has been by far the most common diagnosis made for the past 3 years and disease is most common in the summer months. In ponds that experience outbreaks, mortalities can be very high (up to 100% loss); fish typically lack evidence of concurrent pathogens, indicating that this strain of bacteria is very different from previous catfish isolates.

The gross and microscopic manifestations of disease have special and useful identifying characteristics that have not been reported. As is common to MAS infections, there is severe hemorrhagic dermatitis with ulceration, as well as abdominal hyperemia, petechiation, and mild ascites. Characteristically present was marked hemorrhage and edema in the submucosa and muscularis of the stomach, with lymphangitis and few bacteria. Also the spleens were large, bloody, and friable with infarctions. Splenic ellipsoid necrosis with macrophages and few bacteria were typical, however large infarcts filled with bacteria were also present. Livers, kidneys, and intestines were less affected. The findings in the stomach have not been reported in MAS in farmed catfish, and the splenic changes were highly characteristic. These changes may have significant bearing on the etiology of this disease in farm raised catfish.

November 7, 2017
4:10 PM – 4:20 PM

RIBONUCLEIC ACID (RNA) DECAY AND THE ESTIMATION OF THE POSTMORTEM INTERVAL (PMI) IN HORSES (2016 ACVP/AAVLD TRAVEL AWARD WINNER)

Nanny Wenzlow, Dan Neal, Maureen Long

Introduction: The goal of this study was to investigate the RNA decay in equine tissues in order to determine the feasibility of this data for the aid in estimating the PMI in horses and to determine the morphological changes of autolysis in the same equine tissues during the first 72 h after death. Currently, no field applied methods exist to accurately estimate the PMI in animal or humans. The PMI determination capability would be of central importance for forensic investigations of suspicious death in horses. The hypothesis investigated is that RNA degrades in a predictable and step-wise fashion in post-mortem tissues and provides a decay profile for the estimation of the PMI in horses.

Material and Methods: Brain, liver, and skeletal muscle from 12 freshly euthanized horses, were held at 22°C and 8 C. The RNA decay was assessed at T0h, T1h, T2h, T4h, T6h, T12h, T24h, T36h, T48h, T60h, and T72h. The RNA degradation was determined by microfluidic analysis and the decay of the mRNA (cDNA) of β-actin, histone and β-tubulin was assessed by conventional PCR. Histologically, criteria for autolysis were evaluated for each tissue at each time point.

Results: In liver tissue, RIN (RNA integrity number) and 28S showed the most predictable decay over time with significant differences for temperature. Muscle RIN and 28S were the most stable and brain showed the most unpredictable decay rates and was the only tissue affected by the storage time. The decay of β-actin mRNA (cDNA) was the most representative in all tissues and the most predictable in liver. Hepatocyte individualization and the separation of bile duct epithelial cells from the basement membrane were the most field represented criteria for autolysis in liver followed by disruption of myofiber continuity, increased eosinophilia and loss of striation in muscle tissue.

Conclusion: Horse liver tissue showed the most predictable decay rate over 72h after death by all three methods. Results for horse liver RIN or 28S taken together with results from conventional PCR for liver β-actin and the extend of hepatocyte dissociation and bile duct separation grouped into a clinical index could estimate the PMI in horses.

November 7, 2017
4:10 PM – 4:20 PM

LARGE WHALE UNUSUAL MORTALITY EVENT, EASTERN NORTH PACIFIC

Stephen Raverty, Paul Cottrell, Kathy Burek Huntington, Heindrich Snyman, Katharine Savage

Between May 2015 and September 2016, 17 finback and 50 humpback whales were reported dead around Kodiak Island through the western Gulf of Alaska, central British Columbia coast and west coast of Vancouver Island. The increased number of large cetacean mortalities and short time and distance between strandings was unprecedented and prompted an enhanced effort to attend and determine possible causes of the stranding. Prime differentials included sonar/seismic testing, radiation, ship strike, infectious disease, predation, and oceanographic changes leading to algal toxin exposure or starvation due to shifts in prey species and distribution. Complete necropsies were impeded in regions due to site access, safety and state of decomposition. Of 11 animals examined, 4 had evidence of blunt or sharp force trauma (ship strike), 5 were in suboptimal nutritional condition, 1 was autolyzed and 1 had no significant lesions. The stranding coincided with anomalous warming of the sea surface water layer and algal blooms. Of 14 animals screened for HABs toxins, 6 had detectable levels of Paralytic Shellfish Poisoning (PSP) and Domoic acid (DA), 3 had PSP only, 3 had DA and 2 were below detectable limits. The contribution of these HABs to morbidity or mortality remains unknown and no specific cause of death was linked to the loss of these whales. Enhanced efforts to monitor large cetacean strandings are ongoing with field observations to assess body condition, prey shifts, and thorough necropsies which may provide baseline information for future stranding anomalies.

November 7, 2017
4:20 PM – 4:30 PM

TiLV: A NOVEL ORTHOMYXOVIRUS CAUSING MASSIVE DIE–OFFS IN TILAPIA FISH

Avi Eldar, Eran Bacharach, Ian W. Lipkin

Tilapias are increasingly important to global food security, with global production estimated at 4.5 million metric tonnes. We have previously described disease outbreaks in wild and on-growing farmed fish in Israel, and demonstrated that the causative agent was a novel virus named Tilapia Lake Virus (TiLV) that could be isolated from brains of diseases fish (Eyngor et al., 2014). Ferguson and coworkers described a disease in farmed tilapia in South America that affected fingerlings and differed in that pathology focused in the liver rather than the central nervous system (Ferguson et al., 2014). Collaborative work revealed the presence of a similar virus in both instances.

High throughput sequencing (UHTS), northern hybridization, mass spectrometry and in situ hybridization studies indicate that TiLV is segmented, negative-sense RNA virus. TiLV contains 10 genome segments, each with an open reading frame (ORF). Nine of the segments have no recognizable homology to other known sequences; one segment predicts a protein with weak homology to the PB1 subunit of the influenza C virus. Our findings suggest that TiLV represents a novel orthomyxo-like virus and confirm that it poses a global threat to tilapine aquaculture.

November 7, 2017
4:30 PM – 4:40 PM

GAMMAHERPESVIRUS-5 REPLICATES IN PULMONARY ALVEOLAR MACROPHAGES AND IS ASSOCIATED WITH A CD3-DRIVEN MAC387+ HISTIOCYTIC RESPONSE IN EQUINE MULTINODULAR PULMONARY FIBROSIS

Brieuc G.A. Cossic, Matthew R. Pennington, Amy L. Glaser, Tarin Rathbone, Gerlinde Van de Walle, Gerald E. Duhamel

Introduction: An association between equine multinodular pulmonary fibrosis (EMPF) and equine gammaherpesvirus-5 (EHV-5) is well-established; however, the cellular target of EHV-5 within the lung parenchyma and associated host inflammatory and adaptive immune cellular responses are incompletely characterized. Therefore, we sought to identify the cellular target of EHV-5 replication in the lungs of horses with naturally-occurring EMPF. Additionally, we compared the degree and distribution of collagen deposition, myofibroblast activation and type II pneumocyte hyperplasia in the lungs of horses with EMPF, equine idiopathic pulmonary fibrosis (EIPF) and healthy horses.

Materials and Methods: Infection with EHV-5 was confirmed by PCR assays, while EHV-5 replication in tissues was demonstrated by using a virus-specific *in situ* hybridization (ISH) assay. Key inflammatory and immune cells, and activation of myofibroblasts and type II pneumocytes were characterized by immunohistochemical staining, while collagen deposition was determined by Masson's trichrome histochemical staining.

Results: When compared with healthy controls, horses with EMPF had prominent myofibroblast activation and type II pneumocyte hyperplasia together with a predominantly CD3+ T lymphocyte and monocyte/macrophage (Iba-1, pan-macrophage/dendritic cell; MAC387, recently infiltrating M1-like pro-inflammatory monocyte/macrophage) cellular infiltrate. When compared with EIPF, horses with EMPF had a two to three folds increase in MAC387 and EHV-5 replicating exclusively within MAC387- pulmonary alveolar macrophages (PAM), presumably M2-like anti-inflammatory macrophages.

Conclusions: These findings delineate EHV-5 tropism for long-lived, self-renewal PAM accompanied with a T lymphocyte-driven histiocytic response in horses with EMPF that will assist in the development of improved protocols for the diagnosis and control of this disease.

November 7, 2017
4:40 PM – 4:50 PM

GROSS AND HISTOPATHOLOGIC CLASSIFICATION OF HEPATOCELLULAR CARCINOMAS IN FENNEC FOXES (VULPES ZERDA)

Colleen F. Monahan, Michael M. Garner, Matti Kiupel

Hepatocellular carcinomas (HCC) are the most common primary hepatic neoplasm of dogs and occur at a high prevalence in fennec foxes. Of 54 fennec fox submissions to Northwest ZooPath, 15 (28%) were diagnosed with HCC. This group consisted of 9 males, 3 females, and 2 of unspecified gender. Affected foxes ranged in age from 8 to 13 years (mean 10.3 years). Clinical signs included weight loss/inappetence (2), lethargy/weakness/ataxia (3), and seizures/shaking/tremors (4). Clinical pathologic abnormalities included elevated liver enzymes (5), hypoglycemia (3), anemia (2), and elevated renal analytes (3). In 8 cases, a palpable or visible hepatic mass was identified. HCC was confirmed microscopically and the human World Health Organization classification of hepatic neoplasms was applied to all cases. Gross morphologic patterns for 13 HCC were massive (8), nodular (4), and diffuse (1). The majority of HCC (12) had a mixed histomorphologic pattern and 2 HCC were trabecular and one compact. Trabecular (11), pseudoglandular (7), compact (6), and scirrhous (1) patterns were observed in the 12 mixed HCC. All HCC were well-differentiated, but 7 cases had moderate anisokaryosis. In 7 HCC, a significant portion of neoplastic cells were identified as clear cells and 6 cases had a pelioid pattern. Necrosis was detected in 6 HCC where it encompassed between 5-40% of the neoplasm. While no outcome data are available, based on morphologic similarities to canine HCCs, complete surgical removal of the affected liver lobe is suspected to be associated with a favorable prognosis.

Diagnostic Pathology Focused Group Poster Session

D-01: A TONGUE OSTEOLIPOMA IN A HOWLER MONKEY (ALOUATTA PALLIATA)

Alexis Berrocal, Fernando Alegre

Lipomas are very common benign tumors located in any part of the body where fat is normally present. Several variants of lipomas have been described, depending on the type of tissue present, among them: fibrolipoma, angiolipoma and chondrolipoma. Osteolipomas are a less frequent subtype, reported only in humans.

A 2-month-old howler monkey (Alouatta palliata) was found orphaned in the wild and raised in a rescue center for 5 months until he died from a digestive problem. The necropsy was performed at the rescue center. Samples of liver, lung, stomach, kidney and the tongue, with two nodules of 3mm and 4mm in the dorsal base, were sent. Tissues were processed for histopathological analysis and stained with H&E and Masson's trichrome. The tongue showed a circumscribed mass surrounded by collagenous tissue and pre-existing smooth muscular fibers. It had a thick wall composed primary of trabecular bone with bone marrow and cartilage differentiation. The rest of the mass was formed by mature adipocytes. Based on the histopathological morphology the tongue nodule was diagnosed as an osteolipoma with cartilage and bone marrow formation.

In the literature, there are only seven cases of osteolipomas from the oral cavity described in humans, and to our knowledge this is first case reported in veterinary medicine. Because of the age of the animal, we believe this tumor was congenital.

D-02: ABERRANT EXPRESSION AND NUCLEAR TRANSLOCATION OF 14-3-3-SIGMA PROTEIN AND ABERRANT E-CADHERIN CYTOPLASMIC UPREGULATION IN A METASTATIC HEPATOCELLULAR CARCINOMA IN A VERVET MONKEY (CHLOROCEBUS PYGERYTHRUS)

Alejandro Suarez-Bonnet, Elena Suarez-Bonnet, Rachel Pittaway, Simon Lawrence Priestnall

Background: The vervet monkey (*Chlorocebus pygerythrus*), an old world monkey of the family Cercopithecidae, is widely used in biomedical research. While hepatocellular carcinoma (HCC) has been experimentally induced in non-human primates, spontaneous liver tumors are uncommonly reported and knowledge regarding the expression of tumorigenic proteins is lacking. This is the first description of HCC in the vervet monkey and the first immunohistochemical analysis of proteins that are directly implicated in human HCC.

Objective: To describe for the first time histologically and immunohistochemically, a spontaneous hepatocellular carcinoma in a vervet monkey and to provide evidence that spontaneous tumors in this species can be a useful model of human HCC.

Methods: Formalin-fixed tissues from several organs of a 26-year-old vervet monkey were submitted for histopathological examination. The sections were additionally immunohistochemically labelled with the following antibodies; pan-keratin, vimentin, 14-3-3σ, E-cadherin, cox-2, CD117, CD31 and P63.

Results: The liver, lung and small intestine contained a poorly-demarcated, infiltrative, highly pleomorphic neoplasm morphologically consistent with a metastatic HCC. Neoplastic cells were positive for pan-keratin. Aberrant cytoplasmic and nuclear expression of 14-3-3σ and cytoplasmic E-cadherin expression was observed in primary and metastatic lesions.

Conclusion: 14-3-3σ protein neo-expression and nuclear translocation may be associated with more aggressive behavior as demonstrated in various human cancers. Cytoplasmic E-cadherin expression may indicate a cell-reprograming event facilitating invasiveness and metastasis. This is the first study to demonstrate 14-3-3σ expression in the vervet monkey and its neo-expression in HCC may indicate a useful spontaneous animal model of this disease.

D-03: CONCURRENT LEUKODYSTROPHY AND CEREBELLAR DEGENERATION IN A JUVENILE LABRADOR RETRIEVER

Brieuc G.A. Cossic, Randall J. Cross, Andrew D. Miller

A 7-month-old male intact Labrador retriever presented with ahistory of epilepsy that progressed with cerebellar signs consistent with aneuroanatomic diagnosis of diffuse. Magnetic resonance imaging revealed symmetric hyperintensities in the forebrain and circumferentially around the midbrain and pons. Histologic evaluation of the central nervous system (CNS) revealed widespread bilaterally symmetrical white matter vacuolation with loss of myelin and retention of unaffected axons throughout the entire brain including the mixed white and gray matters sections of brainstem. Loss of myelin was confirmed via Luxol fast blue and Bielschowsky silver stain confirmed the lack of axonal pathology. In the white matter, blood vessels were cuffed by concentric rings of astrocyte processes that were strongly positive for glial fibrillary acidic protein. In addition, Purkinje cells in the cerebellum were commonly degenerate with hypereosinophilic cytoplasm, shrunken borders, and pyknotic nuclei. Immunohistochemistry for microtubule-associated protein 2 and synaptophysin confirmed the Purkinje cell loss. Secondary degeneration was noted in the cerebellar nuclei. This case is unusual due to the simultaneous occurrence of a leukodystrophy and Purkinje cell degeneration which has not been reported in the veterinary literature and is equally rare in human neuropathology.

D-04: ASSOCIATION OF EQUINE GAMMAHERPESVIRUS-5 WITH LYMPHOHISTIOCYTIC INTERFACE DERMATITIS ON THE MUZZLE OF TWO HORSES IN THE UNITED STATES

Brieuc G.A. Cossic, Amy L. Glaser, Gerald E. Duhamel, Jeanine Peters-Kennedy

Skin biopsy specimens taken from a 15-year-old Haflinger mix gelding and a 9-year-old Thoroughbred mix gelding with non-painful, non-pruritic, scaly, annular to irregular lesions on the muzzle were submitted for histological evaluation. Histopathologically, there was a moderate lymphohistiocytic interface dermatitis with mild multifocal hydropic degeneration of the keratinocytes, pigmentary incontinence and moderate to marked compact orthokeratotic to parakeratotic hyperkeratosis. Throughout the epidermis, but predominantly in the stratum basale, there were small numbers of apoptotic keratinocytes, some with satellitosis. These changes were suggestive of discoid lupus erythematosus-like disease. Additionally, small numbers of keratinocytes within the upper stratum spinosum and stratum granulosum had glassy basophilic intranuclear inclusion bodies that marginated the chromatin, suggestive of herpesvirus infection. PCR assays of DNA extracted from FFPE confirmed the presence of equine gammaherpesvirus-5 (EHV-5). EHV-5 replication within clusters of stratum corneum keratinocytes was further demonstrated by using a virus-specific in situ hybridization method. EHV-5 has a worldwide distribution, and a prevalence ranging between 3 and 100 percent depending on the sampling method and age group. The clinical and pathological features of our cases were similar to those reported by Herder et al., (2012) in a 9-year-old Holsteiner stallion from south Germany with EHV-5-associated erythema multiforme lesions. These cases provide further evidence of EHV-5 association with dermatitis in horses in the United States, and consideration as a differential diagnosis for facial interface dermatitis in horses. The identification of EHV-5 will assist with the development of improved protocols for the diagnosis of skin diseases in horses.

D-05: INTESTINAL HISTOPLASMOSIS IN A CAPTIVE REINDEER (RANGIFER TARANDUS), MISSOURI, USA

Jessica S. Fortin, Michael J. Calcutt, Dusty W. Nagy, Keiichi Kuroki

A 4-year-old castrated male reindeer (*Rangifer tarandus*) (caribou) was brought to the Veterinary Medical Diagnostic Laboratory, University of Missouri, for post-mortem examination. The animal had persistent diarrhea for at least one week prior dead. At necropsy, fecal staining was seen around the perineum. The mucosa of the small intestines and proximal colon was dark red and thickened. Microscopically the intestinal mucosa was diffusely expanded by abundant epithelioid macrophages with numerous intrahistiocytic round to oval, 2-4 µm yeast forms with basophilic centers and peripheral clear zones surrounded by an indistinct outer cell wall. Yeast forms stained positively with Gomori methenamine silver (GMS) and periodic acid-Schiff (PAS). Antibody against *Histoplasma capsulatum* was used for recognition of H and M antigens. Numerous yeast forms stained positively for *H. capsulatum* M and H antigens by immunohistochemistry. The pathologic evaluations were summarized as histiocytic enterocolitis with intrahistiocytic yeast, etiology of *H. capsulatum*. It can be speculated that the animal succumbed to malabsorption caused by intestinal histoplasmosis. The "at risk" population for enteric histoplasmosis in reindeer is probably quite limited. In contrast, white-tailed deer (*Odocoileus virginianus*) are ubiquitous in the regions of endemic histoplasmosis in the Midwestern United States. To our knowledge, histoplasmosis in Cervidae has not been reported previously. We report an occurrence of intestinal histoplasmosis in a farmed reindeer in an endemic area for histoplasmosis.

D-06: INTESTINAL GLANDULAR INCLUSIONS (GLANDULAR CHORISTOMA) IN THE T-SEPTAL LYMPH NODE OF A FLORIDA MANATEE (TRICHECHUS MANATUS LATIROSTRIS)

David S. Rotstein, Martine de Wit, Kane J. Rigney, Gina L Lonati, Donna J. Szemer, Brittany D. Barbeau, Anna L. Panike

The carcass of an immature female manatee that had been rehabilitated for cold stress syndrome two years prior was recovered from an area with a red tide bloom. Lesions included ulcerative skin lesions and emaciation. Gross findings included multi-organ congestion, pulmonary abscesses, and small intestinal thickening. Brevetoxin was detected in liver, urine, and gastric contents at 15 ng/g, 1.5 ng/mL, and 366.4 ng/g, respectively. Histopathologic findings included enteric trematodiasis, ulcerative dermatomyositis, and suppurative lymphadenitis. In the T-Septal lymph node, there were random individual acini lined by goblet cells and cuboidal epithelium consistent with intestinal glandular inclusions. Intestinal glandular inclusion is a glandular choristoma which has been reported rarely in the lymph node of domestic cattle and goats. There are three types including epithelial, nevomelanocytic, and decidual. All are non-neoplastic. The finding was incidental, but provides additional information on the natural history of this species.

D-07: GLOMUS TUMOR OF THE TONGUE IN A CAT

Hao-Che Yen, Victor Fei Pang, Hui-Wen Chang, Chian-Ren Jeng, Fun-In Wang, Chen-Hsuan Liu

Glomus tumors arise from the modified smooth muscle cells of glomus body, which is located in the wall of specialized arteriovenous anastomosis and involves temperature regulation. They are categorized as perivascular tumors in human medicine, and mostly develop in the subungual region of the digit. Glomus tumors are rare in animals, only reported in the dog, cat, non-human primate, cattle, and horse. A 12-year-old female spayed mixbred cat developed a mass on the ventral aspect of the tip of the tongue, which enlarged gradually and bled while eating. Partial glossectomy followed by pathological examination was performed. The tip of the tongue contained an approximately 1.2 cm, distorted, protruding, circumscribed, pale, indistinctly lobulated mass. Microscopically, it comprised varying numbers of irregular, branching vascular channels interspersed by sheets, trabeculae, and whirling and wavy swarms of plump spindloid neoplastic cells supported by some fibrous stroma. The neoplastic cells had scant to a moderate amount of pale eosinophilic cytoplasm with sharply defined cell borders, which were enhanced by periodic acid-Schiff staining. Immunohistochemically, the neoplastic cells were positive for smooth muscle actin, E-cadherin and vimentin, but were negative for cytokeratin, desmin, S-100, neuron specific enolase, and factor VIII-related antigen. Although glomus tumors usually occur in the peripheral soft tissue, they also occasionally develop in the internal organs, such as gastrointestinal tract, urogenital tract, mesentery, mediastinum, liver, pancreas lung, ear, bone etc. This is the first case report of a glomus tumor arising from the tongue in both man and animal.

D-08: RHODOCOCCUS EQUI INFECTION IN GOATS: CHARACTERIZATION OF VIRULENCE PLASMID PHENOTYPE

Lauren W. Stranahan, Quinci Plumlee, Laura K. Bryan

Background: *Rhodococcus equi* is an opportunistic, facultative intracellular bacterium most commonly associated with pulmonary abscesses in foals. Virulence associated proteins (vap) are encoded on plasmids, are essential for bacterial survival within macrophages, and consist of VapA (equine-associated), VapB (porcine-associated), and the recently described VapN (bovine-associated). Cases of disseminated *R. equi* infection have been described in goats worldwide but without full characterization of the virulence plasmid profile.

Objective: The goal of this study was to characterize the virulence plasmid genes, if any, carried by *Rhodococcus equi* strains isolated from goats.

Methods: Clinical history, necropsy reports, and histopathology were reviewed for all cases, when available. Isolates of *R. equi* isolated from five infected goats were analyzed by polymerase chain reaction for *Rhodococcus* virulence-associated plasmid (vap) genes.

Results: Two goats exhibited disseminated infection with caseous granulomas affecting liver, jejunum, spleen, and visceral lymph nodes in one case and the lungs and mesenteric lymph nodes in another. The remaining three cases presented with a subcuticular abscess, suppurative bronchopneumonia, and vertebral osteomyelitis, respectively. Four isolates of *R. equi* from goats carried the VapN plasmid while one isolate lacked virulence plasmids and was classified as avirulent. Possible predisposing factors to infection included endoparasitism and copper deficiency.

Conclusions: This is the first report of the VapN plasmid in isolates of *Rhodococcus equi* from goats. One strain was classified as avirulent, indicating that *R. equi* infection in goats, as in humans and companion animals, may be associated with more than one virulence profile and is likely opportunistic.

D-09: MYCOTIC HYPOPHITIS DUE TO MORTIERELLA WOLFII IN AN ADULT COW

Alycia P. Fratzke, Caitlin E. Older, John F. Edwards

A 3-year-old, Charolais cow presented to the Texas A&M University Veterinary Medical Teaching Hospital for progressive neurologic signs that developed shortly after giving birth to a normal calf. Physical and neurologic examination revealed mildly depressed mentation, bilateral proptosis, lateral strabismus of the left eye, and vertical strabismus of the right eye. The cow was humanely euthanized and submitted for necropsy. On gross examination, the pituitary gland and circle of Willis were incorporated into a 2 cm-thick, firm, pale yellow to tan mass firmly adhered to the underlying bone. On cross section, the mass contained multifocal areas of edema, necrosis, and a tan exudate. Histology of the pituitary gland and surrounding tissue consisted of necrotizing vasculitis with thrombosis, pyogranulomatous inflammation, coagulative necrosis, and fibrosis. Vessel walls contained numerous non-parallel walled, pauci-septate, rarely branching fungal hyphae. The ITS-2 region was amplified from DNA extracted from formalin-fixed paraffin-embedded tissue. Results from sequencing of the PCR product matched *Mortierella wolfii*. *M. wolfii* is a common cause of mycotic placentitis and abortion in New Zealand cattle, but it is rarely reported in other parts of the world. Rarely, systemic infection with *M. wolfii* in adult cattle has been reported secondary to abortion or metritis causing lesions within the lungs, uterus, liver, kidneys, and brain. In this case, the uterus, lungs, liver, and kidneys were grossly and histologically normal. This report shows a unique case of mycotic hypophysitis without evidence of systemic infection.

D-10: ULTRASTRUCTURAL FEATURES OF CANINE NEUROAXONAL DYSTROPHY IN A PAPILLON

Miyuu Tanaka, Shinobu Yamaguchi, Hideo Akiyoshi, Takeshi Izawa, Jyoji Yamate, Mitsuru Kuwamura

Background: Neuroaxonal dystrophy (NAD) is a rare neurodegenerative disease characterized by severe axonal swelling (spheroids) throughout the nervous system. In dogs, NAD has been reported in several breeds including Papillon dogs. Recently, a missense mutation in *PLA2G6* gene (a causative gene for infantile neuroaxonal dystrophy in human) was identified in the Papillon dog NAD.

Objective: We performed ultrastructural analysis on the central nervous system (CNS) to clarify the detailed ultrastructural features of the dystrophic axons in a Papillondog with NAD.

Methods: Formalin-fixed spinal cords and brain from a NAD-affected Papillon dog (4 months old, male) were embedded in epoxy resin. This dog had a mutation in *PLA2G6* gene. Ultrathin sections were stained with uranyl acetate and lead citrate and examined in an electron microscope.

Results: Dystrophic swollen axons were found throughout the CNS, many spheroids were predominantly localized in the spinal cord dorsal horn and medulla oblongata including nuclei cuneatus, nuclei gracilis, nuclei olivalis, nuclei spinalis nervi trigemini, and lemniscus medialis. Dystrophic axons consisted of accumulation of filamentous and granular materials, and densely packed mitochondria. Numerous swollen edematous mitochondria with degenerated inner membranes and abnormal vesicles were often observed in the CNS.

Conclusions: Similar degenerative mitochondria are reported in *Pla2g6* knockout mice. Mitochondrial degeneration may be related with the pathogenesis of NAD in Papillon dogs.

D-11: ORAL PAPILLARY SQUAMOUS CELL CARCINOMA IN A 4-MONTH-OLD PUPPY

Wade Won, Tzushan Sharon Yang, Alicia Olivier

Oral papillary squamous cell carcinoma, a subtype of oral squamous cell carcinoma, is often a pattern of squamous cell carcinoma diagnosed in young dogs but also occurs in adult dogs. In humans, papillomaviruses are often associated with the development of oral squamous cell carcinoma. However, this relationship is rarely definitively confirmed in dogs. In the present report, a 4-month-old female Labrador retriever puppy had a rapidly growing oral mass along the rostral mandible. Computed tomography of the head revealed a large (5.3 x 3.5 x 3.8 cm), invasive, expansive, mixed soft tissue mass, resulting in marked cortical lysis of the left rostral mandible with displacement and deviation of the dentition. Histologic morphology of the mass was consistent with a papillary squamous cell carcinoma with effacement of bone and periodontal tissue. Real-time PCR for canine papillomavirus (CPV 1-8) and immunohistochemistry for canine papillomavirus were negative. Although involvement of other strains or a novel papillomavirus cannot be ruled out, the current case does not support the hypothesis of papillomavirus induced oral squamous cell carcinoma in this young dog. Necropsy examination revealed no evidence of metastasis.

D-12: SPONTANEOUS TERATOMA IN A WISTAR RAT

Abelardo A. Morales, Manuel M. Moya, Yetzy Y. Moreno, Emilio E. Suniaga

Background: The aim of this study was to describe a spontaneous teratoma in a Wistar Rat.

Methods: A spontaneous tumor was found in the left adrenal gland of a 35-day-old male Wistarrat which had undergone no experimental treatment died from enlargement of the abdomen, with weight loss syndrome.

Results: At necropsy, a spherical mass measuring about 2.5 X 3cm in diameter in the left adrenal gland. Histologically, the tumor consisted of various tissues derived from all three embryonic germ layers, i.e. the ectoderm (nervous tissue and skin), mesoderm (striated muscle, bone and cartilage), and endoderm (glandular tissue). The tumor also contained loosely or compactly arranged embryonic connective tissues.

Conclusion: In conclusion this from these findings, the tumor was diagnosed asteratoma.

D-13: HISTOLOGICAL AND IMMUNOHISTOCHEMICAL FEATURES OF SMALL INTESTINAL LYMPHATIC HYPOPLASIA IN DOGS PRESENTED WITH PROTEIN-LOSING ENTEROPATHY

Joseph M. Malatos, Gerald E Duhamel

Intestinal lymphatic hypoplasia (ILH) is a rare, but well-documented cause of protein-losing enteropathy (PLE) in human infants. To our knowledge, this condition has not been previously reported in veterinary medicine. Here we present the clinical and small intestinal histopathological findings in three dogs presented with clinical signs of PLE. The onset of PLE was early in an 18-month-old Great Pyrenees, while the other two dogs, a Pug, and a Tibetan Terrier had a late onset at 4- and 12-year-old, respectively. The histological features of intestinal lymphatic and blood vessels were assessed by using immunohistochemical staining with antibodies specific for human prospero homeobox 1 (prox-1), a lymphatic endothelial nuclear transcription factor, and human von Willebrand factor, a marker of vascular endothelial cells. Intestinal biopsies taken from each dog showed similar changes consisting of severe mucosal edema together with a lack of lamina propria lymphatic vessels, dilated and tortuous blood capillaries, and small to moderate numbers of mixed inflammatory cells. Other histological features of ILH were variable amongst the cases and included club-shaped villous tips inconsistently lined by low cuboidal epithelial cells with cytoplasmic microvesicles, thin and inconspicuous villous longitudinal smooth muscles, and extrusion zone epithelial inversion. While ILH is an uncommon diagnosis, it should be considered as a differential in cases with clinical evidence of PLE. Diagnosis of ILH on endoscopic or full-thickness biopsies can be difficult; however, the use of a specific immunohistochemical marker of lymphatic endothelial cells was essential for making this diagnosis.

D-14: FIBROUS OSTEODYSTROPHY AND WIDESPREAD METASTATIC MINERALIZATION IN A GRAY MOUSE LEMUR (MICROCEBUS MURINUS)

Kerriann M. Casey, Caitlin J Karanewsky, Jozeph L. Pendleton, Mark R. Krasnow, Megan A. Albertelli

A 9-year-old, intact male, gray mouse lemur (*Microcebus murinus*) was submitted to necropsy for impaired resolution of an oblique fracture of the right tibia. Pre-mortem serum chemistry analysis showed elevated BUN (221 mg/dL), creatinine (1.68 mg/dL), and phosphorus (15.5 mg/dl). Calcium levels were not evaluable due to limited blood sample volume. Radiographically, there was severe, generalized reduction in bone density and multiple areas of radiolucency (“moth-eaten” appearance). Grossly, appendicular fractures were present in the right tibia and right radius. Calvarial bones were pliable. The ascending/descending aorta and pulmonary arteries were segmentally dilated and hard. Bilaterally, the kidneys were symmetrically enlarged and mottled pale-tan to white. Parathyroid glands were grossly unremarkable. Histologically, all bones (appendicular, vertebral, calvarial, mandibular/maxillary) had replacement of cortical and trabecular bone by highly-cellular fibrous connective tissue surrounding poorly-mineralized islands of osteoblast-lined immature woven bone. Intratrabecular resorption cavities and Howship’s lacunae were associated with high numbers of plump osteoclasts and active osteoclastic osteolysis. Non-union fractures were present in the right tibia, radius/ulna, and scapula where they were associated with fibrocartilaginous calluses and/or bony sequestrae. Mineralization was found in all examined arteries (tunica media and/or tunica intima), select basement membranes (renal tubules, glomeruli, gastric mucosa, pulmonary alveoli, trachea, parathyroid gland), and cardiomyocytes. Additional renal lesions included glomerulonephritis, acute tubular necrosis, tubular ectasia and proteinosis, and tubulointerstitial nephritis. Although serum parathyroid hormone (PTH) and calcium levels could not be confirmed, widespread metastatic mineralization and fibrous osteodystrophy were considered direct sequelae to pre-existing renal disease and altered serum phosphorus levels.

D-15: A NEW TYPE TAUOPATHY IN AN OLD CAT

Mutsumi Yamazaki, Natsuho Nishiki, Koji Nishida, Fuyuki Kametani, Nobutaka Arai, Kinji Shirota, Yumi Une

An 18-year-old female mixed-breed cat showed decreased physical activity and neurological signs including an asymmetric pupillary reflex (right: mydriasis, left: miosis), poor balance on the left body side, and involuntary movement of the right leg. “Decreased emotion” had also been noted in this cat. Pathological examination of the brain showed moderate atrophy, increased cellularity, and diffuse severe gliosis and neuronal degeneration in the cerebral cortex and midbrain. Eosinophilic small granular structures consistent with hyperphosphorylated tau (HpT) accumulations were found not only in the neurons, but also in the microglia, and the HpT was found to include two isoforms of tau, 3-repeat and 4-repeat taus, on immunohistochemical examination. Neurofibrillary tangles were also frequently seen in these areas. However, the beta amyloid deposits were only detected in part of the cerebral cortex of the parietal lobe. Hyperphosphorylated tau accumulations in aged cats have been reported as a pathology similar to Alzheimer’s Disease in humans, but they have been observed mainly in neurons of the hippocampus. In this case, however, HpT accumulations were detected throughout the cerebral cortex and midbrain, and they were characteristically detected in the neurons, as well as microglia. The distribution and isoform components of the accumulated HpT in the brain with significant gliosis might be unique to this cat. To the best of our knowledge, there have been no cases of tauopathies of cats and humans similar to this cat case, suggesting that this is a new type of feline tauopathy.

D-16: CONCURRENT MULTIPLE MYELOMA AND ACUTE MYELOID LEUKEMIA IN A CAT

Emma Borkowski, Kimberly Hooi, Dorothee Bienzle, Janet Beeler-Marfisi

A 10-year-old male neutered domestic shorthair cat was presented for evaluation of ptyalism. The cat had marked hyperproteinemia despite euhydration, and diffuse ulcerated oral lesions. Serum protein electrophoresis showed a monoclonal gammopathy in the beta-2 region. Further diagnostic tests and therapy were declined. The cat represented 6 months later for evaluation of continuing ptyalism. A mild nonregenerative anemia, worsened hyperglobulinemia, persistent oral ulceration, circulating atypical plasma cells and immature granulocytes were present. Biopsies of bone marrow for cytology and histopathology, and of the oral lesions for histopathology, were submitted.

Bone marrow films had two populations of atypical cells. The first consisted of ~30% large myeloblasts/promyelocytes with prominent nucleoli and frequent mitotic figures. The second population was ~15% atypical plasma cells. On histopathology the bone marrow was ~80% cellular, contained similar abnormal cell populations, and plasma cells occurred in clusters and included multinucleated forms. The oral lesion consisted of ulcerated buccal mucosa with numerous inflammatory cells, including plasma cells. Interpretations were acute myeloid leukemia (AML), multiple myeloma and chronic ulcerative stomatitis. Immunohistochemistry with antibody to MUM-1 confirmed plasma cells in bone marrow. Treatment with melphalan was initiated, but hyperglobulinemia and anemia persisted following 2 months of treatment.

In people, AML has been reported to develop during therapy for multiple myeloma, but synchronous diagnosis of both appears very rare. This cat had untreated myeloma for at least 6 months prior to diagnosis of AML. Possible causes for synchronous development of multiple hematopoietic neoplasms are shared chromosomally unstable regions and driver mutations.

D-17: AN ATYPICAL VACCINE-INDUCED CANINE DISTEMPER IN A LITTER OF GERMAN QUAIL DOGS

Marjukka Anttila, Veera Karkamo, Riikka Holopainen, Tiina Nokireki, Katri Vainio-Siukola

Six puppies from a litter of ten German Quail dogs developed clinical signs after a routine vaccination with a DHPPi vaccine at the age of 12 weeks. Four puppies developed a severe disease with depression, vomiting, diarrhea and fever and three of them died and one puppy was euthanized six days later. Two puppies that developed a milder disease survived. The four puppies that were vaccinated with a different vaccine did not develop clinical symptoms.

Complete necropsy was performed for three puppies and samples were collected from all major organs. Immunohistochemical staining for canine distemper virus, adenovirus and parvovirus was performed on sections of liver, lung, lymph nodes and small intestine. RT-PCR for canine distemper virus was done using RNA extracted from paraffinized sections of small intestine followed by sequencing of the PCR product.

Macroscopically the lymph nodes, thymus and spleen were enlarged and oedematous. In the ileum there were multifocal hemorrhagic ulcers. Histologically there was marked lymphocytolysis and hemorrhagic necrosis in the thymus, spleen, lymph nodes and lymphoid tissue of the ileum. There were no specific lesions in any other organs including the brain. The Peyer's patches were strongly positive for canine distemper virus in IHC and subsequently with RT-PCR. The sequence of the PCR product was identical with the vaccine strain.

Morbilliviruses are known to cause lymphopenia and immunosuppression in highly susceptible hosts. Vaccine virus induced fatal lymphocytolysis has not been reported previously. The unexpected reaction to a vaccine virus probably has a genetic basis in this litter.

D-18: 360-TORSION OF THE GALL BLADDER AND QUADRATE LIVER LOBE IN A GERMAN SHEPHERD DOG

Russell S. Fraser, Lillian Cousto, Monica Jensen, Alex zur Linden, Brandon L. Plattner

Background: Gall bladder torsion in dogs is rare and potentially fatal, and all three cases reported in the literature involved German shepherd dogs. A 10-year old female spayed German shepherd presented with tachycardia (150 bpm), mucus membrane pallor and in lateral recumbency, following a 48-hour history of vomiting. Abdominal radiographs and ultrasound examination identified a large fluid filled organ in the cranial abdomen. Based on these data, gastric dilatation with volvulus were suspected. Due to a poor prognosis, euthanasia was elected and the dog was submitted for post-mortem examination.

Results: There was a 360 degree torsion of the gall bladder and quadrate lobe of the liver. The dark grey to black gall bladder measured 19x6x6 cm, and contained abundant dark brown, moderately granular fluid. The quadrate lobe of the liver was soft, dark purple to black, and was partially adhered to the gall bladder. Histopathology revealed coagulative necrosis affecting 95 percent of the gall bladder mucosa and in some areas, transmural coagulative necrosis of the gall bladder wall. Within the quadrate lobe of the liver, there was complete loss of hepatic architecture with replacement by hemorrhage amid a few scattered remnants of portal triads. No predisposing cause was identified.

Conclusions: We report a case of gall bladder and quadrate liver lobe torsion in a German shepherd dog. German shepherd dogs appear to be predisposed to gall bladder torsion, and this condition should be considered as a differential diagnosis when imaging is equivocal for GDV.

D-19: CD117 (KIT PROTEIN) EXPRESSION IN CANINE LYMPHOMA

Ian K. Hawkins, Marcia R.S. Ilha, Moges W. Woldemeskel, Lisa W. Whittington

Background: CD117 (KIT protein) is a transmembrane tyrosine kinase receptor that is expressed in a variety of canine neoplasms, including lymphomas. To the best of our knowledge, no studies have been undertaken to assess the prevalence of CD117 expression in canine lymphoma. Determining this prevalence may lead to the development of novel treatment strategies (e.g. tyrosine kinase inhibitor drugs) for lymphoma.

Objective: To determine the prevalence of CD117 expression in canine lymphoma.

Methods: Histopathology from 87 previously diagnosed cases of canine lymphoma were retrieved from the Tifton Veterinary Diagnostic Laboratory archive and reviewed to confirm diagnosis. Immunohistochemistry for CD3 and CD20 was performed on the cases to categorize each lymphoma according to the World Health Organization's Classification system for canine lymphoma. CD117 immunohistochemistry was then performed on all lymphomas. CD117 immunostaining was assessed for percentage of positive cells per section, staining strength, and staining distribution.

Results: Following immunohistochemistry, 67.8% (59/87) of the lymphomas were B cell lymphomas and 32.2% (28/87) were T cell lymphomas. Immunohistochemistry for CD117 revealed only two (2.3%) lymphomas that expressed the immunomarker in the majority of neoplastic cells. Both CD117 positive lymphomas were T cell lymphomas.

Conclusions: The number of lymphomas expressing CD117 in this study was lower than anticipated, and if representative, indicates a low prevalence of CD117 expression in canine lymphoma. However for those expressing CD117 in this cohort, the majority of neoplastic cells exhibited immunoreactivity, consequently such lymphomas may be sensitive to tyrosine kinase inhibitor treatment.

D-20: HISTOLOGIC EVALUATION OF PITUITARY ADENOMAS IN CANINE TRANSSPHENOIDAL HYPOPHYSECTOMY SPECIMENS

Margaret Allan Miller, Tina J. Owen, David S. Bruyette, Catharine Scott-Moncrieff, Jose' A. Ramos-Vara, Annie V. Chen-Allen, Linda G. Martin, Andrea L. Vanderpool, Deidre M. DuSold

Background: Histopathology of canine pituitary adenomas has been predominantly a postmortem endeavor.

Objective: Standardized histologic evaluation of 16 canine hypophysectomy specimens.

Methods: Pituitary height/brain area (P/B) ratio was calculated from magnetic resonance images. Histochemistry included reticulin, periodic acid-Schiff (PAS), and immunohistochemistry for adrenocorticotrophic hormone (ACTH), melanocyte stimulating hormone (MSH), growth hormone (GH), and Ki-67.

Results: Four female and 12 male dogs were 4-13 years old; median, 9.8. Seven were brachycephalic breeds; 9 had hyperadrenocorticism. Adenoma, with loss of reticulin framework, was diagnosed in all specimens. All dogs had pituitary enlargement (P/B ratio, 0.47-1.31). One nonfunctional plurihormonal adenoma expressed ACTH and GH. Nine corticotroph adenomas were sparsely to densely granulated, chromophobic to basophilic, PAS- and ACTH-positive. Six melanotroph adenomas resembled corticotroph adenomas, but also expressed MSH. Five dogs with corticotroph adenoma and four with melanotroph adenoma had hyperadrenocorticism. Six corticotroph adenomas and 1 melanotroph adenoma were in brachycephalic breeds. Median P/B ratio for corticotroph adenomas was 1.07; for melanotroph adenomas, 0.68. Median survival for dogs with corticotroph adenoma was 300 days; for melanotroph adenoma, 894 days. Median Ki-67 proliferation index in corticotroph adenomas was 8.6%; in melanotroph adenomas, 1.2%. Median survival of dogs with Ki-67 index >4.1% (overall median) was 260 days compared to 592 days for Ki-67 index <4.1%.

Conclusions: Canine pituitary adenomas usually develop from corticotrophs or melanotrophs; either can be associated with hyperadrenocorticism. Melanotroph adenomas, presumably of pars intermedia origin, tended to be smaller and less proliferative; affected dogs survived longer and were less likely to be brachycephalic.

D-21: XANTHINE NEPHROLITHIASIS IN A GOAT

Krystal J. Vail, Tasha Likavec, Philippa Gibbons, Raquel Rech

A two-year-old, female, mixed breed goat presented to Texas A&M University Veterinary Teaching Hospital for a 24-hour history of anorexia and one week of severe weight loss. The doe's diet consisted of coastal hay or pasture supplemented with commercial goat pellets. On physical examination, the doe was dull, unable to stand and emaciated. Serum chemistry revealed an elevated BUN (223 mg/dl) and creatinine (4.65 mg/dl). Abdominal ultrasound showed bilaterally decreased renal corticomedullary distinction, poor visualization of the renal pelvis and a dilated ureter. On necropsy, the kidneys were small, pale, firm and gritty. The pelves were severely dilated by variably-sized, gravel-like, yellow nephroliths. Numerous fine, granular, yellow nephroliths were embedded within the markedly atrophied inner medulla. Histologically, the majority of cortical and medullary tubules were distended by yellow-brown, multilayered crystals. The remaining tubules were either ruptured or atrophied. The interstitium was replaced by abundant fibrous connective tissue infiltrated by lymphocytes, plasma cells, fewer macrophages and multinucleated giant cells. Samples submitted to the Minnesota Urolith Center were evaluated by infrared spectroscopy and identified as 100% xanthine. Xanthine urolithiasis is rare in domestic species and is typically associated with use of xanthine oxidase inhibitors or an inborn metabolic disorder. In the absence of a history of xanthine oxidase inhibitor use, it is likely that xanthine nephrolithiasisin this case was caused by a genetic mutation.

D-22: METASTATIC DISEASE IN A DOG WITH A WELL-DIFFERENTIATED PERIANAL GLAND TUMOR

Maggie McCourt, Greg Levine, Melanie Breshears, James Meinkoth

Fine needle aspirates from a perianal mass on an 8-year-old, intact male, miniature poodle presenting for tenesmus showed a uniform population of well-differentiated hepatoid cells with no notable criteria of malignancy. Cytologic diagnosis was a perianal gland tumor, with adenoma likely given the cytomorphology. Abdominal ultrasound revealed multiple, markedly enlarged, intra-abdominal lymph nodes. Aspirates of these also showed well-differentiated polygonal, hepatoid cells displaying no notable criteria of malignancy. The presence of metastasis led to the interpretation of a well-differentiated, malignant perianal gland tumor despite the benign cellular appearance. Histopathology of the surgically excised perianal mass and one enlarged lymph node revealed lobules of uniform polygonal hepatoid cells arranged in organized islands and trabeculae surrounded by a single layer of uniform reserve cells. Few mitotic figures were present. The only indication of malignancy within the primary mass was the presence of small islands of well differentiated hepatoid cells seen infiltrating into adjacent skeletal muscle and adipose tissue. Histologic diagnosis was perianal gland adenocarcinoma. Most textbooks describe perianal gland carcinomas as showing increased atypia including pleomorphism, disorganization of hepatoid cells, and increased numbers of pleomorphic reserve cells with mitotic figures. This case is an example of the uncommon occurrence of a well-differentiated perianal gland tumor with metastasis and highlights the importance of realizing that with these tumors, a benign cytologic and histologic appearance may not correlate with biologic behavior.

D-23: OSSIFYING FIBROMA ARISING FROM THE SPHENOID BONE IN AN ACUTELY BLIND HORSE

Molly C. Friedemann, Rodolfo G. Madrigal, Jessica M. Vallone, Lucien V. Vallone, Todd Laughrey, Michelle Coleman, Raquel R. Rech

A 5-year-old Quarter Horse gelding presented to Texas A&M University Veterinary Medical Teaching Hospital for acute bilateral blindness of three weeks duration. On ophthalmic examination, the left eye displayed an absent menace response, fixed mydriasis, and marked atrophy of the optic disc. The right eye displayed an inconsistent menace response, sluggish pupillary light reflex, and slight optic disc pallor. No further neurologic abnormalities were noted. An MRI and a CT showed a large, well-circumscribed, multilobulated, soft tissue and mineral attenuating mass that originated from the caudoventral aspect of the left ethmoid labyrinth, invaded the ventral aspects of both the left and right ethmoid turbinates, effaced both the left and right sphenopalatine sinuses, and asymmetrically compressed both optic nerves just rostral to the optic chiasm. Due to poor prognosis, euthanasia was elected. Postmortem examination revealed a large, bony mass arising from the sphenoid bone. The mass effaced endoturbinates II and III, the sphenopalatine sinuses bilaterally, and compressed the cranioventral cranial cavity. Histology of the mass was compatible with an ossifying fibroma, with a variably dense fibrovascular stroma containing many bony trabeculae lined by osteoblasts. Microscopic analysis of the optic tracts revealed axonal degeneration, digestion chambers, and gitter cells. Ossifying fibromas are classified as benign, proliferative, fibro-osseous lesions that most frequently occur in the rostral mandible of young horses, and must be differentiated from other similar entities, including osteoma, fibrous dysplasia, and osteosarcoma. This case represents an unusual location and presentation for an ossifying fibroma.

D-24: A CASE OF SUSPECT PALYTOXICOSIS (PALYTHOA TOXICA POISONING) IN A HOUSEHOLD INVOLVING HUMANS AND A DOG

David S. Rotstein, Aurimar Ayala, Jennifer Jones, Jake Guag, Olgica Ceric, Sarah Nemser, Renate Reimschuessel

Palytoxin is a highly potent biotoxin produced by zoanthid corals. The biotoxin binds to the sodium pump resulting in a loss of the cellular ion gradient, which can lead to a range of serious health effects in humans and animals, including respiratory symptoms and death. In August 2015, one household reported sudden onset of a series of symptoms of unknown etiology among all human and animal household members. Six family members reported headache, coughing, chest pain, throat pain, weakness, and a metallic taste, and sought medical care. All 2 cats and 3 dogs living in the household also showed signs of illness. All dogs had oral white foam and the cats had difficulty breathing. A 3 YO NM Chihuahua belonging to the family acutely died. Household environmental assessments and case interviews resulted in the identification of palytoxin as the suspect causative agent. Symptoms were consistent with palytoxin and a possible source for the toxin was identified. The family kept 4 fish tanks and had recently purchased used sand online from a tank containing corals. The sand was introduced to the home two days before onset of symptoms in humans and the death of the dog. A necropsy on this dog showed gross multi-organ congestion. Histopathological findings included multi-organ congestion and a non-specific acute neutrophilic and exudative pneumonia. No infectious agents were observed. Sand, water, and corals from the tank were negative for palytoxin. While palytoxicosis was not confirmed, it remains suspect. Consumer education could raise awareness and limit exposures.

D-25: ECTOPIC THYROID CARCINOMA AND THYROID ADENOMA IN A GUINEA PIG (CAVIA PORCELLUS)

Hirotaka Kondo, Iori Koizumi, Narumi Yamamoto, Hisashi Shibuya

A 56-month-old male guinea pig (*Cavia porcellus*) presented with gradual weight loss from 757 g to 691 g for the past month. On physical examination, decreased cardiac contractility, moderate cardiomegaly, pericardial effusion and pulmonary edema were detected. After a month, the case died due to respiratory failure. On necropsy, the left ventricle and interventricular septum were thickened, with moderate amounts of serosanguineous pericardial fluid. On the base of heart, there was a 5 x 3 x 3 mm, tan, smooth nodule. The right thyroid gland was enlarged and measured 8 x 5 x 5 mm. Histologically, the nodule on the base of heart comprised moderately cellular neoplastic tissue composed of cuboidal cells arranged in variably-sized follicles. Most follicles were filled with eosinophilic homogeneous material (colloid). Anisocytosis and anisokaryosis were moderate. There were two mitoses per 10 400x fields. Within the neoplastic tissue, there were foci of interstitial osseous metaplasia with mature bone marrow tissue. These gross and histologic findings were consistent with ectopic thyroid carcinoma. Other significant histologic lesions included a thyroid adenoma of the right thyroid gland, myocardial hypertrophy and degeneration, suppurative bronchopneumonia, and centrilobular hepatocellular degeneration. In guinea pigs, reports on endocrine tumors including thyroid tumors are increasing; however, to the authors' best knowledge, this is the first report of ectopic thyroid carcinoma in this species. Although serum total T4 concentration was not evaluated, the tumor in this case was considered functional because concurrent lesions related to hyperthyroidism were observed.

D-26: IMMUNOHISTOCHEMICAL CHARACTERIZATION OF INFLAMMATION IN GLIOMATOSIS CEREBRI IN FIVE DOGS

Elizabeth C. Alloway, Keith E. Linder, Debra A. Tokarz

Gliomatosis cerebri (GC) is a rare neoplasm characterized by diffuse infiltration of neoplastic glial cells affecting at least three cerebral hemispheric lobes or the spinal cord. For human brain tumors, a growing body of literature supports a role for tumor-infiltrating immune cells in prognosis, immune modulation, and potential therapeutics. Although lymphocytic perivascular inflammation and microgliosis have been noted in several reported cases of canine GC, a detailed report of the inflammatory cell population in canine GC is lacking. To investigate this, immunohistochemical markers for glial cells (GFAP, Olig2) and immune cells (Iba-1, CD3, CD20) were applied to 5 cases of canine GC. The neoplastic cells, characterized by enlarged elongated nuclei, in 4/5 cases were largely immunopositive for Olig-2, with high numbers of GFAP-positive and IBA-1-positive cells present throughout the tumor interpreted as reactive astrocytes and microglia, respectively. In these cases, there were variable but generally very low numbers of CD3-positive cells throughout the tumor and CD20-positive cells mainly present in perivascular cuffs. In the fifth case, only 5-10% of the tumor cells were immunopositive for Olig-2, while approximately 50% of the cells were immunopositive for IBA-1. As in the other cases, GFAP-positive cells with a reactive astrocyte morphology were present throughout the tumor. This case notably featured higher numbers of CD3- and CD20-positive cells throughout the tumor and perivascularly. These results demonstrate variation in inflammatory cell infiltration and Olig2 immunoreactivity among canine GC and suggest that tumor-infiltrating immune cells may hold prognostic and therapeutic relevance for this tumor.

D-27: SUBCUTANEOUS RHABDOMYOSARCOMA IN AN OLD PET RABBIT

Chunho Park, Chikage Nakajima, Kazunori Kimitsuki, Nozomi Shiwa, Yasuhiko Tsuchida

Background: In rabbits, rhabdomyosarcomas are extremely rare neoplasms. We report the case of rhabdomyosarcoma with various histopathological appearances, and the differential diagnosis of this unusual tumor is discussed.

Material and Methods: An 11-year-old castrated male cross-breed pet rabbit was presented to the local veterinary hospital with a rapidly growing subcutaneous mass in the crural regions of the right hind limb. The mass was excised and fixed in 10% neutral-buffered formalin solution. The mass measured 4.7 × 3.2 × 2.5 cm, and the cut surface had a varied appearance, including greyish solid areas and whitish myxoid areas. The sections were stained with hematoxylin and eosin, Alcian Blue (pH 2.5) and phosphotungstic acid-hematoxylin (PTAH). Immunohistochemistry was performed using the desmin, vimentin, α-smooth muscle actin, cytokeratin AE1/AE3, adipophilin and S-100 protein. For electron microscopic examination, the formalin-fixed specimen was cut into 1-mm blocks, fixed in 1% buffered osmium tetroxide and embedded in epoxy resin.

Results and Conclusions: The mass comprised solid and myxoid areas. Solid areas were characterized by a storiform or interlacing pattern of spindle cells, strap cells, multinucleated giant cells and round cells with eccentrically located nuclei, whereas the myxoid areas were composed predominantly of elongated fusiform cells with hyperchromatic nuclei embedded in Alcian Blue-positive myxoid stroma. Immunohistochemically, tumor cells in both areas were positive for desmin and vimentin. Ultrastructurally, the tumor cells in the solid areas had abundant myofilaments with electron dense Z-band structures. Based on these pathological findings, this case was diagnosed as rhabdomyosarcoma in a rabbit.

D-28: RUPTURE OF A CLINICALLY SILENT MALIGNANT PHEOCHROMOCYTOMA COMPLICATED BY CARDIORESPIRATORY FAILURE IN A DOG DURING THE ANESTHESIA

Cheng-Shun Hsueh, Chian-Ren Jeng, Victor Fei Pang, Hui-Wen Chang, Fun-In Wang

Rupture of pheochromocytomas is a rare and lethal condition due to overwhelming clinical consequences related to excessive release of catecholamine and massive hemorrhage. An 11 year-old, female spayed, schnauzer was attended for vomiting consultation, due to the presence of a foreign body on the pylorus found in radiography. The physical and laboratory findings included grade IV heart murmur and neutrophilia with left shift, monocytosis and increased serum alanine aminotransferase, alkaline phosphatase concentration, respectively. Ten minutes after the beginning of gastrotomy to remove the foreign body, the patient presented with tachypnea and eventually collapsed despite of the first aid. Necropsy revealed abdominal bruise, generalized hemoabdomen with a ruptured liver. Retroperitoneal hemorrhage was accompanied with a large sanguineous mass at the rightadrenal gland. The left cardiac ventricular wall was diffusely beige discolored. The lungs were diffusely wet and heavy. The stomach had a worn one-dollar coin. Histologically, the adrenal mass is a malignant pheochromocytoma, as confirmed by positive staining for the neuron specific enolase and synaptophysin and negative for Melan-A. The left ventricular cardiomyocytes show contraction band necrosis accompanied with mild lymphohistiocytic inflammation and severe pulmonary edema. The cause of death is thus considered thecardiorespiratory failure due to massive release of catecholamines from the ruptured pheochromocytoma. Hereinwe report arare case of silent, preoperatively undiagnosed pheochromocytoma, which was activated during surgical operation and resulted in typical pathological consequences.

D-29: CD204-EXPRESSING TUMOR ASSOCIATED MACROPHAGES ARE ASSOCIATED WITH MALIGNANT, HIGH-GRADE, AND HORMONE RECEPTOR-NEGATIVE CANINE MAMMARY GLAND TUMORS

Byung-joon Seung, Seung-Hee Cho, Soo-hyeon Kim, Jung-hyang Sur

Background: Tumor-associated macrophages (TAMs) are an important component of leukocyte infiltration in tumors. TAMs can be classified into M1 and M2 phenotypes. High numbers of M2-polarized TAMs is associated with worse clinical course in various tumors.

Objective: The objectives of the present study were (1) to compare the expression of CD204 (a M2-polarized macrophage receptor) -possitve TAMs between benign and malignant tumors; (2) to compare the expression of CD204-positive TAMs between tumors with different clinicopathological characteristics such as histological grade, hormone status, lymphatic invasion and subtype; and (3) to determine the relation ship between CD204-positive TAMs and clincial obesity index.

Methods: In the present study, the expression of CD204, a M2-polarized macrophage receptor, was investigated using immunohistochemistry in the area surrounding TAMs in 101 cases of canine mammary tumors (CMTs).

Results: The mean number of CD204-positive macrophages was significantly higher in malignant CMTs than in benign CMTs (P = 0.000). The mean number of CD204-positive macrophages was significantly different between three histologic grade classes (P = 0.000) and significantly higher in grade III than in grades I and II. Moreover, the mean number of CD204-positive macrophages was significantly higher in HR-negative malignant CMTs (P = 0.035) and malignant CMTs with lymphatic invasion (P = 0.000) than in HR-positive malignant CMTs and malignant CMTs without lymphatic invasion.

Conclusions: These findings suggested that CD204-positive macrophages affect the development and behavior of CMTs and highlight the potential of CD204 as a prognostic factor.

D-30: MICROPHTHALMIA WITH MULTIPLE CONGENITAL OCULAR DEFECTS IN PORTUGUESE WATER DOGS: 17 CASES

May P.Y. Tse, Gillian C. Shaw, Andrew D. Miller

Portuguese water dog breeders and veterinary ophthalmologists recognize microphthalmia with multiple congenital ocular anomalies in the breed; however, no microscopic descriptions exist of the condition. Sixteen Portuguese water dog puppies (<8 weeks; 9 females, 7 males) and one adult (5 years; castrated male) with microphthalmia submitted for necropsy were examined (one globe was excluded due to level of artifact). The two most common abnormalities were microphthalmia (32/33 eyes; 97%) and congenital cataract (26/33 eyes; 79%). Other common histologic lesions include failure of development of the filtration apparatus and drainage structures (16/33 eyes; 48%), anterior uveal abnormalities, narrowing/collapse of the anterior chamber, anterior segment dysgenesis (each in 15/33 eyes; 45%), Descemet's membrane absence and/or doubling (11/33 eyes; 33%), posterior synechiae, lens capsule discontinuity and/or lens protein leakage (each in 9/33 eyes; 27%), descemetization of the anterior iris surface (8/33 eyes; 24%), aphakia (6/33 eyes; 18%), anterior synechiae, retinal atrophy (both in 5/33 eyes; 15%), retinal detachment with retinal pigmented epithelial cell hypertrophy (4/33 eyes; 12%) and corneal hypoplasia (2/33 eyes; 6%). The spectrum of severity varied from localized to generalized ocular dysgenesis, which mostly affected the anterior segment. The cases with axial lesions including lens capsule discontinuity, posterior synechiae and Descemet's membrane abnormalities are compatible with features described in anterior segment dysgenesis in humans. Our findings further elucidate the clinical findings in affected Portuguese water dogs and implicate a genetic defect that causes extensive and variable ocular dysgenesis early in embryogenesis.

D-31: EVALUATION OF OPTICAL COHERENCE TOMOGRAPHY FOR SURGICAL MARGIN ASSESSMENT IN RESECTED CANINE SOFT TISSUE SARCOMAS

Jonathan P. Samuelson, Laura E. Selmic, Jennifer K. Reagan, Kelly J. Mesa, Elizabeth A. Driskell, Joanne Li, Marina Marjanovic, Stephen A. Boppart

Background: In veterinary medicine, the standard of care for surgical margin assessment is histopathology. The majority of surgical margins from excised lesions are examined using radial sectioning. This method assesses only a small percentage of the total surgical margin, and histopathology usually takes a few days to complete. Optical Coherence Tomography (OCT) is a potential diagnostic tool to deliver real-time and accurate surgical margin evaluation.

Objectives: To correlate histologic features with OCT images from excised canine soft tissue sarcomas (STS), and to establish image evaluation criteria for identifying incomplete surgical margins.

Methods: Fourteen canine specimens were examined. Immediately following resection of the STS, two to four areas of the margin were imaged with OCT. Imaged areas were then examined histologically. Histopathology and OCT images were then evaluated for correlation, and appearances of different tissue types on OCT imaging were characterized.

Results: Adipose tissue exhibited a relatively low scattering and a hole-filled texture pattern on OCT. Skeletal muscle and sarcoma tissue were both highly scattering and dense. Sarcoma tissue did not have a defined structure, while alignment patterns of muscle fibers were typically present in skeletal muscle.

Conclusions: Surgical margin images can be acquired with non-invasive OCT technology. Tissue types have different appearances which closely correlate with low power histology images. Optical coherence tomography may be used to image surgical margins following resection of STS, but further research is needed to assess the diagnostic accuracy of this imaging modality.

D-32: PREECLAMPSIA IN A RHESUS MACAQUE (MACACA MULATTA) WITH FETAL INTRAUTERINE GROWTH RESTRICTION

Lois Colgin, Kamm Prongay, Anne Lewis

An 8-year-old, multiparous female rhesus macaque presented from an outdoor corral in the third trimester of pregnancy with possible dystocia and generalized edema affecting the limbs, neck and back. She had previously delivered three live infants. A fetal heartbeat was detected on abdominal ultrasound. Key laboratory findings included panhypoproteinemia, azotemia, a neutrophilic leukocytosis with a left shift and proteinuria. A presumptive diagnosis of preeclampsia was made and an emergency cesarean section elected. Blood pressure prior to surgery was 134/80 mmHg but normalized following delivery of a moderately undersized infant for gestational age. The poor prognosis given the premature birth, low body weight, abnormal neurologic signs and underdevelopment of the respiratory tract prompted euthanasia. At necropsy, there were multiple petechiae affecting the brain and asymmetrical marginal atrophy and necrosis of the monodiscoidal placenta and multiple placental infarcts. Relevant microscopic lesions in the placenta included mural fibrinoid necrosis of decidual vessels, persistence of vascular smooth muscle cells, thrombosis, infarction with associated acute inflammation, syncytial knotting, attenuation of terminal villi and prominent stromal calcification. Mural fibrinoid necrosis was also present in the decidua of the fetal membranes. The clinical signs, laboratory findings and gross and microscopic lesions are consistent with preeclampsia, a pregnancy-related disorder responsible for maternal morbidity and mortality in humans. Intrauterine growth restriction is a fetal complication of preeclampsia. Although spontaneous disease is rarely reported in nonhuman primates, preeclampsia may be an under-recognized cause of adverse pregnancy outcomes in nonhuman primates.

D-33: POTENTIAL HAZARD TO HUMAN AND ANIMAL HEALTH IN EGYPT: II. DISEASES OF EGYPTIAN CROW (CORVUX CORNIX).

Sahar S. Abd El-Rahman, M. Afify, Alaa M. Ali, Dalia A. Hamza, A. Zanaty, Hafez M. Hafez

Background: Free-living birds, including migratory and non-migratory species, constitute an issue of major interest to many scientists and public because they can act as long-distance vectors for a wide range of microorganisms that could be transmissible to humans, animals, and poultry.

Objective: The present study aimed to give a spotlight on the pathological affections that could be found in Egyptian (hooded) crows *(Corvus cornix)* and their potential role in transmitting viral, bacterial and parasitic pathogens to various livestock in Egypt as well as to illustrate the zoonotic importance of this species of birds.

Methods: The study was conducting on fifty apparently healthy crows collected from non-urban areas (6th of October city and 10th of Ramadan city) in Egypt during the period of the hunting season from September, 2015 to July, 2016. Birds were kept for 24hrs in the laboratory during which thorough clinical examination was carried out. After scarification; tissue specimens were collected from various internal organs for histopathological, bacteriological and virological examinations.

Results: The observed pathological lesions were numerous with variable incidences in different body organs; the prevalence and scoring of which were allocated. Bacteriological examination revealed isolation of *E. coli O157,* Klebsiella, Enterobacter, Citrobacter, Shigella, Pseudomonas and Salmonella as a single or as a mixed infection. While virological examination revealed isolation of H9N2 and PPMV-1 from all birds.

Conclusion: It was concluded that Egyptian crow could play an important role in transmitting some bacterial and viral pathogens to human, poultry farms and other live stocks.

D-34: CHARACTERIZATION OF A SPONTANEOUS ASTROCYTOMA IN AN AFRICAN HEDGEHOG (ATELERIX ALBIVENTRIS)

Melissa A. Roy, Anibal G. Armien

This case report outlines the diagnostic approach established to describe a spontaneous brain tumor in an African hedgehog (Atelerix albiventris). The whole body of an adult male captive African hedgehog was submitted for necropsy following a several month history of a swollen neck. On gross examination, the cervical and mandibular lymph nodes were enlarged, and histologic evaluation revealed a myeloproliferative disorder in the lymph nodes and spleen, as well as chronic, marked spongiform changes to the white matter (wobbly hedgehog syndrome). In addition, histologic evaluation revealed a moderately well demarcated 3mm mass at the level of the hypothalamus. The mass was expanding the right lateral ventricle, compressing the hippocampus ventrally and the surrounding white matter dorsally. Immunohistochemical assays revealed the tumor cells were reactive to GFAP and S100.Examination of the cells' ultrastructure revealed numerous thick cell processes, and cytoplasm rich in extensive networks of abundant intermediate filaments.Based on these findings, in addition tothe histologic appearance of the cells, the tumorwas classified as a low grade astrocytomaof gemistocytic origin. Astrocytomas are well described in humans and dogs; however only two individual case reports exist describing astrocytomas in hedgehogs. This case report demonstrates how different diagnostic techniques can compliment one another to achieve a final diagnosis, and also serves to broaden the understanding of neurological tumors in non-traditional veterinary species.

D-35: DEVELOPMENT OF AN IMMUNOHISTOCHEMISTRY PANEL IN DIAGNOSING CANINE PLEURAL DISEASES

Kendra Marie Andrie, Stephen Pannone, Kelly S. Santangelo

Differentiating malignant mesothelioma (MM), mesothelial hyperplasia (MH), and adenocarcinoma (ACA) in all species can pose a diagnostic challenge for pathologists. In conjunction with routine histopathology the development of immunohistochemistry panels has been useful in human medicine for distinguishing these entities. In people, the application of immunobiomarkers including epithelial membrane antigen (EMA/Muc-1), p53, desmin, calretinin, glucose transporter 1 (GLUT-1), and insulin-like growth factor II messenger RNA-binding protein 3 (IMP3) has proven most helpful when applied as a panel. Our teams' ultimate goal is to identify and validate efficacious ancillary immunohistochemistry and immunocytochemistry tests in dogs to accurately diagnose ambiguous cases. In this study, sixty archived formalin-fixed paraffin imbedded tissues with the aforementioned diagnoses (n=20 per group) were collected and evaluated for expression of calretinin, desmin, and EMA/Muc-1 with the aim of developing a novel diagnostic immunohistochemistry panel to aid in better characterizing these confounding cases. Our work thus far has revealed that the pattern of desmin and calretinin immunolabeling in canine MM, MH, and ACA does not mimic that described in the human literature. Specifically, desmin serves as a nonspecific marker for canine mesothelial tissue, hyperplastic and neoplastic. Further, the diagnostic utility of calretinin has not proven sufficient in our cohort as all cases have been immunonegative for this marker. Currently, our group is working on validating EMA/Muc-1 and GLUT-1 expression. Our findings emphasize the perplexing immunohistochemical profile of canine pleural diseases and the need for further investigation of novel biomarkers and/or new diagnostic techniques for accurate diagnosis of these entities.

D-36: ILEOCECAL INTUSSUSCEPTION WITH CONCURRENT CHRONIC LYMPHOCYTIC LEUKEMIA IN A DACHSHUND DOG

Chien-Hao Chen, Hui-Wen Chang, Fei Pang, Cho-Hua Wan, Chian-Ren Jeng

Background: The exact cause of intussusceptions in both human and veterinary medicine is unknown. Any lesion in the bowel wall or irritant in the lumen that alters the normal peristaltic pattern may initiate invagination. In human medicine, ileocolic or ileocecal intussusception has been reported as a complication of lymphocytic leukemia. Herein we report a counterpart case in canine with systemic metastasis of chronic lymphocytic leukemia.

Case Presentation: A 10-year-old, male castrated dachshund showed melena, hematochezia, anemia with leukocytosis and was euthanized. At necropsy, enlargement of multiple lymph nodes and ileocecal intussusception were observed. The bone marrow of femoral diaphysis was rather soft, fleshy, and yellowish on transverse sections.

Results: Microscopically, the biopsy tissue of bone marrow was effaced by diffuse infiltration of uniform neoplastic cells, majorly the small mature lymphocytes, characterized by high nucleus to cytoplasm ratio with minimal cytoplasm and nuclear homogeneity. The intussuscepted intestinal segment exhibited inside-out structures with dense infiltrations of neoplastic cells aforementioned. The results of immunohistochemistry were positive for CD3 but negative for CD79a. The result of flow cytometry was marked by CD3+, CD21- and CD34-.

Conclusions: This may be the first report of ileocecal intussusception with chronic lymphocytic leukemia infiltration in animals. In human medicine, the pathologic features most commonly is leukemic infiltration in the intestine, which might share similar pathogenesis in the present case.

D-37: LEISHMANIA SPP. INFECTION IN FIVE HORSES FROM COSTA RICA

Alexis Berrocal, Maria Garcia-Ortega, Javier Salguero

Leishmaniasis is one of the most prevalent parasitic public health problems worldwide. In the present study, autochthonous cases of equine cutaneous leishmaniosis due to *Leishmania spp*in Costa Rica are described. The clinical features and gross pathology of the 5 cases presented were typical of a recurrent chronic alopecic and ulcerative lesion within the skin from ears or head and neck areas. A histopathological examination of biopsied lesions pointed out the possibility of Leishmania spp. infection with the presence of a granulomatous inflammatory reaction in the dermis with multinucleated giant cells(MNGCs), macrophages, lymphocytes and few neutrophils and eosinophils. Necrotic areas were occasionally observed. To rule out some differential diagnoses among them mycotic, an immunohistochemical detection of Leishmania spp. was conducted in paraffin tissue section using a rabbit polyclonal antibody raised against Leishmania donovani, L. infantum and L. amazonensis and a established protocol for diagnostics at the School of Veterinary Medicine, University of Surrey, United Kindgdom. Strong positive reaction was observed within the cytoplasm of macrophages and MNGCs where amastigotes where clearly identified. Tissue sections were obtained for laser-capture microdissection followed by PCR and sequence analysis to identify the species of Leishmania spp. This report shows the importance of Leishmania spp. as a causative agent of equine cutaneous disease in the new world and the importance as a possible emerging pathogen. Furthermore, the awareness of considering the Leishmaniasis in the differential diagnosis of equine dermatitis.

D-38: FIRST CONFIRMED MUCOPOLYSACCHARIDOSIS TYPE VII IN A 15-WEEK-OLD BOSTON TERRIER

Mason C. Jager, Gerald E. Duhamel

A 15-week-old, intact female Boston Terrier puppy was presented to Cornell University Hospital for Animals with dull mentation, non-ambulatory tetraparesis, severe joint laxity, and poor body condition. Radiographs and magnetic resonance imaging revealed multiple skeletal abnormalities. Cytology of a cerebrospinal fluid and blood samples showed leukocytes with metachromatic cytoplasmic granules consistent with Alder-Reilly type bodies. At necropsy, several deformities were found including shortened vertebral bodies, incomplete ossification of the vertebral end plates with subsequent widening of the intervertebral disc spaces, deformed epiphyses of the appendicular skeleton, facial dysmorphism, nodular thickening of the atrioventricular valves, meningeal thickening and opacity, and bilateral corneal opacities. Microscopically, chondrocytes, macrophages, and other stromal cells in the vertebral column, Purkinje cells of the cerebellum, stromal cells and macrophages of the aorta and trachea, proximal convoluted tubular epithelial cells, and synovium of appendicular joints had prominent cytoplasmic vacuolation. The combined clinicopathologic, gross, and histologic findings of this puppy were strongly suggestive of mucopolysaccharidosis (MPS) with features similar to previous reports of MPS in dogs. The diagnosis of MPS was confirmed on the basis of a urinary Berry MPS spot test (PennGenn Laboratories, University of Pennsylvania), and a severe deficiency of serum β-glucuronidase. These results are consistent with Type VII MPS, also known as Sly syndrome in humans, which to our knowledge has not been reported in Boston Terriers previously.

D-39: A CASE OF FELINE PROGRESSIVE HISTIOCYTOSIS WITH PULMONARY METASTASIS

Michael Zinn, Keiichi Kuroki

Multiple biopsies of nodular skin foci from a 10 year-old, female spayed domestic short hair cat were examined. The cat had a history of several month duration of 0.5 – 2 cm sized plaque-like to nodular, often ulcerated skin lesions on face, trunk and extremities. A previous cytology evaluation was pyogranulomatous inflammation. Fungal culture yielded no growth. Microscopically, the dermis is infiltrated, replaced and expanded by sheets of round to polygonal, markedly pleomorphic cells. Mitoses average 3 per hpf. Epidermis and follicular epithelium are multifocally infiltrated by these neoplastic cells. The neoplastic cells stained positively for IBA-1 and CD18 and stained negative for CD3 and CD20 by immunohistochemistry. The lesions were diagnosed as feline progressive histiocytosis. The cat was euthanized after histologic diagnosis. Beside the skin lesions, dozens of similar plaques and nodules were found in the lungs at necropsy. Multiple lymph nodes including inguinal nodes, popliteal nodes, brachial nodes and hilar nodes were enlarged. The skin, lung and lymph node lesions were all composed of proliferation of pleomorphic histiocytic cells. The purpose of report is to describe a case of feline progressive histiocytosis and to provide an overview of this rare histiocytic proliferative condition in cats.

D-40: MULTIPLE OSTEOCHONDROMATOSIS AFFECTING BOTH LONG AND FLAT BONES IN A FREE-RANGING WHITE-TAILED DEER

Martha Frances Dalton, Heather Fenton, Mark G. Ruder, James Crum, Elizabeth W. Uhl

An emaciated adult, pregnant female, white-tailed deer from West Virginia was euthanized and submitted for evaluation because of numerous hard masses protruding from the skull, mandible, scapula, ribs, vertebrae and pelvis. The masses were exophytic proliferations of cancellous bone ranging in size from <1 cm nodules to a very large (34 x 16 x 12 cm) mass that obscured the right side of the face. Histologically, the growths consisted of bone forming from disorderly endochondral ossification beneath a cartilaginous cap or extending from the periosteum. While osteochondromatosis has been previously described in white-tailed deer, this is the first report of multiple osteochondromatosis affecting both the flat and long bones in a free-ranging animal. In humans, horses and likely dogs, multiple osteochondromatosis results from autosomal dominant genetic mutations. The condition is characterized by osseous outgrowths arising from bones formed by endochondral ossification, most commonly the vertebrae, ribs, and long bones of the limbs. The microscopic lesions resemble growth plates and stop expanding after the animal reaches skeletal maturity, however, malignant transformation is also reported. Multiple osteochondromatosis in cervids is more similar to feline osteochondromatosis in that the lesions continue to progress, are randomly distributed and juxtacortical, and are on the bones of the skull as well as the long bones. The term 'cervid osteochondromatosis' is proposed to indicate that it is a distinct condition with similarities to feline osteochrondromatosis, but likely induced by a pathogenesis specific to cervids, as based on previous cases, it appears to be more common in females.

D-41: DISSEMINATED HISTOPLASMOSIS IN AN ADULT CAT

Ji-Hang Yin, Jessica S. Fortin

A 12-year-old, castrated male, American Domestic Shorthair cat was presented to the Veterinary Medical Diagnostic Laboratory, University of Missouri, for post-mortem examination with a history of weight loss, lethargy and inappetence for three weeks. In the radiological examination of the lung, metastatic neoplasia or granulomatous inflammation was suspected. At necropsy, all lobes of the lungs contained multifocal to coalescing tan to white, round, firm nodules, ranging from 2-10 mm in size. Multifocal dark red nodules, measuring 1 cm in diameter, were present in the hepatic parenchyma. The spleen also contained similar nodules. Microscopically, organs including lungs, liver, spleen and bone marrow were affected by multifocal granulomatous inflammation. The foci of inflammation were comprised of moderate to large numbers of epithelioid macrophages containing multiple round to oval (approximately 2-4 mm in diameter) intracytoplasmic yeast with a basophilic center surrounded by a peripheral clear zone and enclosed with an indistinct outer cell wall. Yeast forms stained positively with Gomori methenamine silver (GMS) and periodic acid-Schiff (PAS). Antibody against *Histoplasma capsulatum* was used for immunohistochemical recognition of H and M antigens. Numerous yeast forms stained positively for *H. capsulatum* M and H antigens. Therefore, the evaluations in this case were summarized as a disseminated granulomatous inflammation with intrahistiocytic yeast, consistent with *H. capsulatum.* Pathological findings which were not merely confined to the lungs but other organs in spleen, liver and bone marrow were an indicative of a dissemination of *H.capsulatum.*

D-42: OSTEOGENESIS AND DENTINOGENESIS IMPERFECTA IN A 4-MONTH-OLD ENGLISH MASTIFF

Randi Gold, Erin Edwards

Osteogenesis imperfecta, also known as “brittle bone disease,” is an inherited connective tissue disorder that causes defects in type 1 collagen, the predominant collagen type in bone, dentin, ligaments, tendons, and the ocular sclera. The disease results in low bone mass and reduced bone strength, often manifesting as multiple intrauterine fractures, skeletal abnormalities, and either stillbirth or perinatal death. A 4-month-old, female intact, English Mastiff dog presented to the Texas A&M University Veterinary Medical Teaching Hospital for multiple limb fractures. Due to a poor prognosis euthanasia was elected. Gross examination revealed diffuse osteopenia with multiple chronic and acute skeletal fractures affecting three long bones and two ribs. Bones broke easily with moderate pressure. All adult teeth were undersized and opalescent, and multiple deciduous incisors were retained. Histopathology of the long bones demonstrated severe, diffuse osteopenia with retention of unossified cartilage spicules from the primary spongiosa. The incisor teeth had multifocal odontoblast and ameloblast disorganization and piling (dysplasia) with dentin hypoplasia. A diagnosis of osteogenesis imperfecta and dentinogenesis imperfecta was made. Most reports of osteogenesis imperfecta in domestic animals are in calves and lambs; however, rare cases in various breeds of dogs and cats have been reported. To our knowledge, this is the first report in a Mastiff dog. Osteogenesis imperfecta should be considered as a differential diagnosis for any young domestic animal species presenting with multiple fractures without a history of trauma.

D-43: HISTOPATHOLOGIC AND ELECTRON MICROSCOPIC CHARACTERIZATION OF UVEITIS PRESENT IN EQUINE RECURRENT UVEITIS

K. Paige Carmichael, Shannon G. Kirejczyk, Silvia G. Pryor, Kathryn Myrna

Background: A 19-year old Appaloosa gelding and a 14-year old Quarter horse gelding were enucleated due to unilateral corneal ulceration and blindness secondary to chronic intraocular inflammation, and from severe uveitis refractory to therapy, respectively.

Methods: Affected eyes were surgically removed, fixed intact in Davidson's solution and examined histologically. In addition, tissues were submitted for electron microscopy.

Results: Microscopically, the uvea of both globes examined have diffuse infiltrations of lymphocytes, and anterior uveal lymphonodular aggregates. Both cases contained abundant homogenous eosinophilic, material on the surface of and replacing the non-pigmented epithelium of the ciliary body and processes. Both samples were congophilic and exhibited apple-green birefringence under polarized light. EM of the non-pigmented epithelium of both horses revealed massive extracellular deposits of non-branching fibrils (7-15nm in diameter), consistent with amyloid.

Conclusion: The uveal eosinophilic material seen in cases of ERU has been definitively characterized as amyloid using routine and specially stained histopathology sections, and electron microscopy. A recent case report (Ostevik et al) that detected amyloid in ERU using anti-amyloid immunohistochemistry further supports our conclusion. The electron microscopic appearance of amyloid in cases of ERU has not been previously described.

D-44: INFILTRATIVE LYMPHOCYTES SUBPOPULATIONS AND MAST CELLS IN CANINE INFLAMMATORY BOWEL DISEASE

Andres A. Espinoza, Carlos M. Gonzalez, Ivan A. Contreras

Background: Inflammatory bowel disease (IBD) is a group of gastrointestinal diseases characterized by chronic gastrointestinal symptoms and histologic evidence of inflammatory cell infiltration. Lymphocytes and mast cells may play a role in the development and maintenance of IBD in dogs, considering they are an important cell component in lesions associated to this disease.

Objective: Our objective was to quantify and evaluate changes in lymphocytes and mast cells populations in intestinal samples of dogs with IBD.

Methods: Lymphocyte (CD3 and CD79a) and mast cell (CD117) immunophenotype identification was performed in formalin-fixed paraffin embedded sections of small intestine with IBD and controls. Also, Toluidine blue staining was used to detect mast cells by routine conditions. Computer assisted morphometric analysis was performed. Wilcoxon test was applied to evaluate the difference in the number of the cells populations.

Results: CD3+ lymphocytes and CD117+ mast cells were significantly higher ($p<0.05$) in the lamina propia of the intestinal tissue from dogs with an IBD diagnoses when compared with healthy dogs. Also found a higher detection in the number of mast cells when we used CD117 immunophenotyping compared with Toluidine blue staining.

Conclusions: A higher number of infiltrative CD3+ lymphocytes in the lamina propia of the intestinal wall correlates with an increase in mast cells suggesting an altered Th1 immune response in dogs with IBD. The higher detection of mast cells by of CD177+ in comparison to Toluidine blue can be explained by the degranulation of the mast cells occurring during the lapse of this disease.

D-45: EQUINE MULTINODULAR PULMONARY FIBROSIS IN A THOROUGHBRED HORSE IN JAPAN

Akihiro Ochi, Koji Tsujimura, Takanori Ueno, Maki Sekiguchi, Yoshinari Katayama

Background: Equine multinodular pulmonary fibrosis (EMPF) is a progressive fibrosing interstitial lung disease, which has been associated with equine herpesvirus type 5 (EHV-5). EMPF has been reported in the United States, Europe and Australia, but reports of the disease in Japan are lacking.

Case Presentation: A 14-year-old Thoroughbred gelding presented a one-month history of fever, innappetance, lethargy and persistent elevation of serum amyloid A (SAA), an inflammatory marker. Antibiotic treatment was initiated, but the animalshowed a poor response. Therefore, the horse waseuthanized for animal welfare reasons.

Results: At necropsy, there were multifocal to coalescing,tan to white nodules that were up to 3 cm in diameter were scattered throughoutthe parenchyma. The other findings were chronic cellulitis in the right hind limb, and fibrin accumulation in the hock joint of the limb. Histological examination of the lung revealed a marked interstitial fibrosis and hyperplasia of type II pneumocyte with the infiltration of inflammatory cells.Eosinophilic intranuclear inclusion bodies were occasionally observed in macrophages. The lung was positive for EHV-5 gene by PCR, and EHV-5 RNA was identified in the macrophages by *in situ* hybridization.

Conclusions: The gross and histological findings were consistent with EMPF. In addition, EHV-5 was detected by PCR and *in situ* hybridization. This report describes EMPF in a horse in Japan.

D-46: ERYTHROLEUKEMIA IN A RETROVIRUS-NEGATIVE CAT: A NECROPSY CASE

Satoshi Suzuki, Naotaka Ogino, Ikki Mitsui, Hiroyuki Ito, Takuro Kariya

Erythroleukemia in cats is most frequently associated with FeLV infection and is often preceded by myelodysplastic syndromes (MDS). Only one previous report describes acute erythroleukemia in a retrovirus-negative cat. The mechanism inducing erythroleukemia in retrovirus-negative cats has not been clarified yet. This study describes the clinicopathologic findings in a case of erythroleukemia that progressed from MDS. A four-year-old spayed female Scottish Fold cat was presented for lethargy, anorexia and fever. CBC revealed severe anemia, mild thrombocytopenia and mild leukocytosis. Immature erythrocytes were observed in the peripheral blood smear. The results of FeLV/FIV snap test, PCR test for proFeLV/proFIV, and IDEXX RealPCR Test for feline vector-borne disease panels were negative. Abdominal ultrasound revealed an enlarged spleen and splenic lymph node, and FNA showed numerous immature erythrocytes. Erythroid hyperplasia was seen in bone marrow smears. The ratio of erythroid blasts was less than 20%. Based on these findings, the patient was diagnosed with MDS, possibly the early acute phase of acute myeloid leukemia (AML) M6. Despite treatment, the disease progressed and the patient died on day 118. Postmortem examination revealed marked splenomegaly and hepatomegaly, pulmonary edema, cardiomegaly, jaundice, enlarged pancreaticoduodenal lymph node and mild ascites. Histopathologically, neoplastic erythroblasts infiltrated the bone marrow and invaded other organs.Immunohistochemically, these cells were negative for CD3, CD20 and Granzyme B. Based on these findings, histopathological diagnosis was determined as AML. The cause of death was acute pulmonary alveolar damage This report helps understand the pathology and progression of erythroleukemia.

D-47: CUTANEOUS SQUAMOUS CELL CARCINOMA WITH VISCERAL METASTASES IN A RAINBOW BOA (EPICRATES CENCHRIA)

Rachel Teixeira-Neto, Miranda Vieson, Matthew Allender

A 19-year-old, intact male, captive Rainbow Boa was euthanized after being found unresponsive in its enclosure. The animal presented for necropsy examination with a previously biopsied and diagnosed cutaneous squamous cell carcinoma (SCC). Grossly, the snake had multiple sites of irregular, firm, ulcerated, and crusty integument lesions at the dorsal caudal and mid-body. Scattered throughout the liver parenchyma were a few, 2-4 mm diameter, raised, white nodules. Histopathology of the cutaneous lesions revealed an unencapsulated, highly invasive, squamous epithelial cell neoplasm with features of SCC arising from the epidermis, effacing the dermis and extending into the underlying skeletal musculature.. The hepatic parenchyma and one adrenal gland were multifocally replaced by discrete masses of neoplastic keratinizing squamous epithelium, consistent with metastatic SCC. Given the histopathologic findings, a diagnosis of multifocal primary invasive cutaneous SCC with hepatic and adrenal gland metastases was assigned for this snake. Among reptiles, snakes appear to have the highest incidence of neoplasms. However, SCC in ophidian species are uncommonly reported, being documented in the oral cavity, cloacal and pharyngeal regions and skin. Although vascular invasion was already evidenced in one case of ophidian cutaneous SCC, reports of visceral metastasis are lacking to date.

D-48: DUCTAL PLATE MALFORMATIONS IN CAPTIVE SNAKES

Olivia M. Swartley, Brigid V. Troan, John M. Cullen

The formation of intrahepatic bile ducts is initiated when a subset of bipotential hepatoblasts immediately adjacent to the portal mesenchyme differentiate into biliary precursor cells and form a cytokeratin 7 positive single-layer termed the ductal plate. The ductal plate then becomes bi-layered and subsequently remodels into an array of tubules. Most undergo apoptosis, but the remaining tubules form bile ducts lined by functional cholangiocytes and are incorporated into the portal tract connective tissue. If not completed properly, this process results in abnormalities known as ductal plate malformations (DPMs). DPMs are recognized in humans and various species of domestic animals; however, to our knowledge, no reports have been documented in snakes. We investigated the occurrence and described the histopathologic features of DPMs in snakes submitted to North Carolina State University College of Veterinary Medicine from 2008 to 2016. Malformations were identified in 14 snakes: 7 colubrids, 6 vipers, and 1 boid. There was no sex predilection and the mean age was 17 years. All livers had focal to multifocal, well-demarcated, DPMs characterized by minimally to mildly dilated, irregular, small bile ducts embedded within a robust fibrous stroma. In five cases, DPMs progressed to large cystic structures. In one case, there was malignant transformation into a cholangiocellular carcinoma. Pan-cytokeratin immunohistochemical stains confirmed the irregular bile ducts are lined by biliary epithelial cells and Gomori's trichrome stain highlighted the associated fibrous connective tissue stroma. In conclusion, DPMs similar to those described in mammals are also present in lower vertebrates such as snakes.

D-49: MYENTERIC GANGLIONEURITIS AND GENERALIZED LYMPHOPLASMOCYTIC MYOCARDITIS IN A KEEL-BILLED TOUCAN (RAMPHASTOS SULFURATUS): FIRST CASE REPORT.

Alonso Reyes-Matute, José-Alberto Mercado-Saldaña, Vanessa Hernández-Urraca, Vicente Ávila-Reyes, Luz-María Vargas-Soto

Myenteric ganglioneuritis and non-suppurative myocarditis are uncommon lesions in pet birds, which have been associated with the Bornavirus that causes proventricular dilation disease. This disease usually affects psittacids and to a lesser extent other birds. A 1 year-old toucan (*Ramphastos sulfuratus*), was presented with anorexia, lethargy, cardiac arrhythmias and difficulty to fly. The bird was held together with another toucan, a *Psittacus erithacus*, an *Ara macao*, an *Amazana auropaliata*, three *Melopsittacus undulatus*, five *Agapornis sp*, and numerous passerines. Radiographs from the abdomen were taken where hepatomegaly was diagnosed. Iron storage disease was the presumptive diagnosis. The toucan did not show improvement despite treatment and later died. The postmortem study revealed a severe ganglioneuritis and lymphoplasmocytic perivasculitis in the crop, gizzard, proventriculus and small intestine, as well as a lymphoplasmocytic myocarditis surrounding the conducting cardiomyocytes, which showed moderate degeneration. Other findings were a moderate hepatomegaly due to hemosiderosis, related to iron storage disease, proventricular candidiasis and a granulomatous pododermatitis. The myocarditis and ganglioneurits are consistent with those caused by the avian Bornavirus. Although there were no changes in the gastrointestinal tract, or clinical signs associated with it, it is likely that the severe myocarditis present in the specimen has led to death. In psittacids, the presentation of arrhythmias and alterations in the blood pressure in the birds affected by the avian Bornavirus is documented. Cardiac lesions may result in sudden death associated with the involvement of conducting cardiomyocytes. There are no reports of similar lesions in this species of toucan.

D-50: CLINICAL FEATURES AND PATHOLOGY OF ATYPICAL CANINE PERIPHERAL ODONTOGENIC FIBROMAS

Benjamin E. Curtis, Paula Schaffer, Chad B. Frank

Peripheral odontogenic fibromas (POF) are considered benign neoplasms of canine periodontal ligament. These tumors commonly displace teeth, but do not invade bone or metastasize. A subset of POFs with increased mitotic figures, overall hypercellularity, and increased nuclear and cellular pleomorphism were retrospectively identified over a 4 year period. Data collected from the record included signalment, degree of excision, and location of the mass, when noted. Histologic features including mitotic index, presence of hard substance, presence of odontogenic rests, degree of cellularity, and bone reaction/invasion were noted. Follow-up questionnaires were distributed to referring veterinarians in an effort to better understand the behavior of these atypical lesions. Questions included presence/absence of recurrent disease, spread of disease (and to which organs), follow-up measures (radiation, chemotherapy, surgery), survival time, and cause of death/euthanasia if applicable. The goal of this study is to identify if atypical features are associated with more aggressive behavior, and if further clinical action is warranted in cases of atypical POF.

D-51: SYSTEMIC ASPERGILLOSIS MANIFESTING AS NEUROLOGIC DISEASE IN AN EQUINE

Amanda J. Anderson, Tatiane T.N. Watanabe, Rudy Bauer

A 2.5-year-old Quarter Horse gelding acutely presented with fever and colic that rapidly progressed to neurologic abnormalities including reluctance to move, head pressing, mild forelimb hypermetria, urine dribbling, muscle fasciculations, obtunded mentation, and intermittent blindness. Despite initial improvement after several days of supportive care, the gelding died one week after the initial onset of clinical signs and was submitted to the Louisiana Animal Disease Diagnostic Laboratory for postmortem evaluation. Gross necropsy examination revealeda 3.0 cm spherical, red area of malaciawithin the midbrain/thalamus, primarily involving the right side. Multifocal tan-green plaques were present on the mucosal surface of the medial nasal septum and turbinates. Lung lobes and myocardium contained multifocal round, white nodules with occasional targetoid appearance. Rare, ulcerative, transmural, ovoid, black to tan, poorly-demarcated foci were present on the serosal surface throughout the intestines. Histologically, several regions of the brain contained large foci of malacia, most prominent in the thalamus, with intralesional fungal hyphae, vasculitis, and thrombosis. The nasal septum had ulceration and pyogranulomatous to mixed inflammation with similar Gridley-positive intralesional fungal hyphae. The lesions within the lungs, myocardium, and intestines had similar histologic findings. Positive A*spergillus* sp. cultures were obtained from the brain, lung, and nasal turbinates. *Salmonella* sp. was isolated from the large colon. The final diagnosis was severe, multi-systemic mycosis involving multiple organs including the brain, which would account for the neurologic clinical signs.

D-52: INVESTIGATING THE ASSOCIATION OF BARTONELLA SPP. IN CANINE NODULAR PYOGRANULOMATOUS PANNICULITIS

Joseph M. Malatos, Kathleen Kelly, Jeanine Peters-Kennedy

Canine sterile (idiopathic) nodular pyogranulomatous and granulomatous panniculitis is a routine diagnosis representing approximately 0.9% of canine haired skin biopsies from the Cornell University Surgical Pathology Service; however, the cause is currently unknown. As *Bartonella* spp. is an emerging zoonotic pathogen that induces granulomatous inflammation including rare case reports of pyogranulomatous dermatitis and panniculitis in dogs, we hypothesized that canine idiopathic panniculitis cases would be associated with the detection of *Bartonella* spp. DNA. To evaluate the association of *Bartonella* spp. with pyogranulomatous or granulomatous nodular panniculitis in dogs, a retrospective study was performed on canine skin biopsies from the NYS pathology archive from 2014-2015. DNA was extracted from formalin fixed paraffin embedded canine skin samples from twenty-one cases with a histologic diagnosis of pyogranulomatous or granulomatous nodular panniculitis and twenty-one controls with a diagnosis of subcutaneous lipoma. No organisms were identified with histochemical stains (Fite-Faraco, Ziehl-Neelson, Grocott's methenamine silver, Periodic acid-Schiff stains, and a gram stain). Polymerase chain reaction was performed on extracted DNA using primers targeting the *Bartonella* 16S-23S internal transcribed spacer region. *Bartonella* DNA was not amplified from the panniculitis cases (0/21) or from lipoma controls (0/21). These results suggest that *Bartonella* spp. is not frequently associated with pyogranulomatous or granulomatous nodular panniculitis in the northeastern United States. Ruling out infectious etiologies is important as the mainstay of treatment for an idiopathic sterile nodular panniculitis often includes immunosuppression.

D-53: IMPROVING DETECTION OF CETACEAN MORBILLIVIRUS IN SOUTH ATLANTIC CETACEANS

Kátia Regina Groch, Eva Sierra, Josué Díaz-Delgado, Antonio Fernandez, José Luiz Catão-Dias

Background: *Morbillivirus* (*Paramyxoviridae*) is a re-emergent pathogen associated with severe epizootic outbreaks in cetaceans. Recently, the virus was associated with fatality involving a stranded Guiana-dolphin (*Sotalia guianensis*) in Brazil. Partial sequence of the viral phosphoprotein gene and phylogenic analysis showed that the Guiana-dolphin morbillivirus (GDMV) shared only 79% of nucleotide and 58% of amino acid identities with other cetacean morbillivirus (CeMV), suggesting that the virus might represent a new lineage of CeMV.

Objective: To develop an effective real-time RT-PCR for screening of morbillivirus, capable of detecting the three main variants of CeMV of the Atlantic Ocean: GDMV, dolphin morbillivirus (DMV) and pilot-whale morbillivirus (PWMV).

Methods: Viral RNA was extracted from frozen tissues of a GDMV-positive Guiana-dolphin, a DMV-positive bottlenose-dolphin (*Tursiops truncatus*), a PWMV-positive short-finned-pilot-whale (*Globicephala macrorhynchus*) and a canine distemper virus (CDV)-positive dog. Sequences of DMV, PWMV and GDMV were aligned, most conserved regions of the phosphoprotein gene were selected, primer sets were generated using the Primer 3 software and tested by computer simulations using Primer BLAST (NCBI). The final three primer sets were selected to test field samples by real-time RT-PCR method using SYBR® Green. The methods developed were standardized at the Institute for Animal Health (Canary Islands, Spain), validated through genetic sequencing and sensitivity was studied performing serial dilutions.

Results: The selected primer sets effectively amplified morbillivirus variants from cetaceans and from a CDV-positive dog.

Conclusions: This method may be suitable for screening studies of known and possible novel morbillivirus variants in South Atlantic cetaceans.

D-54: SAVANNAH MONITOR (VARANUS EXANTHEMATICUS) WITH PANCREATIC ENDOCRINE CARCINOMA, HEPATIC ADENOCARCINOMA, AND MULTIPLE METABOLIC DISORDERS

Ethan R. Biswell, Abigail C. Durkes

A 7-year-old female Savannah monitor, *Varanus exanthematicus*, was found dead in its enclosure with a history of intermittent anorexia of approximately 2 weeks duration. Plasma biochemical analysis from the annual checkup (one month before death) indicated mild hyperglycemia and moderate hypercalcemia. Gross and histopathologic evaluations revealed marked serosanguinous effusion of the coelomic and retroperitoneal cavities, fibrinous coelomitis, visceral gout (multiple organs), diffuse hepatic lipidosis, hepatic adenocarcinoma with ductal hyperplasia, cholestasis, encapsulated hepatic hematoma, end stage kidney disease with urate tophi, mild heterophilic enteritis with ulceration, and multiple pancreatic endocrine carcinomas. The cause of death is attributed to one or more of these abnormalities. Varanids are found to have increased predilections to neoplastic disorders compared to other lizard groups, as well as have a high incidence of prior neoplasia or simultaneous second primary neoplasm of a different organ system. Pancreatic endocrine tumors in lizards have been found by IHC to produce one or more hormones (i.e. insulin, glucagon, somatostatin, pancreatic polypeptide, or gastrin). In some cases, pancreatic endocrine tumors are believed to cause metabolic disorders, mainly diabetes mellitus, and would explain the hyperglycemia and hepatic lipidosis in this case. Very few reports of pancreatic neoplasia in Varanids have been reported. The few reported in this species tend to be accompanied by another primary tumor (majority hepatic). Endocrine cell tumors should be considered a differential diagnosis with mild hyperglycemia, intermittent anorexia, and/or sudden death in monitor lizards.

D-55: CONGENITAL ANOMALIES IN DOMESTIC ANIMALS IN GRENADA, WEST INDIES: A RETROSPECTIVE STUDY OF 18 CASES (2001-2016)

Alfred Chibeka Chikweto, Claude De Allie, Keshaw Prasad Tiwari, Ravindra Nath Sharma, Muhammad Iqbal Bhaiyat

Background: Congenital defects are common in both animals and humans. The etiologic agents involved can range from genetic to environmental in nature. A retrospective study of 18 cases of congenital anomalies that presented at the St. George's University was carried out.

Objective: Our objective was to report on congenital anomalies in domestic animals in Grenada.

Methods: The data were collected from necropsy records of carcasses submitted to the Pathology Laboratory of the School of Veterinary Medicine and covered the period between January 2001 and December 2016.

Results: During this period, a total of 2,334 animals (1,963 dogs, 148 goats, 109 chickens, 72 sheep, 36 pigs and 6 cattle) were necropsied. Congenital anomalies diagnosed were as follows: cyclopia in a lamb; polymelia and atrioventricular septal defects in a lamb; arthrogryposis in a lamb; cleft palate and polydactyly in 2 goat kids; patent foramen ovale in a pig; scrotal hernia in a pig, dysplastic atrioventricular valves with biventricular hypertrophy in a calf; supernumerary feet and duplication of cloaca in a broiler chicken; persistent right aortic arch with congenital megaesophagus in a puppy, cleft palate in 2 puppies; ovarian teratoma in a bitch; cryptorchidism in 2 dogs; unilateral renal hypoplasia in a dog; bilateral renal dysplasia in a dog; congenital diaphragmatic tear in a dog and true hermaphroditism in a dog.

Conclusions: Increased congenital anomalies in food animals can impact negatively on productivity and food security for the island. Although these congenital anomalies have been reported worldwide, this is the first report from Grenada.

D-56: INFLAMMATORY EYE LESIONS IN DOGS WITH LEISHMANIASIS

Gisele F. Machado, Livia C. Bregano, Valeria M. F. Lima, Tatiane T. N. Watanabe, Ingeborg M. Langohr

Visceral canine leishmaniasis (VCL) caused by *Leishmania infantum* is a chronic disease that affects dogs. Ophthalmic manifestations of VCL are very common, usually observed together with periorbital alopecia and exfoliative dermatitis. In endemic areas, leishmaniasis should therefore be considered in the differential diagnosis of ocular adnexal and anterior segment inflammatory lesions in dogs. Thirteen dogs with clinical VCL (confirmed by serum ELISA and spleen PCR) were submitted to routine gross examinations. One eye of each dog was enucleated, fixed in 10% formalin (pH 7.4), and routinely processed for histopathology. Aqueous humor was collected from the contralateral eye for anti-*Leishmania* IgG detection (ELISA). Grossly, one dog had uveitis. Histologically, all the examined eyes had inflammatory lesions varying from mild to severe. These alterations were most frequent in the conjunctiva (72%), ciliary body (60%), sclera (38%), and limbus (37%). Other ocular structures such as the retina and choroid (7.7%), periorbital nerves (38%) and muscles (23%) were less affected. The inflammatory cell distribution was perivascular and focal to multifocal coalescing. In some eyes, the infiltrate was typically granulomatous; in others, it was predominantly lymphoplasmacytic. In some cases, the cell infiltrate also included eosinophils. Abundant amastigote forms were observed in one case with granlomatous inflammation. Anti-*Leishmania* IgG was detected in the aqueous humor of all tested dogs. In dogs with VCL, ocular inflammatory lesions are therefore present even without gross evidence of ocular disease and, in routinely stained eye sections, without histologic evidence of the parasite.

D-57: OVER-EXPRESSION OF RAD51 CORRELATES WITH MALIGNANCY OF CANINE INTESTINAL NEOPLASIA

Seung Hee Cho, Byung Joon Seung, Soo Hyeon Kim, Jung Hyang Sur

Background: RAD51 assists in the repair of DNA double strand breaks, a process that is dependent on BRCA2. Therefore, it has great relevance in terms of genomic stability. Numerous studies report that RAD51 is overexpressed in different cancers, including human colorectal tumors. Canine intestinal neoplasia (CIN) includes neoplasms that develop in both the small intestine and colon, and the relationship between CIN and RAD51 or other related biomarkers has not been studied.

Objective: To investigate the expression of RAD51 in different canine intestinal tumors, and confirm the role of RAD51 in CIN. Antibodies to RAD51 were used to establish the association of their expression with the pathological characteristics of the canine intestinal tumor using immunohistochemical methods.

Methods: Immunohistochemical analysis for RAD51 was carried out in formalin-fixed, paraffin-embedded sections of different tumors from the small intestine and colon of 22 dogs, using a standard peroxidase-antiperoxidase method. Expression of this protein in benign (polyps and adenomas, n = 8) and malignant (adenocarcinomas, n = 14) tumors was compared. Immunohistochemical analysis was also done for other biomarkers (BRCA1, TGF beta, and PTEN).

Results: RAD51 expression was significantly higher in malignant tumors than in benign lesions (p=0.023). There were no significant differences in BRCA1, TGF beta, or PTEN expression.

Conclusions: This study suggests that overexpression of RAD51 is a predictor of clinical outcome in CIN, and indicates the promise of future studies using RAD51 as a prognostic marker and therapeutic target.

D-58: REDUCED EXPRESSION OF P21 IN CANINE HEPATOCELLULAR CARCINOMA

Soo Hyeon Kim, Byung Joon Seung, Seung Hee Cho, Jung Hyang Sur

Background: p21, one of the cyclin-dependent kinase inhibitor, is a major cell cycle regulator and widely known as tumor suppressor. In human, p21 has revealed to be decreased in hepatocellular carcinoma. Additionally, p21 is upregulated by Snail, zinc-finger transciption factor and facilitated by Survivin, which belongs to inhibitor of apoptosis family. Complex interactions of these molecules were reported to regulate the proliferation of cells and disrupted expression of these molecules induce tumor formation.

Objective: The study was designed to find the different expression rate of p21 in canine normal liver tissues and hepatocellular carcinomas. Expressions of p21-related molecules, Snail and Survivin werealso analyzed for identifying the tumorigenicity in canine hepatocellular carcinomas.

Methods: Slides were examined by pathologists and classified as normal liver tissues (n=10) and hepatocellular carcinomas (n=10). Immunohistochemistry was conducted using p21, Snail and Survivin polyclonal antibodies. The results were evaluated by different criteria based on previous studies.

Results: Expressions of p21, Survivin were observed mostly in cytoplasm and Snail showed both cytoplasmic and nuclear staining patterns. Expression of p21 was decreased in hepatocellular carcinomas than in normal liver tissues (P=0.015). However, expressions of Snail and Survivin did not show significant difference between in hepatocellular carcinomas and normal liver tissues.

Conclusions: Progression of canine hepatocellular carcinoma seems to be associated with decreased p21 expression, but Survivin and Snail are not likely to be the key factor of the tumorigenicity in canine hepatocellular carcinoma.

D-59: DERMATITIS WITH ASSOCIATED SARCOPTES SCABEI, CORYNEBACTERIUM SP., AND STAPHYLOCOCCUS HYICUS INFECTION IN A MINIATURE POT-BELLIED PIG (SUS SCROFA DOMESTICUS)

Elizabeth A. Kieran, Michael J. Dark, Heather S. Walden, Elizabeth A. Nelson

A 4 month old, male miniature pot-bellied pig (Sus scrofa domesticus) presented to the University of Florida Veterinary Medical Center for sudden onset dyspnea and unresponsiveness. Initial exam findings included obtunded mentation, cyanosis, bradycardia, bradypnea, hypothermia, hypoglycemia, poor body condition, and severe exudative dermatitis with lichenification; despite supportive therapy spontaneous death occurred 36 hours after hospital admission. Post-mortem gross evaluation of the skin identified widespread, marked to severe lichenification, multifocal to coalescing erosions, numerous tan crusts forming broad superficial mats, serous to greasy exudate oozing from all skin surfaces, and patchy reddening of the ventral body wall. Histologic evaluation of the skin identified marked to severe, chronic dermatitis with marked orthokeratotic and parakeratotic hyperkeratosis with multifocal intracorneal pustules, numerous 300 x 400 micrometer intracorneal mites (morphology consistent with Sarcoptes scabiei), and numerous intracorneal coccobacilli. Aerobic culture of the skin yielded heavy growth of predominately Corynebacterium sp.; Staphylococcus hyicus was also cultured. A skin sample was also submitted for parasite identification, confirming Sarcoptes scabiei infestation. S. scabiei infestation is a highly contagious disease of economic importance in production swine operations. The mite causes mechanical injury to the skin in addition to irritant effects induced by secretions and excreta and hypersensitivity reactions to mite antigens. A recent study exploring scabies infection in a porcine model has demonstrated scabies impacts host skin microbiota, with a shift from normal to pathogenic bacterial populations and notably identified Corynebacterium sp. infection associated with crusted skin lesions and associated mite microbiota.

D-60: GLYCOGEN BRANCHING ENZYME DEFICIENCY IN A QUARTER HORSE FOAL (EQUUS CABALLUS)
Elizabeth A. Kieran, Jeffrey R. Abbott

Aone day-old Quarter horse colt (Equus caballus) initially presented to the University of Florida Veterinary Medical Center for inability to stand and nurse. The foal was born to a primiparous mare with reported premature rupture of the umbilical cord during parturition. Initial findings on evaluation included obtunded mentation, bradycardia, injected mucous membranes, hypoxemia, and hypoglycemia.The foal's status initially improved with supportive therapy and was subsequently discharged. Mild diarrhea developed during ongoing care at home. At seven days age the colt suddenly died while playing in a paddock. Post-mortem gross evaluation identified mild serosanguinous effusion and minimal to mild enterocolitis. Histologic evaluation identified numerous PAS-positive, diastase-resistant sarcoplasmic inclusions within cardiacmyocytes. Similar intracytoplasmic and extra cellular inclusions were identified within the neurons and neuropil of the brain, respectively. DNA testing of frozen tissue confirmed the colt was homozygous for the GBE1 genetic defect underlying glycogen branching enzyme deficiency (GBED)of Quarter horses. GBED is a glycogen storage disease type IV and is an autosomal recessive, fatal hereditary condition in Quarter horses that may cause sudden death in utero or during early stages of life. The condition is caused by a nonsense mutation in codon 34 of the GBE1 gene.

D-61: AMEOBIC ENCEPHALITIS IN A CHIMPANZEE
Amit Kumar, Dodd Sledge

A 32-year-old male chimpanzee that was part of a zoological collection died after a <72 hour history of circling, abnormal mentation, and prolonged seizures. On necropsy, the cerebral vessels were diffusely congested, the leptomeninges were mildly edematous, and there was a focal region of hemorrhage and necrosis in the frontal lobe of the right cerebral hemisphere. Within this region of hemorrhage and necrosis, there was common fibrinoid vasculitis and vascular thrombosis. Scattered throughout the necrotic neuropil and concentrated around vessels, there were numerous 35-75 um in diameter, variably shaped amoebal trophozoites that had abundant stippled eosinophilic cytoplasm and small eosinophilic nuclei with a single large karyosome. Intermixed with trophozoites in central regions of hemorrhage, there were rare cysts that were spherical, 20-40 um in diameter, and that had a thick refractile wall. Amebic encephalitis is uncommon in humans and has been rarely reported in other primates. The amoeba associated with encephalitis are typicallyopportunistic pathogens that are naturally free-living in the environmental soil or water sources. The exact source of infection in this case is unclear. Other chimpanzees in the exhibit were not affected.

D-62: FIRST DESCRIPTION OF MALIGNANT GASTROINTESTINAL STROMAL TUMOR WITH TRANSPLANTATION METASTASIS IN A HORSE

Tatiane Terumi Negrao Watanabe, Frank Andrews, Rudy Bauer

Background: Gastrointestinal stromal tumor (GIST) has been described in humans, non-human primates, dogs, horses, ponies, donkeys, and rats. In horses, GIST has been reported as having a benign behavior without criteria of malignancy and most commonly reported in older female horses with unclear breed predisposition.

Case Report: A 15-year-old, bay, Quarter horse mare (*Equus caballus*), weighing 376.0 kg had 1-year history of clinical signs of progressive weight loss and muscle wasting. Clinical differential diagnosis included possible abdominal abscess. Due to the poor prognosis and the chronicity of the case, humane euthanasia was elected by the owners.

Results: Necropsy examination revealed numerous nodules and masses occupying nearly 70% of the entire abdominal cavity. The left lateral lobe of the liver had a large, multinodular, cavitated, mottled dark to pale tan mass (weighing 68.0 kg) that markedly effaced and compressed the parenchyma. On cut surface, the mass was mottled dark red to yellow/tan and gelatinous. Smaller similar nodules were observed throughout the abdominal cavity, spleen and serosal surface of the small and large intestines. Histologically, the neoplastic masses were composed of multinodular, fairly well-demarcated, markedly expansile loose proliferations of spindle cells in streaming bundles separated by abundant myxomatous matrix frequently forming large mucinous lakes. Immunohistochemistry for CD 117 (c-kit) revealed strong immunopositive cytoplasmic labeling.

Conclusions: Based on the histologic findings associated with the positive immunohistochemistry labeling for c-kit, the diagnosis of GIST with transplantation metastasis was made. To our knowledge, this is the first report of malignant GIST in horses.

D-63: PRIMARY CEREBRAL HISTIOCYTIC SARCOMA IN A DOG

Lori E. Bedient, Josh Ramsay

A 5 year old castrated male Pembroke Welsh Corgi with a history of acute progressive tetraparesis, decreased right eye menace and obtunded mental state presented deceased on arrival to the Washington State University Veterinary Teaching Hospital. Gross necropsy revealed a 2.0 cm diameter round to oval, soft, slightly raised, pink left parietal lobe mass, and a presumptive diagnosis of a glial cell tumor was made. Further dissection of the brain revealed an extensive, well-demarcated cortical mass effacing the gray and white matter from the rostral frontal lobe to the level of the amygdala. Histopathology identified sheets and whorls of round to polygonal cells with marked anisocytosis and anisokaryosis and a low mitotic rate. Abundant eosinophils, neutrophils, lymphocytes and plasma cells, the presence of giant cells and frequent emperipolesis were key features of the neoplasm. Immunohistochemistry for the CD18 antigen and Glial Fibrillary Acidic Protein showed positive and negative membrane immunoreactivity respectively, indicating cells of leukocytic origin. A diagnosis of primary cerebral histiocytic sarcoma was made based on IHC results, cellular features and recruitment of other leukocytes into the mass. Evidence of neoplasia was not identified grossly or histologically in other organs. Primary cerebral histiocytic sarcoma is rare in dogs, comprising only 2% of all primary brain tumors in one retrospective study. This tumor is most commonly diagnosed in Pembroke Welsh Corgis, but has been occasionally reported in other breeds. This case highlights the need for inclusion of this uncommon tumor in differentials for primary brain neoplasms in this breed.

D-64: EXTRAGONADAL INTRACOELOMIC TERATOMA IN A COCKATIEL

Franziska Sebastian, Shelley J. Newman, Michael P. Jones

A 14 year-old, female intact Cockatiel was presented for a 2-week history of lethargy, hyporexia and falling off its perch. A large hard coelomic mass was palpated and radiographs and ultrasound determined the mass to be mineralized and surrounded by fluid. On gross examination the body was in poor nutritional and muscular condition. The mass was approximately 52 x 41 x 29mm, hard, gray, yellow and dark brown, filled most of the coelom and compressed internal organs caudally or cranially. It was not connected to the oviduct, ovary or adrenals. Superficially, moderate amounts of clotted blood covered the mass. On cut-section, there was mostly bone tissue, with variably cystic structures.

Histologic examination revealed a pleomorphic neoplasm composed of approximately 80% mature bone trabeculae embedded in haphazardly arranged streams of interlacing bundles of elongate and highly pleomorphic mesenchymal cells with occasional glands lined by cuboidal epithelial cells, which were rarely ciliated, and areas, which resembled skeletal muscle, as well as large multinucleated cells with up to 20 nuclei. There is marked anisokaryosis and anisocytosis, with karyomegaly. The presence of two or three, such as in this case, germ cell lines forming a mass lesion is indicative of a teratoma.

While extragonadal teratomas are described as single reports in bald eagles, fantail pigeons and domestic turkeys, this is the first report of an extragonadal teratoma in a Cockatiel.

D-65: CLINICOPATHOLOGICAL FEATURES OF SALMONELLA DUBLIN INFECTION IN ONTARIO CATTLE

Andrew Brooks, Murray Hazlett, Janet Shapiro, Maria Spinato, Margaret Stalker, Josepha DeLay, Beverly McEwen, Andrew Vince, Durda Slavic

Salmonella enterica ssp. *enterica* serovar Dublin has emerged as an important bovine pathogen in Ontario. *Salmonella* Dublin was known to cause disease in cattle in Quebec and the northeastern United States and in 2012 the first case was detected in Ontario by the Animal Health Laboratory (AHL). Since then, the AHL bacteriology laboratory has screened all bovine lung samples submitted for culture for *Salmonella* Dublin by an enrichment method. In this report we characterize the major clinicopathological features of *Salmonella* Dublin infection based on 51 culture-positive submissions to the AHL from 2012 to 2016. *Salmonella* Dublin predominantly caused disease in young Holstein calves, between 1-6 months of age, in veal or dairy operations. The major clinical problems associated with infection were respiratory disease, diarrhea, increased mortality, sudden death and fever. Of the 30 submissions which included gross postmortem and/or histopathology the most frequent pathological diagnoses were septicemia, characterized by systemic necro-inflammatory lesions, followed by pneumonia and occasionally enterocolitis. In cases of pneumonia, *Salmonella* Dublin was often accompanied by other respiratory pathogens such as *Mannheimia haemolytica* and *Histophilus somni*. *Salmonella* Dublin was most frequently isolated from the lung or other filtering organs such as the spleen and liver. *Salmonella* Dublin now seems established in Ontario and a current surveillance project, supported by the Ontario Animal Health Network, aims to raise awareness of this pathogen among veterinarians and producers and to identify risk factors associated with herd infection.

D-66: LYMPHOID NEOPLASIA IN AN ATLANTIC STINGRAY (DASYATIS SABINA)

Rita McManamon, Alvin C. Camus, Trevor T. Zachariah

A wild-caught, zoo-housed, approximately 15-year-old, female, Atlantic stingray (*Dasyatis sabina*) with a history of progressive emaciation, hepatomegaly, and an enlarging right liver lobe mass, developed tachypnea, anorexia, and lethargy. Previous history included goiter, resolved with iodine supplementation, and unilateral (O.D.) enucleation due to a melting corneal ulcer. Euthanasia was elected due to the grave prognosis. Gross necropsy findings showed severe emaciation, ventral erythema, splenomegaly, and hepatomegaly with a normal left lobe and a firm, yellow, multilobulated mass (6.7 cm X 8.7 cm) in the right lobe. Histopathology revealed neoplastic round cells, compatible with lymphoid origin, widely disseminated in vascular lumens, as well as aggregates and sheets that severely affected the gills, spleen, and liver. The intestines, coelomic adipose, meninges, epigonal organ, ovary, uterus, and dermis also showed significant infiltration, expansion or effacement. Neoplastic cells filled multiple vascular lumens in brain, left eye, pancreas, stomach, renal sinusoids, thyroid, and skeletal muscle. The neoplastic cells were monomorphic, with distinct borders, high nuclear: cytoplasmic ratios (9:1), and scant pale eosinophilic cytoplasm. Anisocytosis and anisokaryosis were mild. Nuclei were round to oval, with stippled to clumped, hyperchromatic chromatin. No mitotic figures were observed in ten 40xHPF. Electron microscopic and immunohistochemical studies to confirm a lymphoid origin and characterize the cells are in progress. While the reported incidence of neoplasia in elasmobranch fish is low, likely due to limited surveillance and reporting, a growing list of tumors are being described. However, lymphoid neoplasia has only been reported once in an elasmobranch species, the bonnethead shark.

D-67: CEREBRAL PERIVASCULAR WALL TUMOR IN A DOG

Adam W. Stern, Ian T. Sprandel, Marejka H. Shaevitz

Background: Perivascular wall tumors (PWT) are part of the spectrum of soft tissues sarcomas in dogs. PWTs are most commonly reported in the subcutaneous tissue. Metastasis of PWTs to the lung, chest, joint and lymph node have been reported. There is a single report of a primary hemangiopericytoma within the brain of a dog.

Methods: The current report describes the gross, histopathological and immunohistchemical findings of a cerebral PWT in 12-year-old, castrated male, Golden Retriever.

Results: AT necropsy within the left frontal lobe is a 1.2 cm in diameter, well-demarcated, tan to pink, and relatively firm mass. Histologically, neoplastic cells are spindloid and are arranged in bundles and frequently whorl around capillaries. Neoplastic cells exhibit positive cytoplasmic immunoreactivity for vimentin. Neoplastic cells lacked expression for GFAP, S100, smooth muscle actin (SMA), and desmin.

Conclusions: Histopathological characteristics are consistent with a PWT with the major feature of perivascular whorling. In humans, perivascular whorling is diagnostic for the myopericytoma. Recently the spectrum of PWTs was morphologically and phenotypically characterized in the dog and whorling was not reported in canine hemangiopericytoma but was reported in myopericytomas. Immunohistochemically, the neoplastic cells in this report lacked expression for desmin and SMA. Although most canine myopericytomas express both desmin and SMA, several myopericytomas lacked expression of one or both of these markers; whereas, hemangiopericytomas express SMA. The neoplasm in this case is morphologically consistent with a myofibroma. Although rare, PWTs should be considered a differential for neoplasia in the brain.

D-68: IMMUNOHISTOCHEMICAL DIAGNOSIS OF ENTEROPATHIES CAUSED BY CORONAVIRUS AND ROTAVIRUS IN CALVES

Alfredo Pérez-Guiot, Carlos-Gerardo Salas-Garrido, Mario-Adán Bedolla-Alva

A study was conducted from 2008-2012 in which 50 intestine samples from calves from the "*Centro de Enseñanza y Diagnóstico de Enfermedades de Bovinos-UNAM*" located in Tizayuca, Hidalgo, Mexico were analyzed. The samples were taken from calves of 1 to 3 weeks of age that showed clinical signs of diarrhea, as well as macroscopic and/or microscopic lesions suggestive of viral enteritis. Intestinal tissues were fixed in formalin and were evaluated by histopathology and immunohistochemistry for rotavirus and/or coronavirus antigens. The most representative histopathological lesions (>60%) were: degeneration of enterocytes of the villi, degeneration of enterocytes of the crypt, villus atrophy, and shedding of enterocytes. Fifty cases (100%) were immunopositive for coronavirus and 47 cases (94%) were positive for rotavirus. In addition, a method was established to correlate the degree of microscopic lesions with the immunopositivity. Of the 50 cases reviewed, 22 cases (45.8%) had severe injuries (5-6) and high immunopositivity for coronavirus, one case (2%) had severe injuries and high immunopositivity for rotavirus, and 13 cases (27%) had severe injuries and high immunopositivity for both coronavirus and rotavirus. Hence, immunohistochemistry is an important tool for rotavirus and coronavirus diagnosis since they are the most prevalent in neonatal calf diarrhea.

D-69: HISTOPATHOLOGY AND IMMUNOHISTOCHEMESTRY OF PROLIFERATIVE LESIONS IN THE UTERUS OF AFRICAN HEDGEHOGS (ATELERIX ALBIVENTRIS)

Luz-María Vargas-Soto, Adriana Méndez-Bernal, Alonso Reyes-Matute, Elizabeth Morales-Salinas

Neoplasia is a common finding in African hedgehogs (Atelerix albiventris) and represents one of the main causes of mortality in this species. Most uterine tumors in women and domestic animals such as dogs, cats, horses, cows, goats, pigs and rabbits develop from endometrial epithelium and myometrial smooth muscle fibers. Proliferative lesions in uterus of African hedgehogs have not been well characterized. This study aims to report the microscopic and immunohistochemical features of proliferative lesions in the uterus of fourteen African hedgehogs submitted to the Department of Pathology at the FMVZ-UNAM. Immunohistochemestry was performed in these tumors, where immunopositivity to desmin, cytokeratin (AE1 / AE3), and smooth muscle actin (SMA) was assessed. The proliferative lesions in the uterus were classified as neoplastic and non-neoplastic. Within the non-neoplastic, two endometrial polyps were diagnosed. The neoplastic lesions were divided into primary and metastatic. The primary neoplasms of the uterus were those originating from the endometrial stroma: adenosarcoma (3), carcinoleiomyosarcoma (1), adenoleiomyosarcoma (1), endometrial stromal sarcoma (3), leiomyoma (1) and lipoma (1). In all tumors the neoplastic endometrial epithelium expressed AE1/AE3 cytokeratin. The stromal component of the tumors and the leiomyoma were strongly positive to SMA and moderately positive to desmin. The neoplasms that metastasized to the uterus were a fusocellular sarcoma and a germ cell like tumor. The present study demonstrates the complex origin of neoplasia arising from the uterus in African hedgehogs compared to other domestic species and may serve as a tool for their prognosis and treatment.

D-70: DIFFUSE MALIGNANT MELANOMA OF UVEAL TRACT IN A FELINE

Lucia A. Garcia-Camacho, Dulce M. Badillo-Leal, Ignacio C. Rangel-Rodriguez

In dogs and cats, ophthalmic melanocytic tumors of the uveal tract are common, showing an expansive growth and invasion of adjacent structures. A 11-year-old male domestic catwas presented to the teaching hospital with the complaint of a bulged mass on the left eye. At cytology an epithelioid malignant melanoma was diagnosed. After surgical enucleation, theglobewas distorted by a pigmented 3.6 X 2.9 cm mass which was extended from the iridocorneal angles with bilateral spreading into the limbic sclera toward the optic nerve. The anterior and posterior chambers were occluded and effaced, displacing the crystalline lens. The tumor anterior portion was markedly dark-brown, showing gradual discoloration to posterior parts. Microscopically, the uveal tract and limbic sclera wereeffaced by strikingly pleomorphic predominantly round and fusiform melanocyte proliferation, which invade the sclera and extend out to surround the ocular globe. In addition, the retina was fully replaced by neoplastic cells, and the blood vessels of the retina and choroid displayed hyaline walls, indicating chronic secondary glaucoma. The tumor cells display marked anisocytosis and anisokaryosis, scant to abundant cytoplasm with a variable amount of melanin, ovoid to irregular nuclei with vesicular to fine granular chromatin and prominent or multiple (2-3). The mitotic index is high (6-8 atypical figures per high power field). A final diagnosis of high grade diffuse malignant melanoma of the uveal tract was made. Since feline diffuse melanoma of the iris is frequent, displaying high potential of metastasis, most likely the present tumor arose from that location.

D-71: POORLY DIFFERENTIATED FIBROSARCOMA IN A CAT WITH SYSTEMIC METASTASIS

Mayra Chávez-Rodriguez, Adriana Méndez-Bernal, Laura Romero-Romero, Malibé Cano-Saldaña, Hortensia Corona-Monjaras

Fibrosarcoma is the most common malignant mesenchymal neoplasm in cats with 3 different etiologies: fibrosarcoma associated with multicentric feline sarcoma virus (FeSV), post-injection fibrosarcoma and solitary fibrosarcoma. The most common form is the post-injection fibrosarcoma. The aim of this study is to report a case of poorly differentiated fibrosarcoma in a 7-yr-old short haired, spayed female, cat. The patient was referred to the Veterinary Hospital (HVE) at the UNAM, for presenting a firm, multilobulated, non-movable and painful mass in the right leg; that was extending to the hip joint, with a total diameter of 23 cm. Fine needle aspiration was performed, in which pleomorphic mesenchymal cells were observed. Advanced imaging (CT) and right leg amputation were decided. The histopathological diagnosis was fusocellular sarcoma; additional tests of electronic microscopy (TEM) were performed in which cells with abundant endoplasmic reticulum, characteristic of fibroblasts were observed, granting a diagnosis of poorly differentiated fibrosarcoma. Seven months later, the patient presented with depression, anorexia and paraplegia. Due to worsening of clinical signs, euthanasia was elected. The postmortem study showed metastasis to various organs such as heart, intestine, lung, lymph nodes and adrenal gland, as well as an abdominal aorta thrombus of 3 cm in length, which completely obliterated the vascular lumen. The gross findings, clinical presentation and metastasis described in this case although reported, are uncommonly observed in fibrosarcomas in cats.

D-72: SUBCUTICULAR PARAVERTEBRAL CALCINOSIS CIRCUMSCRIPTA IN THE NECK OF A CAPTIVE AFRICAN SPURRED TORTOISE (GEOCHELONE SULCATA)

Heindrich N. Snyman, Menita Prassad, Bruce Burton

An approximately 15 year old captive male African spurred tortoise (*Geochelone sulcata*) presented with a focal, 2.5 cm diameter, firm and mobile, subcuticular mass along the dorsal midline of the neck. The mass was monitored over a 6 month period and following progressive enlargement was surgically excised and submitted for further evaluation. Upon receipt the mass was ~ 6.1 cm in diameter, contained a thick outer fibrous capsule, and on cut section was multiloculated with abundant amounts of soft pasty to gritty, pale yellow debris. Bacterial and fungal cultures were taken and the remainder submitted for histopathological evaluation. Histologically the mass contained multiple varisized deeply basophilic granular mineral lakes dissected by intervening fibrous bands with rare peripheral macrophages and heterophils, scattered osseous metaplasia and few peripheral lymphoid follicles. Mineral content was confirmed with Von Kossa and PAS stains. Based on the characteristic histological features and negative bacterial and fungal cultures, a diagnosis of calcinosis circumscripta was made. Calcinosis circumscripta is a well-recognised condition in domestic mammals. It is most common in dogs and thought to be a form of dystrophic mineralization, often occurring at sites of previous trauma or sustained pressure. It is comparably rare in reptiles with a few sporadic reports being limited to aquatic turtles and lizards. Review of this tortoise's life history revealed a thermal burn wound at the same site almost 10 years prior. This report adds to the knowledge on this distinct entity and suggests a similar pathogenesis for its development in reptile species.

D-73: COMPARATIVE ANALYSIS OF HISTOPATHOLOGICAL INDEX OF LESION IN LIVER, GILL AND KIDNEY OF TILAPIAS, SAMPLED FROM THE BILLINGS WATER RESERVOIR IN SÃO PAULO STATE, BRAZIL

Sandy Lorena Pulecio Samtos, José Henrique Hildebrand Grisi Filho, Ivy Tasso Gomes, Ana Carolina Camachos Lopez, Vera Lisa Generosa Paiva Silva, Bárbara Held, Marta Condé Lamparelli, Gilson Alves Quinráglia, Lilian Rose Marques de Sá

Introduction: Billings Dam is one of the most exposed reservoirs to pollution due to the rapid urban and industrial development of the Metropolitan Region of São Paulo. The histopathological index of lesion (HI) is calculated based on the findings of microscopical examination of each individual fish, assigning a value to the severity and distribution of the lesions and can be used as bio-markers to assess the effects of the ecosystem and its pollutants on health of fishes. Tilapia is one of the human consumed fish from the reservoir.

Objective: it is to compare HI of gill, liver and kidney between two periods of 2014 during the water crisis in the state of São Paulo, in order to determine if there is difference among organs and which organ had a higher lesion index in each sampling periods.

Methods: 46 tilapias captured at different control points of the reservoir were analyzed in summer (n=24) and spring (n=22). Necropsy and tissue processing were performed for histopathological evaluation of gills, liver and kidney.

Results: Gills HI were 8.63±3.73 and 7.68±3.81 (p=0.276), liver HI were 3.33±1.86 and 3.73±2.03 (p=0.524) and kidney HI, 5.17±2.68 and 4.23±2.39 (p=0.171) for summer and spring, respectively. Conclusions: Gills were the main organ affected by adverse water conditions in summer and spring of 2014 when water crisis started in São Paulo. The results showed that organs of fishes were differently affected by environment and the water quality between sampling periods did not differ significantly.

Financial support:CNPq-Brasil, FEHIDRO – AT 603

D-74: DIFFUSE-TYPE GASTRIC MUCINOUS ADENOCARCINOMA IN A CALIFORNIA KING SNAKE

Cheng-Shun Hsueh, Chian-Ren Jeng, Victor Fei Victor Fei, Fun-In Wang, Hui-Wen Chang

An adult female California king snake (*Lampropeltis getula californiae*) housed in Taipei Zoo had presented clinical features of anorexia, poor spirit, and abdominal swelling for two weeks. Exploratory laparotomy revealed a gastric mass with mural perforation and extruded caseous material. Gastrectomy was simultaneously performed. Grossly, the gastric wall was diffusely and annularly widened and had multiple warty, mottled white to beige, protrusions along the mucosal surface with two circular perforations. Histologically, the gastric mass reveals invasive, transmural growth with the neoplasm being composed of nests and sheets of lightly stainingepithelial cells, separated by fibrous connective stroma, and frequent large pools of mucinous background. Intra/extracellular mucin is highlighted by an alcian blue stain. The clustered neoplastic cells are round to oval with distinct-borders, and contains light basophilic, foamy or vacuolated cytoplasm, that peripheralizes the nuclei. The peripheralized nuclei are flattened and crescent-shaped, with prominent marginated chromatin and distinct nucleoli. The epithelial origin of these cells was confirmed with positive immunoreactivity to a pan-cytokeratin antibody. Neoplastic cells also manifest marked anisocytosis, anisokaryosis and a high mitotic rate. Based on the WHO histological criteria, a diagnosis of diffuse-type gastric mucinous adenocarcinoma was made. To the authors' knowledge, this is the first report of a diffuse-type gastric mucinous adenocarcinoma in a California king snake.

D-76: CAROLI SYNDROME IN A 6-YEAR-OLD ROTTWEILER

Nathan D. Helgert, Mee-Ja M. Sula

A 6-year-old intact female Rottweiler was submitted for necropsy after several days of lethargy, ascites, and dyspnea following a lifelong history of liver disease. Grossly, the liver was firm, orange-tan with an enhanced reticular pattern and a bosselated surface, and was smaller than expected at 1.9% of body weight. On cut section, the bile ducts were markedly ectatic and contained viscous light green fluid. The gall bladder was distended, diffusely thickened, and white. There was also a single renal cortical cyst. Within the lungs, there were multiple regions of atelectasis with raised light tan, firm circular nodules.

Histologically, the liver had abundant bridging fibrosis that extended through the limiting plate replacing hepatocytes. Within these areas, there were abundant biliary profiles and biliary epithelial cells that did not form lumens and occasionally, were in direct contact with the surrounding hepatocytes. Some bile ducts were markedly ectatic and lined with hyperplastic epithelium. Scattered throughout the kidneys, there was mild interstitial nephritis with fibrosis, multiple ectatic renal tubules, and a single large cyst.

Gross and histologic findings paired with the early onset of liver disease is consistent with Caroli syndrome, a ductal plate malformation caused by persistence of embryonic hepatic and biliary epithelial progenitors. In humans, ductal plate malformations are frequently associated with adult polycystic kidney disease and pancreatic cysts. In veterinary species, ductal plate malformations are frequently, but not necessarily, associated with renal cysts. A renal cyst was present in the current case.

D-77: MYELOID SARCOMA (MS) PRESENTING AS A PARA-OVARIAN MASS IN A JUVENILE DOG, JJ Bailey, JW Koehler

Jessica J. Bailey, Jey W. Koehler

A 15-month-old intact female Dachshund dog with a previous 1-week history of dyspnea, lethargy, and anorexia presented with radiographic evidence of pleural effusion, a diffuse interstitial lung pattern, a mottled spleen, enlarged lymph nodes, and an enlarged right ovary. At necropsy, there was a right para-ovarian mass that surrounded the proximal ureter and adrenal gland, and infiltrated the regional hypaxial muscles. Multiple 1-2-mm flat, round, subpleural foci were scattered throughout the lungs, and there were extensive pleural adhesions. Histologically, the para-ovarian mass was composed of neoplastic round cells, and there was multi-organ infiltration and intrasinusoidal presence of neoplastic cells in the liver, supportive of leukemia. The morphology and immunohistochemical profile (Iba1, myeloperoxidase, and elastase) of the neoplastic cells are most consistent with myelomonocytic differentiation. The final diagnosis was acute myelomonocytic leukemia (AML-M4) with widespread tissue invasion and right para-ovarian tumor formation (myeloid sarcoma). Myeloid sarcoma, also known as granulocytic sarcoma (GS) or chloroma, is described in humans as a focal mass of extramedullary proliferation of myeloid precursors seen either preceding or following a diagnosis of acute myeloid leukemia. However, in domestic animals, myeloid sarcomas are largely incidental findings and not typically associated with the subsequent onset of AML. This case is of interest because there are very few reports of myeloid sarcoma in domestic animals, especially in combination with leukemia, and none describing this anatomic location. A review of the current literature for myeloid sarcomas in conjunction with AML in humans and domestic animals will be discussed.

D-78: PRIMARY GASTROINTESTINAL FOLLICULAR LYMPHOMA IN THREE DOGS

Michael A. Richardson, Tuddow Thaiwong, Matti Kiupel

In humans, primary gastrointestinal lymphomas are most commonly B-cell lymphomas with diffuse large B-cell lymphoma (DLBCL) and mucosa-associated lymphoid tissue lymphoma (MALTOMA) representing the majority of cases. Gastrointestinal follicular lymphomas (FL) are rare primary gastrointestinal neoplasms with an indolent behavior and good response to surgical resection or monoclonal antibody therapy. In dogs, enteropathy-associated T cell lymphomas (EATL) large cell type (type 1) is the most common gastrointestinal form of canine lymphoma; however, DLBCL also occurs as primary gastric and, to a lesser degree, intestinal lymphoma. Primary gastrointestinal FL has not been reported in dogs. We identified three cases of primary rectal FL in dogs with histomorphologic and immunohistochemical features similar to the human entity. The affected dogs were all castrated males, between 9 months to 2 years old and of varying breeds. Expanding the rectal submucosa in all 3 caseswere large, uniform, densely packed follicles that had no polarity, differentiating them from hyperplastic lesions. There were uniform proportions of centrocytes and centroblasts through all follicles with an absence of a dark zone of mantle cuff, and no tingible body macrophages. Immunophenotypically, centrocytes and centroblasts expressed typical B-cell antigens (CD20, CD79a, and PAX5), anti-apoptotic protein BCL2, and lacked expression of the T-cell antigen CD3. PCR for rearrangements of the immunoglobulin heavy chains confirmed a clonal expansion of B-cells in all cases. Surgical excision or chemotherapy appeared curative. Recognition and accurate diagnosis of this distinct manifestation of gastrointestinal lymphoma is important for appropriate therapeutic selection and prognosis.

D-79: FEMALE REPRODUCTIVE PATHOLOGY IN GERIATRIC FALLOW DEER (DAMA DAMA)

Colleen F. Monahan, Wynona Shellabarger, Kimberly A. Thompson, Kirk Suedmeyer, Dalen W. Agnew

Spontaneous reproductive neoplasms in cervids, and ruminants in general, are rare. This study included 15 female fallow deer (*Dama dama*) (age 11 to 17 years) from a single inbred and reproductively senescent population-submitted to the Michigan State University Veterinary Diagnostic Laboratory or the Reproductive Health Surveillance Program (Association of Zoos and Aquariums). These cases represented 72% of the total mortality among females in this population, with complete necropsies performed, from 2007-2017. An additional case was from another zoo. These deer were housed with sterilized males for many years. The objective of this study was to describe the reproductive pathology of a population of aged fallow deer which had been barren for many years. Reproductive tract examination yielded 44 lesions, including 18 neoplastic and 26 non-neoplastic. The most common neoplasm was endometrial carcinoma (9/18), 8 of which had extensive metastasis to multiple organs. Additional neoplasms included leiomyoma (3), leiomyosarcoma (2), endometrial stromal tumor (1), endometrial adenoma (1), uterine squamous cell carcinoma (1), and vaginal adenocarcinoma (1). One leiomyosarcoma had extensive metastasis to multiple organs. The most common non-neoplastic lesion was cystic endometrial hyperplasia (9). Other non-neoplastic reproductive lesions included ovarian subsurface epithelial cysts (3), ovarian follicular atrophy (3), paraovarian cysts (2), ovarian follicular cystic hyperplasia (2), serosal inclusion cysts (2), endometrial polyps (2), adenomyosis (1), endometrial atrophy (1), and vaginal cystic epithelial hyperplasia (1). This case series and a few case reports in the literature suggest that fallow deer are predisposed to reproductive neoplasia and/or that pregnancy may be protective.

D-81: NASAL ADENOCARCINOMA WITH INTESTINAL DIFFERENTIATION IN A BEAGLE

Chi-Fei Kao, Fei Victor Pang, Hui-Wen Chang, Chian-Ren Jeng, Fun-in Wang, Chen-Hsuan Liu

Background: Non-intestinal glandular malignancy with intestinal differentiation has been well documented in humans but not in animals. Aside from microscopic similarity, the neoplastic cells exhibit unique immunohistochemical and molecular characteristics indicating intestinal differentiation. Here we report the first case of a canine nasal adenocarcinoma histologically and immunohistochemically resembling colorectal adenocarcinomas.

Case Description: A 9-year-old, male intact beagle with 1-year unilateral nasal discharge presented with left facial swelling and a maxillary mass. Diagnostic imaging revealed osteolysis of the left maxilla and destruction of the nasal conchae by a soft tissue opacity involving the leftorbital cavity, frontal sinus and Eustachian tube. Thoracic and abdominal cavities appeared unremarkable. Several biopsies of the mass were preserved and processed routinely for histological evaluation.

Results: Microscopically, the mass comprised mucinous pools and tubulopapillary growths reminiscent of neoplasms arising from the intestinal tract with varying numbers of goblet cells. The neoplastic cells were cuboid to columnar, having eosinophilic cytoplasm and mostly basally placed nuclei. They were positive for pan-CK, CDX-2, Villin and CK20 but negative for CK7. Fine needle aspiration of the submandibular lymph node showed clusters of epithelial cells with mild pleomorphism, indicative of metastasis.

Conclusions: Although without necropsy, we would not be able to completely rule out a primary intestinal tumor, the absence of related history, failure to identify other primary neoplasm, and the extreme rarity of nasal metastasis from colorectal adenocarcinomas, made this unlikely. To our knowledge, this should be the first case demonstrating intestinal differentiation in nasal adenocarcinoma in the veterinary literature.

D-82: CO-INFECTION OF MYCOBACTERIUM MARINUM AND M. FORTUITUM IN A CAPTIVE ADULT DIAMONDBACK WATER SNAKE (NERODIA RHOMBIFER) CAUSING DISSEMINATED MYCOBACTERIOSIS WITH AN ACUTE CUTANEOUS MANIFESTATION

Tatiane Terumi Negrao Watanabe, Emi Sasaki, Gordon J. Pirie, Nobuko Wakamatsu

Background: Mycobacteriosis is a sporadic disease in reptiles and potentially zoonotic. *Mycobacterium marinum*, the most common isolate in reptiles, is a slow growing species commonly found in both fresh and salt water and causes cutaneous nodules and papules known as "aquarium/fish tank granuloma" in humans. *Mycobacterium fortuitum* is a ubiquitous, rapid growing species, and the infection has been reported in pet snakes and the owner suffering from lymphadenitis.

Case Description: An adult male captive diamondback water snake (*Nerodia rhombifer*) was found dead after presenting one day history of lethargy with multifocal, acute ulcerative cutaneous lesions throughout the body. The snake ate two sunfish (*Mola* spp) five days prior to death. In addition to the cutaneous lesions, gross examination revealed multifocal to coalescing, white to pale tan nodules in the lung and liver and segmental impactions of the small and large intestines with digesta containing fish's bones. Histopathologic examination confirmed severe granulomatous inflammation with numerous intrahistiocytic acid-fast bacteria in the skin and underlying muscle, lung, liver, and intestines. *M. marinum* and *M. fortuitum* were identified by culture of the hepatic granuloma, followed by PCR and rpoB gene sequencing.

Conclusions: This is a case report of mycobacteriosis in a water snake co-infected with slow and rapid growing mycobacterial species. Ingestion of infected fish/water and direct contact with contaminated water are considered most likely routes of the infection. In addition to forming granulomatous nodules in multiple organs, the mycobacterial infection caused acute ulcerative cutaneous lesions and intestinal impactions due to intestinal mural granulomas.

D-83: SYSTEMIC AMYLOIDOSIS IN A CAPTIVE POPULATION OF PRONGHORN ANTELOPE (ANTILOCAPRA AMERICANA)

Margaret E. Martinez, Katie Seeley, Dawn Zimmerman, Priya Bapodra, Rachel Cianciolo

Background: Fourteen pronghorn antelope (*Antilocapra americana*) from a single captive herd underwent complete or partial autopsies between 1997 and 2016.

Objective: Our goal was to characterize the histologic and ultrastructural changes, as well as the underlying cause of the amyloid.

Methods: Histologic examination of all H&E and congo red stained microscopic slides, as well as transmission electron microscopy and mass spectrometry was performed in two cases. Four banked serum samples from affected pronghorns had serum amyloid

A, haptoglobin, beta and gamma globulin levels measured. Pedigree analysis and retrospective investigation into the clinical histories was performed.

Results: Ten animals had histologic evidence of amyloidosis resulting in a prevalence of 77% of autopsied animals. The majority (90%) of animals had histologic amyloid in the kidneys often causing global to segmental expansion of the glomerular mesangium and capillary wall. Transmission electron microscopy demonstrated glomerular deposits of haphazardly arranged non-branching fibrils consistent with amyloid, as well as a coarser fibril. Mass spectrometry revealed the presence of serum amyloid A and fibronectin. The banked serum amyloid A, beta and gamma globulin levels from affected pronghorns were within normal ranges for healthy domestic cattle. Clinical commonalities between most of the cases included elevated fecal strongyle counts (*Haemonchus* spp.), anemia, hypoproteinemia and azotemia. At post-mortem examination, several (40%) animals had a diagnosis of pneumonia. There was no significant difference between the mean degree of relatedness and presence of amyloidosis.

Conclusions: Therefore, the systemic reactive amyloidosis was likely secondary to underlying chronic inflammation caused by haemonchosis and/or pneumonia.

D-85: LARGE WHALE UNUSUAL MORTALITY EVENT, EASTERN NORTH PACIFIC

Stephen Raverty, Paul Cottrell, Kathy Burek Huntington, Heindrich Snyman, Katharine Savage

Between May 2015 and September 2016, 17 finback and 50 humpback whales were reported dead around Kodiak Island through the western Gulf of Alaska, central British Columbia coast and west coast of Vancouver Island. The increased number of large cetacean mortalities and short time and distance between strandings was unprecedented and prompted an enhanced effort to attend and determine possible causes of the stranding. Prime differentials included sonar/seismic testing, radiation, ship strike, infectious disease, predation, and oceanographic changes leading to algal toxin exposure or starvation due to shifts in prey species and distribution. Complete necropsies were impeded in regions due to site access, safety and state of decomposition. Of 11 animals examined, 4 had evidence of blunt or sharp force trauma (ship strike), 5 were in suboptimal nutritional condition, 1 was autolyzed and 1 had no significant lesions. The stranding coincided with anomalous warming of the sea surface water layer and algal blooms. Of 14 animals screened for HABs toxins, 6 had detectable levels of Paralytic Shellfish Poisoning (PSP) and Domoic acid (DA), 3 had PSP only, 3 had DA and 2 were below detectable limits. The contribution of these HABs to morbidity or mortality remains unknown and no specific cause of death was linked to the loss of these whales. Enhanced efforts to monitor large cetacean strandings are ongoing with field observations to assess body condition, prey shifts, and thorough necropsies which may provide baseline information for future stranding anomalies.

D-86: HISTOPATHOLOGIC CHARACTERIZATION OF FELINE BOWENOID IN SITU CARCINOMA-A CASE SERIES

Education Focused Scientific Session

November 7, 2017 | 8:00 AM – 12:00 PM
Session Chair: M.S. Camus
Committee Members: K.P. Carmichael (Past Chair), M.K. Keel, M. Sula, R. Mo-Peters

November 7, 2017
11:30 AM – 11:45 AM

USING MICROSOFT POWERPOINT TO CREATE INTERACTIVE LEARNING EXERCISES FOR VETERINARY PATHOLOGY STUDENTS

Brian F. Porter

Microsoft Powerpoint® is a program that is capable of more than lecture-style presentations. Using animations and hyperlinks, it can be used to develop highly interactive learning exercises for the pathology classroom or laboratory. Clinical information can be combined with images to produce case studies that help demonstrate the clinical relevance of pathology principles. Animations allow students to search for and identify specific features within gross and histologic images, while hyperlinks let students jump from slide to slide, mimicking a web-based experience. Use of these exercises in laboratories can enhance gross specimen evaluation and potentially replace slide evaluation using microscopes or whole slide scans.

November 7, 2017
11:45 AM – 12:00 PM

DEVELOPMENT OF A VIRTUAL VETERINARY PATHOLOGY ROUNDS I-BOOK TO SUPPLEMENT LEARNING

Oscar G. Illanes, Carmen I. Fuentealba

In most Veterinary Schools, gross post-mortem findings from cases submitted to the pathology services are presented to students and faculty during weekly or bi-weekly necropsy rounds. These activities have an enormous teaching value and foster fruitful discussion and reflection.

At Ross University School of Veterinary Medicine (RUSVM), quality images of the macroscopic findings in animals submitted for post-mortem evaluation are routinely taken and archived in the school's records. The autopsy reports and digital photographs of microscopic findings from cases of high learning potential are also archived. This material is shared with small groups of students rotating through post-mortem services or with the RUSVM Pathology Club. In order to reach a larger audience it was decided to make this material available in an i-Book format as part of the RUSVM iPad-enabled learning program. By using this technology autopsy cases are readily accessible to both students and faculty; their teaching value was enhanced by including literature reviews conducted by selected students. The development of this iBook is a dynamic process and new cases are continuously added.

Surveys indicate that the virtual pathology iBook is an effective method to reach a larger audience and is a valuable teaching tool used by faculty throughout the curriculum. Involvement in this project proved rewarding and motivating for the students participating in the collection and scholarly discussion of the clinical and pathology data.

Veterinary Student Posters

SP-01: PATHOLOGY OF PURPLE ANNULAR PIGMENTATION AND TISSUE LOSS IN COMMON SEA FANS (GORGONIA VENTALINA)

Gina M. Zeitlin, Michelle M. Dennis

Gorgonia vantalina provides a growth substrate and food source for Caribbean reef inhabitants. Purple annular pigmented lesions in sea fans are widely considered specific for *Aspergillus sydowii* infection and yet limited descriptions of the microscopic pathology of such lesions exist. We hypothesized that such lesions may represent a nonspecific response to injury, potentially induced by other etiologies. This study aims to describe the pathology of sea fans affected with purple annular pigmentation resembling aspergillosis, and create a microscopic case definition for sea fan aspergillosis. Twenty *G. ventalina* from two reefs at depths ranging 1.0-2.9m were enrolled in the study. Mean (SE) coral size was 32.9cm (1.37cm). Lesions were circular(n=8) to non-geometric, mean (SE) lesion size was 6.05cm (0.38cm).Of corals sampled, 35% had central areas of tissue loss within pigmented foci. Lesions ranged from 1-4 per colony. Wedge-shaped excisional biopsies were collected from each coral using scissors. Biopsies were fixed in Z-fix and routinely processed for histology. Histopathologic evaluation was conducted on 9 corals. More than one type of micro-organism was seen in the purple pigmented areas in in most corals including fungi morphologically consistent with *Aspergillus spp.* (n=6), algae (n=6), and cyanobacteria (n=3). Failure to consistently demonstrate fungi within these pigmented lesions may indicate a previous infection was cleared, or another etiology instigated the lesion. Additional coral biopsy and histopathology study is required to elucidate the relationship between fungal infection and these pigmented lesions, potentially requiring focus on new or actively growing pigmented lesions.

SP-02: METASTATIC INTESTINAL ADENOCARCINOMA IN A TIGER

Megan Zalek, Dane Rahoi, Kim Newkirk

A 14-year-old, neutered male, Bengal tiger presented to the University of Tennessee Veterinary Medical Center with a three-week history of weight loss. Euthanasia was elected. Necropsy identified a 1 cm circumferential narrowing within the jejunum, which resulted in dilation of the orad intestines. On the mucosal surface at the jejunal narrowing, there was a poorly demarcated firm grey multinodular mass that extended 1.5 cm orad and aboral to the stenotic area. The adjacent mesentery was thickened by coalescing firm, tan nodules. The associated mesenteric lymph node was enlarged, firm and tan. The liver had multifocal firm, tan nodules with umbilicated centers. A thrombus was present in the caudal vena cava. Histopathology revealed a poorly demarcated mass infiltrating and effacing the intestinal wall. The neoplasm was composed of tubules, lined with well-differentiated simple cuboidal to columnar epithelium, separated by collagenous stroma. The liver and mesenteric lymph node were effaced by the same neoplastic population. The neoplastic cells were also present in the caval thrombus. These findings support a diagnosis of metastatic intestinal adenocarcinoma. In addition, this tiger had a gastroesophageal intussusception; this is likely a sequela to the nearly complete obstruction caused by the intestinal adenocarcinoma. Intestinal adenocarcinomas in domestic cats commonly present as an annular stenotic mass, as was seen in this tiger. This is the first report of intestinal adenocarcinoma in a tiger.

SP-03: TERATOCARCINOMA IN AN EMU

Jessica Wild, Amie Perry, Jeann Leal de Araujo, Sharman Hoppes, Raquel Rech

Adenocarcinomas and leiomyomas are the most common neoplasms of the avian female reproductive system. Teratomas are uncommon; most arise from gonads and are composed of at least two of the three germ cell layers: ectoderm, mesoderm, and endoderm. A 17 year-old female emu presented for anorexia and depression and died shortly after arrival at the referral hospital. Gross findings included a large, multinodular mass composed of cystic and solid nodules on the ovary and oviduct with numerous smaller nodules adhered to the mesentery, intestinal serosa, and body wall. Approximately 6 L of fluid were in the coelomic cavity. On histologic evaluation, all three primordial germ cell lines were present. Endodermal elements included both ciliated respiratory and gastrointestinal epitheliums, with goblet cells, often lining a lumen. Mesodermal components included foci of osteoid, hyaline cartilage, connective tissue, lymphoid tissue, and bundles of skeletal muscle and smooth muscle. Ectodermal components included melanocytes and nervous tissue with neurons, glial cells embedded in neuropil or lined by ependymal cells. Based on gross and histologic findings, the diagnosis of teratocarcinoma was made. Immunohistochemistry for cytokeratin (CKA1/CKA3), desmin, and glial fibrillary acidic protein (GFAP) were performed to evaluate the cross-reactivity of antibodies in avian tissues. Cytokeratin was positive in the epithelial cells. Immunolabeling for desmin was observed in the muscle fibers. Glial cells and their processes were positive for GFAP within nervous tissue. The immunohistochemistry for the tested antibodies highlights their cross-reactivity in avian species.

SP-04: SILICATE PNEUMOCONIOSIS IN DOMESTIC AND WILD ANIMALS FROM THE CARIBBEAN ISLAND OF ST. KITTS

Randall Thomas Walker, Pompei F. Bolfa, Oscar Illanes

Silicate pneumoconiosis is a well-known environmentally acquired lung disease in humans characterized by silicate crystal depositions within the pulmonary interstitium. A high incidence of silicate pneumoconiosis has been observed in domestic and wild species on the Caribbean island of St. Kitts. To estimate the prevalence, 219 lung samples, from ten species (chickens, frigate bird, mongoose, cows, pigs, small ruminants, cats, dogs, equines, monkeys), were collected at the abattoir or from previous studies and examined by histopathology and scanning electron microscopy with energy dispersive x-ray analysis (SEM/EDXA). Passive eight hour air samples are being obtained from random geographical areas and animal collection sites throughout the island and analyzed for presence and density of silicates. Microscopic findings, seen in 145/219 (66%) samples, included perivascular and interstitial accumulations of heterogeneous crystalloid particulate material, either free or intracytoplasmic within macrophages. The material was birefringent, acid-fast positive, and largely composed of SiO_4^{4-} based on SEM/EDXA. These lesions were graded based on severity (normal, grade I, grade II, grade III). By assessing the prevalence of pneumoconiosis in the animal population and the composition of locally obtained air samples, we intend to elucidate the silicate source of origin and to correlate environmental exposure with the presence and severity of silicate deposition within the lungs. Results from this investigation my warrant future studies on the prevalence of environmentally acquired pulmonary disease within the human population of St. Kitts.

SP-05: IN VITRO VALIDATION OF PROTEIN GERANYLGERANYLATION AS A PHARMACOLOGICAL TARGET FOR THE TREATMENT OF CANINE MAMMARY GLAND TUMORS

Anne-Laurence Vigneau, Charlène Rico, Derek Boerboom, Marilène Paquet

Canine mammary tumours (CMTs) are the most common neoplasms in intact bitches, and no effective chemotherapeutic options are available for highly invasive and metastatic tumors. Recent studies have shown the potential involvement of dysregulated Hippo signaling in CMT development and progression. Protein geranylgeranylation (GGylation) is an important post-translational modification for many signaling molecules, and blockade of GGylation with statins has been shown to inhibit YAP/TAZ-mediated transcriptional activity via activation of the Hippo pathway. In this study, we sought to determine if protein GGylation represents a valid pharmacological target for the treatment of CMTs. Two CMT cell lines (CMT-9 and CMT-47) were evaluated for their sensitivity to two statins (Fluvastatin and Atorvastatin), the effect of both statins on YAP and TAZ expression levels, and on mRNA levels of YAP/TAZ transcriptional target genes. Results showed Fluvastatin and Atorvastatin to be cytotoxic to CMT-9 and CMT-47 cells, with ED50 values of 0.95 µM and 23.5 µM, respectively. In addition, Fluvastatin (10mM, 24 hours) lowered protein levels of YAP and TAZ in both cell lines. Finally, Atorvastatin and Fluvastatin reduced the mRNA levels of YAP/TAZ-TEAD target genes known to be involved in breast cancer progression and chemoresistance (*CYR61*, *CTGF* and *RHAMM*). Taken together, these findings suggest that the therapeutic targeting of protein GGylation with statins could be a novel approach for the treatment of inoperable canine mammary gland cancers.

SP-06: HYPEREOSINOPHILIC SYNDROME AS A CAUSE OF CHRONIC HEPATITIS/ CHOLANGIOHEPATITIS IN 6 CATS

Linnette Leticia Vasquez, Oscar Illanes, DVM, PhD, DACVP Illanes, Carmen Fuentealba, Taryn Donovan

A retrospective microscopic evaluation of liver biopsies from 24 cats diagnosed with chronic hepatitis at the New York Animal Medical Center over a 15 year span, was performed.Sixof these cats, ranging from 2 to 13 yearsof age, had moderate to severe chronic eosinophilic inflammatory cell infiltration primarily centered within portal areas. Inflammatory cell infiltration resulted in hepatocellular loss and variable degrees of fibrosis. Lesser number of lymphocytes, plasma cells and macrophages were alsopresent. Affected cats had no history ofendoparasitismor any other known cause of eosinophilia. The absence of a recognizable cause of eosinophilia is key in the diagnosis of felinehypereosinophilicsyndrome (FHES);an idiopathic condition characterized by prolonged peripheral eosinophilia and soft tissue infiltration by eosinophils. Tissue damage associated with FHES is thought to be the result of oxidative injury caused by the release of oxygen metabolites by the infiltrating eosinophils. Although liver involvement isconsidered to berelatively rare, findings in this study suggest that felinehypereosinophilicsyndrome should beincludedin the differential diagnosis of chronic hepatitis/cholangiohepatitisin cats.

SP-07: PHENOTYPIC CHARACTERIZATION OF SNTG1 DEFICIENT MICE
Katherine N. Turnbull, Audrey Cleuren, David Ginsburg, Emilee Kotnik

Background: Syntrophin gamma 1 (*Sntg1*) encodes a protein of the syntrophin family, which mediates dystrophin binding and plasma membrane gamma-enolase trafficking. Studies have shown it expressed in the brain, and mutations in this gene have been associated with idiopathic scoliosis.

Objective: Even though this gene is conserved across species, including mammals, amphibians, and fish, little is known about the function of *Sntg1* or the potential effects of *Sntg1* deficiency. Therefore, we generated a *Sntg1* null mouse and characterized the phenotype with a specific focus on histology and hemostatic parameters.

Methods: *Sntg1* knock-out alleles were generated using CRISPR/Cas9 technology in which exon 5 was targeted, resulting in an eleven base-pair deletion causing a frameshift. Mice heterozygous for this deletion were intercrossed to generate null mice and characterized by collecting blood for a complete blood count (CBCs), prothrombin time (PT), activated partial thromboplastin time (APTT), and blood smears. In addition, mice were weighed weekly, and histologic analysis on a subset of organs was performed.

Results: In heterozygous intercrosses, the expected ratio of wild type, null, and heterozygous mice were seen. Preliminary data showed no significant weight differences between null mice and control littermates. Additionally, blood smears, coagulation profiles and CBCs of *Sntg1* null mice were within normal limits and comparable to wild type controls.

Conclusion: Our preliminary data identified no major morphologic or hemostatic alterations associated with *Sntg1* deficiency indicating that *Sntg1* is not an essential gene. Its biologic function remains to be elucidated and will require more extensive phenotypic characterization.

SP-08: TARGETING MINIMAL RESIDUAL DISEASE IN METASTATIC BREAST CANCER

Rebecca K. Tierce, Laura L. Bronsart, Stavros Melemenidis, Marjan Rafat, Yasaman Ahrari, Edward Graves

Breast cancer affects one in eight women in the United States. Metastatic breast cancer comprises a small portion of these cases yet is associated with higher mortality rates. Despite surgical excision of primary tumors, high levels of minimal residual disease and disseminated metastases require prolonged chemotherapy, resulting in complications. Ideal therapies for such patients involve less severe systemic side-effects, while controlling tumor burden and extending life expectancy and quality. *In vitro*, flow cytometry and cytotoxicity assays demonstrated that the combination of local radiotherapy and immunotherapy significantly decreases tumor cell burden compared to either treatment alone. *In vivo*, unilateral irradiation of an orthoptopic tumor in a two-tumor mouse model detected upregulation of ligands associated with human natural killer NK92 cell-mediated apoptosis of MDA-MB-231 human breast cancer cells, quantified via flow cytometry. Unirradiated contralateral tumors also expressed increased levels of these ligands, suggesting combination regional radiotherapy and cellular immunotherapy targets unirradiated minimal residual disease sites. To explore this finding, we first confirmed upregulation of these ligands in the contralateral tumors was due to a physiological radiation response versus intercellular communication between cancer cells. A co-culture assay of irradiated and unirradiated MDA-MB-231 cells was carried out and analyzed via flow cytometry. Following this, abscopal tumor effects and subsequent tumor cell killing were visualized *in vivo* with bioluminescence imaging in a metastatic lung model. A decrease in mouse tumor cell burden following NK92 cellular immunotherapy would support the hypothesis that combination regional radiotherapy and cellular immunotherapy targets sites of unirradiated minimal residual disease.

SP-09: SEASONAL VARIATION OF GREY SQUIRREL (SCIURUS CAROLINENSIS) TESTES IS PRODUCED BY MECHANISMS DISTINCT FROM AGE RELATED TESTICULAR REGRESSION

Lucie Stratton, Abir Mukherjee

Seasonal breeding is seen in many animal species, and is often associated with a seasonal decrease in testicular size and function. Changes likely occur because sperm and testosterone production are energy-intensive processes which are only feasible when reproductive success is likely. In addition to seasonal changes, testicular regression is a common aging change in many species. It is not clear whether the underlying physiological mechanisms in aging are distinct from those in seasonal breeders. Studies reveal anti-aging properties of the gene SIRT1 in the testis. SIRT1 prolongs spermatogenesis and may play a role in regulating testicular size. SOX9, a gene essential for the maintenance of tubular architecture, preventing both feminization and apoptosis of sertoli cells, is reported to decline with aging. Here we investigate the role of SIRT1 and SOX9 in seasonal testicular regression and maintenance of testicular function, as well as changes in proteins associated with proliferation (PCNA) and apoptosis (Cleaved Caspase 3) in a seasonal breeding species. Immunohistochemistry, TUNEL staining and Western Blots were performed on samples of Grey Squirrel (*Sciurus Carolinensis*) testis obtained during different seasons. Winter samples exhibited significantly increased levels of TUNEL (P=0.02) and Cleaved Caspase-3 (P=0.04); these findings indicate higher levels of apoptosis in winter. Changes in PCNA were not significant (P=0.83). There was no significant seasonal difference in levels of SIRT1 (P=0.05), suggesting a distinct mechanism from aging related size regression. Interestingly there were significantly higher levels of SOX9 in winter testis (P=0.03), suggesting a role for this gene in seasonal testicular regression.

SP-10: CANINE CUTANEOUS LYMPHOMA WITH CONCURRENT METHICILLIN RESISTANT STAPHYLOCOCCUS AUREUS (MRSA) INFECTION

Ashley E. Saver, Cory R. Hanks, Marigold E. Ernst, Geoffrey K. Saunders

A nine-year-old, spayed female, English Bulldog presented to the Virginia-Maryland Veterinary Teaching Hospital Emergency service for respiratory distress. The patient had a history of multifocal, progressively worsening skin lesions originally observed six months prior. The patient was previously treated with incomplete courses of multiple antibiotics. Thoracic radiographs revealed a focal pulmonary nodule. A fine needle aspirate was performed on cutaneous nodules and the enlarged right and left popliteal lymph nodes. Lymph node cytology showed a mixed lymphoid population with moderate plasma cells, while skin cytology showed a majority of large lymphoblasts admixed with degenerate neutrophils and macrophages. Numerous doublets of 1 micrometer cocci were observed within neutrophils. Two differentials included cutaneous lymphoma and reactive lymphoid hyperplasia secondary to bacterial dermatitis. Subsequent histopathologic evaluation revealed an unencapsulated, densely cellular neoplasm within the dermis, infiltrating the adjacent subcutis and epidermis. The neoplasm was composed of sheets of large, round cells with round to ovoid nuclei and finely stippled chromatin. In addition to the neoplastic cells, numerous small, reactive lymphocytes, degenerate neutrophils, and histiocytes were present. The majority of the neoplastic round cells were immuno-positive for CD3, giving a definitive diagnosis of cutaneous T cell lymphoma. Bacterial culture and sensitivity yielded significant growth of Methicillin-resistant *Staphylococcus aureus*. Cutaneous lymphoma is uncommon in dogs. Cutaneous T cell lymphoma accounts for approximately 12% of all canine lymphomas and bulldogs are reportedly predisposed. Concurrent inflammation often complicates diagnosis of cutaneous lymphoma, necessitating the use of ancillary tests to provide definitive diagnosis.

SP-12: CLINICOPATHOLOGIC, CYTOLOGIC, AND HISTOLOGIC CHANGES IN AN ALPACA WITH HEPATIC COPPER ACCUMULATION

Shannon J. Reeves, Craig A. Thompson

A 6 year-old, intact female, Alpaca presented to the Purdue Veterinary Teaching Hospital for anemia and weight loss. Hematologic findings included a marked to severe normocytic, hypochromic anemia, marked hypoproteinemia, and mild, mature neutrophilia. Biochemical findings included mild decreases in BUN, Phosphorus,

Calcium, Magnesium, and moderate decreased in Potassium. Biochemistry also revealed a metabolic alkalosis with increased Bicarbonate and decreased Anion Gap. Toxicology revealed a serum Copper level within reference intervals. Approximately two weeks after presentation and several failed transfusions, a liver biopsy was submitted for histopathology with an impression smear submitted for cytology. Histopathology with Hematoxylin and Eosin stain revealed fine, light brown granules in the cytoplasm of hepatocytes, some with mild vacuolation, and aggregates of hematopoietic cells. Special stains were performed which revealed green granules consistent with Copper coinciding with the aforementioned light brown granules. Cytology revealed greater than half of hepatocytes to have dark green cytoplasmic pigment consistent with iron or bile. On cytology there were also noted erythroid precursors which were mostly mature, small and intermediate to large lymphocytes, nondegenerate neutrophils, and macrophages sometimes containing coarse green granules. Hepatopathy and subsequent death with post mortem discovery of changes consistent with Copper accumulation has been documented in camelids. This process is not necessarily accompanied by hemolytic crisis as is described in Copper toxicity in sheep. Disease has been documented in camelids with acceptable Copper levels when the Copper to Molybdenum ratio is not balanced.

SP-13: MUSCLE FIBRE SIZE IN THE DELTAE50-MD CANINE DUCHENNE MUSCULAR DYSTROPHY MODEL

Faye Rawson, John Hildyard, Fran Taylor-Brown, Rachel Harron, Dominic Wells, Richard J. Piercy

Duchenne muscular dystrophy (DMD) is an X-linked recessive disorder that affects 1 in 5000 newborn boys, and which has also been reported in many dog breeds. This progressive myopathy is caused by lack of dystrophin, a protein essential to muscle structure and function. Preclinical testing of potential therapies primarily use the mdx mouse however this model does not wholly replicate the human disease. The deltaE50-MD dog model in contrast shows pathological features more representative of DMD and unlike the other major Golden Retriever dog model (GRMD), the phenotype stems from a mutation in the human dystrophin gene 'hotspot' – a mutation readily accessible to current therapeutic approaches. Comprehensive studies of pathological progression in this model are lacking.

Muscle fibres of dystrophic patients show greater variability in cross sectional area (CSA) and diameter: a hallmark of fibre regeneration and compensatory hypertrophy. Vastus lateralis muscle biopsy samples from a deltaE50-MD dog (DD) and a healthy (WT) littermate were collected at 3-monthly intervals over an 18-month period. Following perlecan immunohistochemistry, analysis of CSA and minimum Feret's diameter (MFD) permitted assessment of age- and disease-related changes and comparisons with human pathology. Coefficient of variation (CV) was used as a measure of relative variability. CV of CSA and MFD were greater at every timepoint in the DD, particularly at 3 months with 2.3- and 2.6-fold increases respectively.

Dystrophin immunohistochemistry revealed isolated revertant fibres in the DD at different ages.

Ongoing analysis of further deltaE50-MD dogs will enable statistical analyses on overall muscle phenotype and progression.

SP-14: ENVIRONMENTAL AWARENESS: RECREATING THE NATURAL TISSUE NICHE IN VITRO FOR CANCER RESEARCH

Brittany Lynn Rasche, Shirisha Chittiboyina, Farzaneh Atrian, Sophie A. Lelièvre

Despite many advances in cancer research, *in vitro* models that accurately represent the environment in which cancer develops have not been established. One example of the environment's role is the link between increased stromal density and increased risk for onset and progression of aggressive forms of breast cancer. Our hypothesis is that stromal stiffness and cells affect the phenotypes of non-neoplastic and preinvasive mammary epithelial cells by influencing the morphology of the cell nucleus. In a first approach, we are building a model that includes cancer cells, non-neoplastic epithelial cells, basement membrane component (laminin), and stromal components (collagen I and fibroblasts). First, non-neoplastic and preinvasive cells were cultured separately in the presence of collagen I matrix of adjustable stiffness. The nuclear morphology was assessed based on DAPI staining (for nuclear areas and circularity) and SC35 immunostaining (for normal differentiation) using *ImageJ*. The non-neoplastic cells showed a significant difference in nuclear area and circularity as well as differentiation, while the preinvasive cells displayed a significant difference in nuclear circularity depending on stiffness. Next, the non-neoplastic and preinvasive cells were cocultured on top of the collagen I matrix. A significant alteration in nuclear area for preinvasive cells was measured as stiffness changed. In future experiments, fibroblasts will be embedded in the collagen matrix underneath the coculture of epithelial cells. The data collected from these experiments will be compared with data from real breast tissue to determine which local environment allows cells to establish a phenotype that most closely resembles the real tissue.

SP-15: DEVELOPING A NOVEL PORCINE MODEL OF LARYNGOPHARYNGEAL REFLUX DISEASE

Betsy A. Pray, Abigail Durkes

Laryngopharyngeal reflux (LPR) disease is believed to involve chronic backflow of gastric refluxate, resulting in damage to laryngopharyngeal epithelium. Gastric refluxate has been shown to contribute to many laryngological conditions, including laryngitis, sore throat, ulcers, and globus pharyngis. Approximately 10% of laryngologic patients and over 50% of patients with voice disorders have symptoms of LPR. Despite the prevalence of LPR, there is no appropriate animal model. The pig provides a unique opportunity to study laryngeal disease as porcine and human vocal folds are similar in terms of architectural, biochemical, neuromuscular, and cellular properties. In this pilot study, a novel in vivo pig model was developed to closely simulate the clinical condition of human LPR by challenging healthy, uninjured laryngeal epithelia with acidified pepsin. An indwelling esophagostomy catheter was placed surgically and positioned near the aryepiglottic folds. The pigs were randomly assigned to a reflux group (n=3) and a sham (n=1). The reflux group a received continuous rate infusion of acidified pepsin solution, and the sham received saline. Necropsies were completed on days 8 and 12. Laryngeal tissues were examined grossly and histologically for differences between treatment groups. Immunohistochemistry for CD3+ T-cells was quantified via digital pathology. Grossly, cloudy fluid was observed in the larynges of the reflux group. Differences between treatment groups in histopathology and CD3+ T-cell quantification and localization were not significant. This pilot study contributes fundamental groundwork necessary for future studies to improve our understanding of the pathophysiology of this complex disease and to eventually improve treatment outcome.

SP-16: EVALUATING THE EFFECT OF WHOLE BODY HZE ION IRRADIATION ON METASTASIS IN MMTV-PYMT-INDUCED MAMMARY TUMORS

Sarah Jane Powers, Elijah Edmondson

High atomic number and energy (HZE) nuclei are components of galactic cosmic rays that pose significant and uncertain health risks for manned missions to interplanetary space. Preliminary carcinogenesis studies in C3H mice have indicated that whole-body exposure to HZE ions increases the risk of hepatocellular carcinoma metastasis. To further investigate this observation, a genetically engineered mouse model (GEMM) of metastasis is utilized. FVB/N-Tg (MMTV-PyMT) mice carry a transgene consisting of a mouse mammary tumor virus promoter controlling the expression of the mouse polyomavirus middle-T antigen, leading to early mammary tumor development and metastasis. Male MMTV-PyMT mice are crossed with female DBA/2J and FVB/N and individuals from each background are exposed to 0.2 Gray of 300 MeV/n Si28 ions (HZE ions), or sham irradiated. Metastases are quantified using area-based measurements to determine metastatic density using digital analysis of whole slide images. Using a GEMM of mammary tumor metastasis, whole body irradiation with HZE ions does not appear to significantly alter metastatic density or incidence and effects of background genetics (FVB vs. DBA) does not appear to modulate susceptibility to metastasis following irradiation.

SP-17: INVESTIGATING AN ETIOLOGIC ROLE FOR FELIS CATUS GAMMAHERPESVIRUS 1 IN HIGH-GRADE, LARGE-CELL LYMPHOMA

Tamsen Polley, Julia Beatty, Amy Durham, Alicia McLuckie, Patricia Pesavento

The first gammaherpesvirus of domestic cats, FcaGHV1, was discovered in 2014. Human gammaherpesviruses are known oncogenic viruses (e.g. Epstein-Barr virus) as in immunosuppressed transplant patients. FcaGHV1 is an endemic lymphotropic virus detected in the blood of 10-19% of cats worldwide. Elucidating the potential epidemiology and role of FcaGHV1 in lymphomagenesis is critical for supporting the global health of companion felines. Whether FcaGHV1 plays a causal role in feline large-cell lymphoma subtypes is the subject of this investigation.

Ninety-one domestic cats from American veterinary hospitals with different subtypes of large-cell lymphomas were assessed for the presence of FcaGHV1 DNA in formalin-fixed, paraffin-embedded tissues by a combination of conventional polymerase chain reaction (cPCR) and in-situ hybridization (ISH). Lymphoma groups included post-transplant associated diffuse B-cell (PT-DLBCL, n=7), gastrointestinal (GI, n=20), T-cell rich B-cell/Hodgkin-like (n=35), nasal (n=20), and DLBCLs from miscellaneous sites (n=9). Three samples were positive for FcaGHV1 DNA by cPCR: PT-DLBCL, GI, and nasal lymphoma. A negative cPCR result is correlated with viral DNA being below the assay detection threshold. Two of the 3 positive cases were analyzed by ISH: PT-DLBCL and GI T-cell lymphoma; virus nucleic acid was detected within the GI T-cell lymphoma. The low cPCR sensitivity may explain the low FcaGHV1 positive tissues, but does not rule out FcaGHV1 involvement.

Our findings demonstrate, for the first time, that FcaGHV1 may be localized to large T-cell neoplastic lymphocytes in the gastrointestinal tract of domestic cats, supporting a causal link between certain subtypes of lymphoma and FcaGHV1 infection in domestic cats.

SP-18: A PILOT RETROSPECTIVE STUDY OF HYPERFERREMIA AND ELEVATED TRANSFERRIN SATURATION IN DOGS: 48 CASES (2017)

Matthew Pate, Eric Fish, Elizabeth Spangler

Background: Reported causes of hyperferremia include increased iron turnover, corticosteroid exposure, and hepatocellular necrosis. The prevalence of hyperferremia and association with other diseases, drugs, and lab parameters are currently unknown.

Objective: Classify cases with hyperferremia and elevated transferrin saturation (TSAT%) to determine associated diseases and if there are any statistically significant differences based on diagnosis, steroids, hemolysis, or evidence of liver damage.

Methods: Laboratory and patient data were retrospectively reviewed from canine patients with a CBC and chemistry performed May 2017 to July 2017. Cases with serum iron >228ug/dL and TSAT% >60% were included. Descriptive and univariate statistics were performed with commercially available software.

Results: Forty-seven of 845 (5.6%) dogs were included. Diagnoses were varied and included neurologic disease (n=11), gastrointestinal disease (n=7), neoplasia (n=6), hemolytic anemia (n=4), hepatitis (n=3), miscellaneous (n=5), multiple diagnoses (n=8), open or no diagnosis listed (n=2), and 1 healthy dog. 20/48 cases (41.7%) received either exogenous corticosteroids (19/20) or had hyperadrenocorticism (1/20). There was no statistically significant difference between steroid exposure groups for serum iron (p=0.102), total iron-binding capacity (TIBC)(p=0.975), or TSAT% (p=0.250). There was a weak, significant correlation between ALT and TSAT% (r=0.394; p=0.005), but not ALT and serum iron (r=0.257; p=0.075) or TIBC (r=0.003; p=0.985).

Conclusions: Hyperferremia was relatively common and associated with more disease processes than previously reported. The weak but significant correlation between ALT and TSAT% suggests hepatocellular injury may contribute to hyperferremia in many cases. There was no statistically significant effect of corticosteroids in this population.

SP-19: AGNOR COUNTS OF CANINE PERIPHERAL BLOOD LYMPHOCYTES: A TOOL FOR PROGNOSTICATION

John Ogunsola, Richard Antia

Objective: A modified agyrophil technique was applied to peripheral blood smears to determine the mean AgNOR counts (MAC) of lymphocytes and ultimately assess the state of the lymphoid system in various clinical conditions of dogs.

Materials and Methods: Fifty dogs, from clinically normal to pets with leukemia, presented to the Veterinary Teaching Hospital, were employed. Blood smears from each dog were stained with routine Romanowsky and modified agyrophil stains. Signalement, clinical diagnoses and hematologic parameters of the dogs were related to the MAC. An AgNOR proliferative index (AgPI) - percentage of lymphocytes with 3 or more AgNORs, was determined and correlated with MAC. Statistical significance was determined at *p less than 0.05.*

Results: MAC ranged from 1.17 in clinically normal patients to 6.00 in leukaemic patients. The MAC was 2.00 in patients (n=26) with lymphocyte counts within reference intervals (900-2400/microlitre); 2.23 in patients (n=4) with lymphopenia; 2.18 in patients with lymphocytosis (n=18) and 4.73 in patients (n=4) with lymphocytic leukemia. Also, the MAC was 2.00 in non-anemic dogs while it was 2.47, 2.49 and 3.06 in patients with mild, moderate and severe anemia respectively. The MAC correlated strongly with AgPI (r=0.91).

Conclusion: The ancillary AgNOR technique provides a cheap, rapid and sensitive tool than routine lymphocyte counts in prognostication and assessing the state of lymphoid activation in a variety of conditions in the dog.

SP-20: ERYTHROCYTE CYTOMORPHOLOGIC CHANGES IN DOGS RECEIVING THE CHOP CHEMOTHERAPY PROTOCOL FOR TREATMENT OF MULTICENTRIC LYMPHOMA

Ariel S. Nenninger, Jeremy Servantes, Shannon Hostetter, Chad Johannes, Austin Viall

Lymphoma is the most common hematopoietic neoplasm of dogs; multicentric canine lymphoma is frequently treated with the modified Madison-Wisconsin CHOP chemotherapy regimen. This protocol consists of an initial treatment of prednisone and four cyclic administrations of vincristine, cyclophosphamide, and doxorubicin over a 26 week period. Serial complete blood cell counts (CBCs) are performed to monitor patients for adverse chemotherapy induced hematologic side-effects. Chemotherapeutics are known to induce morphologic changes in erythrocytes. However, to our knowledge, the erythrocyte morphologic changes experienced by dogs receiving CHOP have not been reported. The purposes of our retrospective study were to: 1) assess differences in erythrocyte morphology between healthy dogs and lymphoma dogs at diagnosis and 2) evaluate changes in erythrocyte cytomorphology as dogs progress through CHOP. Serials CBCs from 28 lymphoma dogs that completed CHOP and CBCs from 26 healthy age/breed matched controls were examined for the presence and severity of erythrocyte morphologic abnormalities. We found no differences in the proportions or severity morphologic abnormalities present in healthy dogs and lymphoma dogs at initial diagnosis. Anisocytosis, polychromasia, macrocytosis, codocytosis, and Howell-Jolly bodies occurred more frequently as dogs advanced through CHOP. However, the severity of these abnormalities did not appear to change as treatment progressed. These findings suggest that dogs receiving CHOP for treatment of multicentric lymphoma experience regenerative erythropoiesis throughout the chemotherapy regimen. Further evaluation is warranted to determine if these changes are due to the chemotherapeutic agents or lymphoma progression, and if there is any prognostic significance to erythrocyte cytomorphology.

SP-21: UREMIC ENCEPHALOPATHY IN A RHESUS MACAQUE (MACACA MULATTA).

Allison Mustonen, Olga Gonzalez, Shyamesh Kumar, Elda Mendoza, Edward J. Dick Jr.

Background: A 14-year-old Specific Pathogen Free female rhesus macaque was admitted for acute onset of obtundation, diarrhea, hypothermia, and hypokalemia. Blood chemistry indicated mild polycythemia; stress leukogram characterized by leukocytosis, mature neutrophilia, lymphopenia, eosinopenia, and monocytosis; severe azotemia; hypercholesterolemia; hyponatremia; hypokalemia; hypochloremia; and a titrational metabolic acidosis with an increased anion gap. Due to deteriorating clinical condition and poor prognosis, the macaque was humanely euthanized and necropsied 30 minutes later.

Methods: A complete necropsy with histology was performed.

Results: Grossly, there were patchy areas of integumentary ecchymosis along the ventral abdomen, medial thigh, and axilla. The gingival and lingual mucosa were multifocally eroded and ulcerated. The kidneys were bilaterally pale and tan. The hippocampus contained bilaterally symmetrical, focally extensive, brown to tan areas. Histologically, both kidneys had tubular epithelial necrosis and multifocal dilated tubules lined by flattened cells. The tubular lumens contained occasional neutrophils, cellular debris, protein, and mineralized concretions. The hippocampus was diffusely necrotic with loss of pyramidal neurons and glial cells, edema, and multifocal hemorrhages. Ulcerative glossitis and stomatitis, gastritis, and multifocal fibrinoid vascular necrosis of the integument, stomach, and brain were identified.

Conclusion: Based upon the clinical history and gross and histologic findings, the lesions in this case are consistent with acute renal failure induced uremic coagulopathy, vasculopathy, and encephalopathy. To the best of our knowledge, this is the first reported case of uremic encephalopathy in a nonhuman primate.

SP-22: IMMUNE-MEDIATED HEMOLYTIC ANEMIA IN A DOG

Stefanie Muller, Lori Rios, Katherine Boes, Sarah Barrett

Canine immune-mediated hemolytic anemia (IMHA) is a severe disease that can be quickly fatal if left untreated. Autoimmune hemolytic anemias occur when self-antibodies attach to red blood cells (RBCs) or erythroid precursors, priming them for destruction through the complement system or through destruction by the mononuclear phagocyte system. This process can be secondary, triggered by dysregulation caused by neoplasia, infection, or drug administration. In contrast, primary canine IMHA is an idiopathic disease that commonly affects middle-aged dogs. We present herein a case of IMHA in an eight-year-old, spayed female, Shih Tzu dog. The dog was presentedfor lethargy of one week duration. Two days after the initial presentation, the dog was started on prednisone following a presumptive diagnosis of IMHA. Blood film examination revealed an inadequately regenerative anemia (PCV = 21%), frequent ghost cells and occasional imperfect spherocytes, with an inflammatory leukocytosis. Abdominal ultrasound revealed mild hepatomegaly and splenomegaly, due to marked extramedullary hematopoiesis observed on cytologic evaluations of fine-needle aspirate preparations. Bone marrow fine-needle aspiration and core biopsy revealed erythroid, myeloid, and megakaryocytic hyperplasia, with frequent erythrophagia. Based on these findings, and the lack of an inciting cause for the hemolysis, a diagnosis of primary canine IMHA was made.

SP-23: RACCOON POLYOMAVIRUS PERSISTENCE AND NEUROGLIAL TUMORS

Alexandra Moskaluk, Kevin Woolard, Patricia Pesavento

Polyomaviruses are widespread in nature, infecting most vertebrate species and capable of long term persistence, without clinical effect, over the lifetime of the host. Rarely, but significantly, these viruses can cause cancer. The trigger for switching from innocuous, low level infection to viral induced oncogenesis, is unknown. In large part, this is because the mechanism for viral persistence is unknown. Some human polyomaviruses have been documented to be shed in urine while others are found more frequently in respiratory secretions, but we do not know specific cell targets of infection in “normal” (non-clinical) naturally infected animals. Moreover, there is a lack of knowledge about when a persistent infection is established, in particular whether or not vertical transmission occurs. This information is vital when considering potential preventative therapeutics such as vaccination. A recently discovered polyomavirus in free-ranging raccoons (RacPyV) in the Western United States can cause neuroglial tumors, and provides an opportunity to study, in a naturally occurring infection, the distribution of virus in animals with or without tumors. We have performed both PCR and in situ hybridization on multiple organ samples of raccoons to interrogate the distribution of the virus and the cell target(s) of infection. This combinational strategy demonstrates that the virus is present at the maternal fetal interface, and in the kidneys of infected raccoons.

SP-24: NEOPLASTIC DISEASES IN CAPTIVE PSITTACINE BIRDS SUBMITTED TO THE ONTARIO VETERINARY COLLEGE

Kai S. Moore, Nicole Nemeth, Thisuri Eagalle, Hugues Beaufrere, Leonardo Susta, Csaba Varga

Background: Psittacines are increasingly popular in households, aviaries and zoological collections, with a growing reliance on pathologists to formulate accurate diagnoses upon which recommendations to improve husbandry and future case outcomes may be based. Comprehensive information about neoplasia in birds, including pathological presentation and grading algorithms, is often limited.

Objective: Our aim was to determine the prevalence of neoplasia, describe pathological presentation, surmise possible risk factors (e.g. signalment, history or age range) and use a ranking scheme to determine the primary and contributing causes of death.

Methods: We conducted a retrospective analysis of diagnostic data from neoplasia cases in psittacine birds submitted to the Ontario Veterinary College (OVC) and Animal Health Laboratory (AHL) at the University of Guelph from 1997 to present.

Results: Neoplasm cases (n=161) were categorized into the following categories: epithelial (e.g. carcinoma), mesenchymal (e.g. sarcoma), round cell (e.g. lymphoma), lipoma/xanthoma, papilloma, cysts, neuroectodermal, benign and uncharacterized tumors. The most affected genus was *Ara* (26; 16.1%), followed by *Nymphicus* and *Melopsittacus* (19; 11.8%, each). The majority of birds affected with neoplasia, regardless of ranking, were mature (75; 46.6%) and geriatric (35; 21.7%) with the remainder immature (7; 4.3%). Carcinoma (62; 38.5%) was the most common neoplasia followed by sarcoma (23; 14.3%). Regardless of tumor type, the most common systems affected were alimentary (42; 26.1%), integument (39; 24.2%) and multi-systemic (27; 16.8%).

Conclusions: A better understanding of these neoplasms will improve diagnostic algorithms and treatment, ultimately improving the health of captive psittacines in Ontario.

SP-25: STING AGONISTS: A POTENTIAL NOVEL IMMUNOTHERAPY FOR CANINE OSTEOSARCOMA

Zachary Millman, Maggie Phillips, Brian Ladle, Dara Kraitchman

Osteosarcoma (OSA) occurs in both humans and dogs with an annual incidence of 450 and 25,000 cases, respectively, making dogs a useful comparative oncology model for the development of immunotherapies for people. The current standard of care treatment for OSA is amputation, or amputation followed by chemotherapy. However, the vast majority of dogs will succumb to metastatic disease. Cancer immunotherapy holds the promise of more effective treatment of the primary OSA tumor as well as distant micrometastases. One such novel immunotherapy is STING (stimulator of interferon genes) agonists, which upregulate inflammatory cytokines such as IFN-β, inducing T cell priming to tumor antigens. STING agonists have been studied extensively in mice and are now in early human clinical trials, but have never been tested in dogs. Unlike humans, where multiple STING alleles influence STING agonist responses, we have found evidence for only a single canine allele. To determine if STING agonists trigger the same inflammatory pathway in dogs, peripheral blood was acquired from 15 canine patients. Peripheral blood mononuclear cells (PBMCs) were isolated and incubated with the STING agonist. Induction of IFN-β (as measured using quantitative reverse transcription PCR) indicated STING agonist activity. When compared with untreated PBMCs, 15/15 dogs responded with an average of 138.3-fold induction of IFN-β expression (range of 10-fold to 395-fold). These data show robust *in vitro* activity of STING agonists in canine samples establishing their potential use as immunotherapies in canine patients with OSA.

SP-26: SPONTANEOUS CERVICAL VERTEBRAL EPIDURAL HEMATOMA IN A QUARTERHORSE WITH NEUROLOGIC SIGNS

Kathryn McCullough, Laura Chen, Nimet Browne, Jennifer Luff

A 1-year-old Quarterhorse gelding presented for acute-onset of neurological signs. Physical examination revealed grade 3 ataxia in the forelimbs and grade 4 ataxia in the hindlimbs. No deficits in mentation or cranial nerve abnormalities were detected. Based upon neurologic examination, the lesion was localized to the C1 - C5 region. The patient's complete blood count and chemistry panel were unremarkable, and serum and CSF titers for *Sarcocystis neurona* were negative. Cerebrospinal fluid evaluation yielded a mild hemodilution suspected to be contamination. Myelogram of the cervical region revealed compression of the cervical vertebrae C6 to C7 and suspected dynamic compression at vertebrae C3 to C4. Due to a poor prognosis, the patient was euthanized. On gross examination, an epidural, circumferential, adherent hematoma was identified filling the spinal canal between vertebrae C6 and C7. Histology confirmed a chronic hematoma with remodeling, and the associated spinal cord segment exhibited mild axonal Wallerian degeneration. Spontaneous cervical vertebral epidural hematoma has been previously but rarely described in the literature, occurring in the absence of reported trauma, coagulopathy, anticoagulant therapy, vascular malformation, neoplasia, or systemic disease. The pathogenesis of this lesion remains unclear. In human medicine, both medical management and surgical spinal cord decompression of cervical vertebral epidural hematomas have been attempted to treat these patients. However, these have had limited success, and permanent neurological damage due to chronic spinal cord compression is likely. A similar clinical course is anticipated for horses, and the few reports of this disease in horses have had a poor prognosis.

SP-27: IMMUNOPHENOTYPING OF LYMPHOCYTES IN A MINIPIG MODEL OF VASCULARIZED COMPOSITE ALLOTRANSPLANTATION

Caitlin E. Mason, Georg Furtmuller, Gerald Brandacher, Sarah E. Beck

Vascularized composite allotransplants (VCAs) are compound transplants that include multiple tissue types, including skin, muscle, and other associated tissues. With rigorous immunosuppressive therapy, successful transplantation can be maintained inanimal modelsand human patients with VCAs with low rates of acute rejection. Clinically, the main target of rejection appears to be the skin, and although the exact immunologic mechanism remains unclear, clinicopathologic monitoring of the skin provides a useful metric for the rejection status of the graft as a whole. A stifle VCA rejection model is achieved with MHC class 1 and 2-mismatched Massachusetts General Hospital (MGH) minipigs, a model with similar dermohistomorphology to humans. However, more detailed studies of the pathology of porcine dermal tissues in acute VCA rejection are necessary. The aims of this project include developing a scoring system based on the human system of the dermatohistpathology of acute rejection in the minipig model and characterizing the lymphocytic infiltrate of the skin in VCA rejection controls using immunohistochemical (IHC) staining. The overarching goals of IHC analysis include optimization of CD3+, CD20+, and FoxP3+ primary antibodies in control lymphoid tissues, and quantification of the amount of CD3+, CD20+, or FoxP3+ inflammation in acute graft rejection. Based on observations in human VCA rejection, we hypothesize that the inflammation will be predominantly CD3+ T cells with significantly fewer CD20+ B cells and FoxP3+ regulatory T cells. These studies will set the groundwork for future immunologic characterization of VCA rejection using the MGH minipig model under a variety of immunomodulatory therapies.

SP-28: ERB-B2 RECEPTOR TYROSINE KINASE 2 (EERB2/HER2) ANTIBODY SPECIFICITY IN CANINE MAMMARY TUMOURS AND OSTEOSARCOMAS

Latasha Ludwig, Emily Brouwer, Courtney Schott, Alicia Viloria-Petit, Geoffrey Wood

There is a general lack of canine-specific antibodies for immunohistochemistry (IHC), so antibodies raised against human proteins are often used. ERRB2 (HER2) is an oncogene amplified in many human cancers, and expression of HER2 is a key prognostic indicator and therapeutic target in breast cancer.The FDA-approved HER2 antibody used for human breast cancerdiagnostics was recently reported to not bind canine HER2. Our lab investigated an alternative anti-human HER2 antibody to determine its specificity in both canine mammary tumours (CMT) andosteosarcomas (OSA).IHC and Western blots (WB) were performed on cell lines and tumoursamples. Human breast cancercell lines (SKBR3, T47D, and MCF-7), a CMT cell line (CF41.Mg), and five canine OSA cell lines were used, along with ten CMT patient tumour samples.IHC was performed for HER2 on cell linesand tumour samples, while WB were performed on protein extracts from the same cell linesand CMT tumour samples. As expected, both IHC and WB were strongly positive for SKBR3 and mildly positive for T47D cell lines. The CMT and OSA cell lines and 8/10 CMT tumour samples were IHC negative, while two CMT tumour samples were positive.However, none of the canine protein extracts had a band at the expected molecular weight of HER2 on WB using this alternative HER2 antibody.These results add to the suspicion that commercial antibodies against human HER2 may not bind canine HER2, and encourage more rigorous testing of antibodies raised against human proteins before use in other species.

SP-29: HUMAN MACROPHAGES SURVIVE AND ADOPT ACTIVATED GENOTYPES IN LIVING ZEBRAFISH

Chaunte Lewis, Colin Paul, Alexus Devine, Kandice Tanner

The inflammatory response, modulated both by tissue resident macrophages and recruited monocytes from peripheral blood, plays a critical role in human diseases such as cancer and neurodegenerative disorders. Here we sought a model to interrogate human immune behavior *in vivo*. We determined that primary human monocytes and macrophages survive in zebrafish for up to two weeks. Flow cytometry revealed that human monocytes cultured at the physiological temperature of the zebrafish survive and differentiate, comparable to cohorts cultured at human physiological temperature. Human cells migrated within multiple tissues at speeds comparable to zebrafish macrophages. Gene expression profiling of *in vivo* educated human macrophages confirmed expression of activated macrophage phenotypes. Here, human cells adopted phenotypes relevant to cancer progression, suggesting that we can define the real time immune modulation of human tumor cells during the establishment of a metastatic lesion in zebrafish.

SP-30: WIDE-SPREAD POST-TREATMENT PULMONARY THROMBOSIS AND RIGHT-SIDED CONGESTIVE HEART FAILURE IN A SHAR PEI DOG WITH HEART WORM DISEASE AND MALIGNANT ROUND CELL TUMORS

Kristal Marie Lebrón, James Fairs, Pompei Bolfa, Oscar Illanes

A 6-year-old male Shar Pei dog, presented with exercise intolerance and early signs of right-sided CHF was diagnosed with heart worm disease and treated with melarsomine. A month later the dog was re-evaluated and tested negative for microfilaria in the modified knott's test. The dog exhibited weight loss and ascites. Abdominocentesis revealed the presence of a modified transudate. An enlarged liver was detected by ultrasound and radiography. Blood biochemistry revealed raised levels of ALT and ALP. Because of the possibility of breed-associated amyloidosis, liver biopsy was declined by the owner. Due to the poor prognosis the dog was euthanized. At autopsy widespread thrombosis and multifocal areas of pleural fibrosis and granulomatous interstitial pneumonia were scattered throughout the lungs. Numerous dead Dirofilaria worms were detected within thrombi obliterating mid-size branches of the pulmonary arteries. In addition, two poorly-demarcated round cell tumors were present within the small intestine and had metastasized to regional lymph nodes, the spleen and liver. Immuno-histochemistry results suggested the possibility of histiocytic cell origin. No amyloid deposition was detected in the liver or kidneys but abundant homogeneous congophilic birefringent extracellular material was found within lymph nodes containing tumor metastases. The findings in this dog highlight the risks associated with heart worm therapy. The severity of the pulmonary lesions was unusual; it is possible that the underlying malignancy and breed predisposition to develop Shar Pei fever, evident by the presence of amyloidosis within lymph nodes, may have had exacerbated thromboembolic disease and the adverse response to melarsormine treatment.

SP-31: THORACIC CORD COMPRESSION DUE TO EXTRAMEDULLARY HEMATOPOIESIS IN A CAT WITH CHRONIC NONREGENERATIVE ANEMIA

Chris R. Larson, Andrea Peda, Mary M. Christopher, Maryanna Thrall, Pompei Bolfa

Extramedullary hematopoiesis (EMH) is thought to be of little clinical significance in animals. Contrary to this belief and similar to what is seen in more than half the cases of EMH in people, we report a case of an FIV-positive cat with chronic nonregenerative anemia that developed hind limb paresis due to an EMH mass lesion that impinged on the caudal thoracic spinal cord.

A 6-year-old, neutered male, DSH cat presented with nonregenerative macrocytic anemia (HCT 16%, MCV 74 fl) of 2 years duration and minimally ambulatory paraparesis. Neurological examination suggested an upper motor neuron lesion within the thoracolumbar spine (T3-L3). The cat had neutropenia and a polyclonal gammopathy and died of a hemolytic crisis; blood film examination was suspect for *Mycoplasma haemofelis*, but PCR was not performed. At gross necropsy, multifocal bilateral dark-red masses were observed on the cranial aspect of the costochondral junctions and in an extradural paraspinal location on the lateral and ventral subpleural surfaces of T4-11. Histologic examination of the masses revealed EMH tissue composed primarily of erythroid precursors and megakaryocytes, with occasional myeloid precursors and blood-filled sinuses. Bone marrow findings supported myelodysplasia as the underlying cause of the hematologic abnormalities.

To our knowledge, this is the first report of an EMH mass lesion resulting in spinal cord compression and neurological signs in a cat.

SP-32: EFFECTS OF DEVELOPMENTAL EXPOSURE TO A MIXTURE OF HYDRAULIC FRACTURING CHEMICALS ON SPERMATOGENESIS IN MICE

Eileen Larsen, Tim J. Evans

Hydraulic fracturing, a practice done by some oil and natural gas operations, can result in the contamination of ground and surface water with endocrine-disrupting chemicals. Previous studies found that male mice exposed *in utero* to a mixture of these chemicals experienced decreased sperm counts and pituitary hormone concentrations, and increased testis weights and serum testosterone concentrations. In this study, sixty-nine 120-day-old male mice were assigned to five treatment groups based on the dosage of an equimass mixture of 23 chemicals used in hydraulic fracturingwith which their dams were orally exposed *ad libitum* in their drinking water for 25 days prior to matingand from gestational day 1 to postnatal day 21. An additional treatment group consisted of the male offspring of dams orally exposed *ad libitum* in their drinking water to the nonsteroidal antiandrogen flutamide. The stage of the cycle of seminiferous epithelium (I-XII) was identified and recorded for 200-300 seminiferous tubules from PAS-stained sections of the testes from each male mouse, and the treatment group means for the relative frequencies of each stage were compared. The frequency of stages I-VII was significantly higher in the 1500 μg/kg/day treatment group and the flutamide group compared to the control group, and the frequency of stages IX-XII was significantly lower in the 1500 μg/kg/day treatment group and the flutamide group compared to the control group. These results suggest that developmental exposure to a mixture of chemicals used in hydraulic fracturing can delay round spermatid development, thereby disrupting murine spermatogenesis.

SP-33: HISTOLOGIC CHARACTERIZATION OF TESTICULAR DEVELOPMENT OF APIS MELLIFERA DRONES DURING SEXUAL MATURATION

Colby Klein, Sophie Derveau, Ivanna Kozii, Sarah Wood, Roman Koziy, Ihor Dyvyluk, Elemir Simko

Background: Current risk assessment procedures for the exposure of pollinators to pesticides are not adequately predicting the detrimental interactions reported in scientific literature.

The "gold standard" for compound risk-assessment relies heavily on histopathology in mammals, but it is not employed in honey bees.

Objective: To characterize the normal morphological changes of testes of honey bee drones during the sexual maturation period in order to establish a normal benchmark for future gonadotoxicity studies.

Methods: At emergence, drones were marked and subsequently sampled each day during the period of sexual maturation (2 weeks). The reproductive tract of each drone was dissected, photographed and analyzed macroscopically and microscopically.

Results: Drones undergo sexual maturation for 2 weeks following emergence. The mature sperm moves from the testes to the seminal vesicles via the vas deferens, while the testes undergo progressive involution and atrophy. The testicular changes during sexual maturation are characterized by several distinct processes: 1) Spermatogenesis continues after emergence for 3 days; 2) By day 8, the entire sperm content is moved from testes to vas deference and seminal vesicles; 3) Testicular parenchyma undergoes degeneration and apoptosis followed by complete parenchymal loss by day 11 – 15 (depending on environmental temperature); 4) The end-stage testes consist of collapsed supporting stroma containing brown pigment and tracheal network. The testicular involution was accelerated by approximately 4 days in drones examined during warmer months.

Our future studies will evaluate effects of insecticides on gonadotoxicity in drones exposed to neonicotinoids during the sexual maturation period.

SP-34: EFFECTS OF INNATE IMMUNE STIMULATION ON EXPERIMENTAL MANNHEIMIA HAEMOLYTICA INFECTION IN CALVES

Emily Isabel Kaufman, Laura Lynn Bassel, Sarah-Nicole Alycia Alsop, Kevin Stinson, Ksenia Vulikh, Laura Rosemary Siracusa, Mary Ellen Clark, Jeff L. Caswell

Bovine respiratory disease is a major cause of death and economic loss in the North American beef industry and a major reason for antibiotic use in beef production. Stress and viral infections are thought to suppress innate immune responses in the lung and thereby contribute to development of bacterial pneumonia. We hypothesized that stimulation of innate immune responses would protect calves against *Mannheimiahaemolytica* pneumonia, and developed a novel model to test this hypothesis. Four pairs of 4-6 week old colostrum-deprived Holstein calves (8 calves total) were administered either aerosolized bacterial lysate orsaline, then 24 hours later all calves were challenged by aerosol with *M.haemolytica*, and euthanized at 3 days post infection. Calves developed depression, inappetance, lethargy, fever, ultrasonographic lung consolidation, and increased blood neutrophils, fibrinogen and haptoglobin. Gross and histologic lesions were of lobular or lobar fibrinous and neutron philicbroncho pneumonia, and *M.haemolytica* was recovered from nasal cavity and lungs .Euthanasia prior to the endpoint was required for 1/4 calves receiving the bacterial lysate and 3/4 control calves. In pairwise comparisons, calves receiving the bacterial lysate had 60 ± 27% lower gross lung lesion scores compared to the corresponding controls. *Mannheimia* loads in the lung were not different among groups.This study reports a novel model of *M.haemolytica* infection using colostrum-deprived calves, which has advantages over existing models of this disease. The findings using this model suggest that stimulation of an innate immune response in the lung partially protects against development of *M.haemolytica* pneumonia.

SP-35: INFECTIOUS CANINE HEPATITIS LIKE LESIONS IN AN ALASKAN BLACK BEAR CUB

Melanie M. Iverson, Kimberlee B. Beckmen, Terry Spraker

An estimated six-month-old female black bear cub was found dead in Juneau, Alaska and submitted for necropsy. Gross necropsy revealed severe jaundice. The gallbladder was bright yellow, severely thickened with prominent vasculature. The liver was swollen with pin-point white foci. The spleen was markedly enlarged, turgid, and had pin-point white foci. Preliminary differentials included canine adenovirus type 1 (CAV-1) known as infectious canine hepatitis, canine distemper virus (CDV), and leptospirosis. Fresh and fixed tissues were initially sent to Colorado State University Veterinary Diagnostic Laboratories. Brain, lung, and spleen tested negative for CDV by real-time polymerase chain reaction (PCR). Kidney tested negative for leptospirosis by PCR. Liver, brain, lung, spleen, and kidney tested negative for canine adenovirus type 1 (CAV-1) and type 2 (CAV-2) by PCR. Histopathologic lesions included severe multifocal necrosis with mineralization and a few intranuclear inclusion bodies within the liver and severe edema within the connective tissues between the gallbladder and liver. These are highly suggestive of infectious canine hepatitis. Previously frozen liver samples were subsequently sent for CAV-1 and CAV-2 PCR testing and immunohistochemical staining at Athens Veterinary Diagnostic Laboratory and CAV-1 and CAV-2 PCR testing at Wisconsin Veterinary Diagnostic Lab. All tests were negative. In October 2015, an Alaskan brown bear found dead tested positive for CAV-1. Bears in Alaska have serum antibody prevalence of CAV-1 but the occurrence of clinical infectious canine hepatitis is unknown. August 2017, a black bear cub with similar gross lesions was submitted for diagnostics.

SP-36: ODONTOGENIC TUMORS WITH INDUCTION IN FOUR DOGS AND A COW: AMELOBLASTIC FIBROMA AND AMELOBLASTIC FIBRO-ODONTOMA

Patrick Huang, Cynthia Bell, Kevin K. Lahmers, Leslie W. Woods, Brian G. Murphy

Ameloblastic fibroma (AF) and ameloblastic fibro-odontoma (AFO) are rare odontogenic tumors with induction (also known as mixed tumors) that have rarely been reported in dogs and cattle. These benign odontogenic tumors occur most commonly in young animals. Although slow growing, they can be locally destructive, destabilizing the adjacent jawbone. Due to the well-demarcated nature of these tumors, complete surgical excision can be curative. These odontogenic tumors are histologically complex and are unique to the oral cavity, often posing a diagnostic conundrum for the examining pathologist. This retrospective study reviews the diagnostic features of AF (4-month-old Labrador retriever dog and an adult dairy cow) and AFO (4-month-old Labrador retriever, 8-month-old Beagle mix, and 9-month-old mixed breed dog). To a degree, odontogenic tumors with induction recapitulate the early stages of tooth development and are comprised of two different types of tissue, neoplastic odontogenic epithelium (OE) and induced basophilic ectomesenchyme (dental pulp). The neoplastic OE is often arranged as thin, branching (plexiform) ribbons, reminiscent of the dental lamina that extends from the oral epithelium during the early stages of tooth development. Importantly, neoplastic OE arranged as plexiform ribbons often lacks most or all of the well-recognized cardinal features of the enamel organ. AFOs are distinguished from AFs by the additional presence of hard dental matrices such as enamel or dentin, typically arranged as undulating ribbons or rings located between the OE and pulp ectomesenchyme. The presence or absence of dental matrices does not affect the prognosis.

SP-37: SYNOVIAL LIPOMAS AND SYNOVIAL LIPOMATOSIS IN DOGS

Chloe Chan Goodwin, Melissa Sanchez, Fabiano Oliviera

Synovial lipoma and synovial lipomatosis (synovial arborescens) are distinct conditions described in humans, but not in dogs. Although they may cause similar clinical symptoms, they differ in their histopathologic characteristics as well as in their etiology. This study describes the clinicopathologic features of two cases of synovial lipomas and one of synovial lipomatosis in dogs.

The primary complaint in all cases was a non-painful, fluctuant mass over a joint. There was no prior history of concurrent illnesses, trauma, or surgeries. No breed predisposition was seen. Masses affected different joints: knee, hock, and interphalangeal. Mean age was seven years. All masses were surgically removed and submitted for histopathology.

Histologically, the case of lipomatosis was similar to human cases and characterized by frond-like projections of the synovium lined by variably hyperplastic synovial cells supported by an expanded subsynovial stroma containing large numbers of mature adipocytes admixed with moderate chronic inflammation. The cases of lipomas were composed of partially encapsulated masses of mature adipocytes lined by a thin layer of synovium.

Synovial lipomatosis is considered a pseudo-tumor and associated with co-morbidities (e.g., trauma and chronic arthritis). Synovial lipomas are considered 'true' neoplasms, not associated with other underlying diseases. The canine synovial lipomas were cured with surgical removal, similar to treatment response in human patients. The case of synovial lipomatosis recurred to its original size three months after surgery. A second surgical resection was not attempted. The recurrence in this patient could have been secondary to persistence of a non-diagnosed joint lesion.

SP-38: SORAFENIB TREATMENT IN A NOVEL MOUSE MODEL OF ACUTE RADIATION-INDUCED DERMATITIS

Kerry Ann Goldin, Jessica Lawrence, Angela M. Craig, Clara Ferreira, Luke Hoeppner, Davis Seelig

Radiation therapy (RT) is prescribed in ~50% of cancer patients in North America. Radiation-induced dermatitis (RID) is a common side effect of RT, affecting up to 95% of patients. The effects of RID range from mild to severe and can lead to pain, disfigurement and may impede completion of the treatment regimen. While the pathophysiology of RID is partially known, a complete understanding is lacking and there are no clear effective preventative strategies. The objective of this study was to characterize the microscopic features of RID in a novel hairless mouse model and to determine the effect of sorafenib, a tyrosine kinase inhibitor, on RID clinically and microscopically. To characterize the progressive pathology of irradiated SKH-1 mice and examine the efficacy of sorafenib, 2 groups (sorafenib+vehicle and vehicle only, n=5 / group) were irradiated. To evaluate the sequential effects of RT +/- treatment, biopsies were collected at 5 time points (2hr, 10d, 12d, 18d, 24d post RT). Tissues were collected for light microscopic and IHC analysis (VEGFR2, CD31, TGFbeta-1). Histologically, RT resulted in progressive epidermal hyperplasia and loss of hair follicles and sebaceous glands. Sorafenib was effective at mitigating epidermal hyperplasia, VEGFR2 and TGF-beta expression, but not the number of dermal vessels. The SKH1-hr1 mice appear to be an adequate model for RID. A single administration of topical sorafenib administered 24 hours prior to RT did not mitigate acute clinical RID but resulted in reduced VEGFR2 expression, reduced TGFbeta-1 expression, and diminished epidermal hyperplasia with no effect on CD31 expression.

SP-39: DETECTION OF ECHINOCOCCUS GRANULOSUS AND ECHINOCOCCUS MULTILOCULARIS POSITIVE CANIDS IN EASTERN OREGON

Kathryn Gaub, Robert Bildfell, Andree Hunkapiller, Donna Mulrooney, Julia Burco

Background: *Echinococcus multilocularis* and *Echinococcus granulosus* are causative agents of disease in humans. In North America, *E. multilocularis* was believed to be restricted to the northern tundra zone of Alaska, USA, and Canada, but has now been reported in canids from several northern states. *Echinococcus granulosus* has not been detected in Oregon cervids for decades, however, re-introduction of wolves from Idaho and Montana has brought infected definitive hosts back onto the landscape.

Objective: To determine the prevalence of *Echinococcus granulosus* and *multilocularis* in Oregon carnivores.

Methods: Intestinal tracts from 67 coyotes, two bobcats, six foxes, and three wolves were collected across Oregon and processed as described by Gesy et al. 2013. Following DNA isolation and extraction, PCR was performed using primers for the NADH dehydrogenase subunit 1 (ND1) mitochondrial gene for both *E. multilocularis* and *granulosus.* PCR products obtained were sequenced using standard Sanger DNA sequencing, with confirmation as *E. granulosus* and/or *E. multilocularis* via BLAST analysis.

Results: *Echinococcus granulosus* was identified in four animals from eastern Oregon: one wolf and three coyotes. *Echinococcus multilocularis* was discovered in six coyotes. One coyote contained both species.

Conclusions: The restriction of *E. granulosus* infection to coyotes in eastern Oregon where wolves have been re-introduced provides evidence of spillover from wolves. The role of coyotes in the dissemination of *Echinococcus* spp. should be further investigated. *Echinococcus multilocularis* infection is now documented in Oregon, where the presence was previously unknown. Public health authorities should be made aware of the potential for these diseases statewide.

SP-40: DOUBLE POSITIVE (CD4+/CD8+) T-CELL LYMPHOMA IN A GUINEA PIG

Miranda Frohlich, Amanda Morphet, Brendan Podell, Kendra Andrie, Gary L. Mason, A. Russell Moore

Case Report: A 3-year-old, male intact, guinea pig (*Cavia porcellus*) presented to the Colorado State University Veterinary Teaching Hospital for a 2-day history of dyspnea, lethargy and anorexia. A syncopal episode, characterized by bradycardia and acute loss of consciousness, was witnessed shortly after presentation. Pericardial effusion was diagnosed via thoracic radiographs and echocardiogram. Twenty-five milliliters of serosanguinous fluid were removed during a therapeutic pericardiocentesis. Cytologically, the fluid contained many large lymphoid cells with variably sized pink cytoplasmic granules, consistent with lymphoma. Flow cytometry was performed on cells isolated from the pericardial effusion, using antibodies targeting guinea pig pan-T cell receptor (clone CT5), CD4 (clone CT7) and CD8 (clone CT6). Neoplastic cells were uniformly triple positive for T cell receptor, CD4 and CD8 by flow cytometry. Clinical signs recurred 7 days later prompting euthanasia with necropsy. Necropsy revealed multicentric lymphoma in the cranial mediastinum, submandibular lymph nodes, heart, lungs, and adrenal glands, and severe, serous, thoracic and pericardial effusions. Anti-human CD3 (clone LN10), and anti-human PAX5 (clone 1EW) did not label the neoplastic cells, but did appropriately label normal lymphocytes within the spleen.

Discussion: The cytologic morphology was suggestive of a T-cell lymphoma. To our knowledge, this is the first reported case using flow cytometry to diagnose lymphoma in pericardial fluid from a guinea pig and represents an unusual morphology with lack of CD3, coexpression of CD4 and CD8, and distinct cytoplasmic granulation. Further investigation into the diagnostic use of flow cytometry in guinea pig lymphoma may be warranted.

SP-41: PATHOGENESIS OF CHIKUNGUNYA VIRUS IN EXPERIMENTALLY INFECTED SOUTHERN TOADS (ANAXYRUS TERRESTRIS) AND LEOPARD FROGS (LITHOBATES PIPIENS)

Michelle Evans, Angela Bosco-Lauth, Sushan Han, Airn Hartwig, Paul Gordy, Richard Bowen

Chikungunya virus (CHIKV) is a mosquito-borne RNA alphavirus that has recently emerged in new regions such as the Caribbean, Europe, and the Americas. While the virus is maintained by an urban human-mosquito cycle, the sylvatic cycle in these regions remains largely unknown. Previous studies have shown that a variety of birds and mammals experimentally infected with CHIKV failed to develop a viremia, indicating that the virus has few species that act as potential reservoirs. However, experimentally infected ectotherms such as frogs, toads, and snakes can develop a viremia theoretically high enough to transmit the virus to mosquitoes. This study examines the pathogenesis of CHIKV infection in southern toads (Anaxyrus terrestris) and leopard frogs (Lithobates pipiens). Six frogs and six toads were inoculated with a SAH strain of CHIKV. Blood was collected on days 1, 3, and 5, while on days 2, 4, and 7 two frogs and two toads were terminally bled, euthanized with an IP injection of pentobarbital, and necropsied to collect tissue for histopathology and virus isolation. Two out of six toads and five out of six frogs developed viremias 1-2 days post infection. Virus was isolated from lung, kidney, and liver in toads, and from the lung, kidney, liver, and skeletal muscle in frogs. None of the viremic animals displayed any histological lesions or inflammatory responses to the virus. This study showed that southern toads and leopard frogs are able to amplify the virus without developing clinical disease, theoretically making these species ideal reservoir hosts.

SP-43: UTILIZING A SCANNER, CELLSENSE STANDARD IMAGE ANALYSIS SOFTWARE AND A CONVENTIONAL LIGHT MICROSCOPE TO ESTABLISH A GRADING SYSTEM FOR EQUINE ARYTENOID CHONDROSIS/CONDROPATHY

Lusan L. DellaGrotte

Background: The pathogenesis of chronic septic chondritis of the equine arytenoid cartilage remains unclear. We are exploring the possibility of altering the parameters of the current surgical therapy to resolve this decline in athletic capacity.

Objective: To develop a grading system that would correlate the size of affected arytenoid cartilage with the degree of loss. These results will provide insight on the current therapy (subtotal arytenoidectomy) and help facilitate a less invasive approach and aim to restore satisfactory athletic capacity.

Methods: 19 diseased arytenoids and 4 controls were sectioned caudal to the corniculate process, scanned (Epson PerfectionV850) and submitted for histology. Cartilage surface area (SA) was calculated using cellSense Standard image analysis software. All were stained with Hematoxylin and Eosin and Safranin O and evaluated using a conventional light microscope (Olympus BX 43) equipped with an Olympus DP 26 digital camera. Safranin O slides were scanned and cartilage SA was calculated. Microsoft Excel and Powerpoint wereused to calculate the percent cartilage SA loss and analyze the disease direction and growth.

Results: 10 tissues were grouped in grade I, 5 in grade II and 4 in grade III, ranging from a surface area of 1.05cm^2-2.00cm^2, 2.10cm^2–2.87cm^2, and 3.24cm^2–9.12cm^2 respectively. The average percent cartilage loss was 53.3%, 50.1%, and 61.8% respectively and the growth of disease had atendencyto migrate towards the mucosal surface.

Conclusions: Through the use of computer technology we propose that size is an accurate predictor of disease severity in equine arytenoid chondropathy helping surgical intervention strategies.

SP-44: CYTOHISTOLOGIC CORRELATION OF A LEFT BODY WALL FIBROSARCOMA IN AN ADULT DUSKY PYGMY RATTLESNAKE

Kara De New, Francisco Conrado, Justin Rosenberg, Sarah Beatty

A 12-year-old, dusky pygmy rattlesnake was presented to the UF Veterinary Hospital for evaluation of a 2.0 cm subcutaneous left body wall mass. The mass was first noted two weeks prior to presentation, when an ultrasound-guided fine needle aspirate was performed. In the aspirates, an atypical mesenchymal population predominated, frequently accompanied by fragments of skeletal muscle with no significant inflammation. These cells were stellate to fusiform, with indistinct cytoplasmic borders, and contained pale basophilic cytoplasm, raising the concern for a mesenchymal neoplasm, primarily a fibrosarcoma. Surgical excision of the mass was pursued, which revealed local body wall invasion with deep margins extending into the coelomic cavity.

Histopathologic evaluation revealed an invasive mass composed of spindle cells that formed streams, swirls, and bundles, supported by loose fibrovascular stroma, supporting the cytologic diagnosis of fibrosarcoma. Fibrosarcomas are one of the most commonly reported neoplasms in snakes. However, they have not been described in a dusky pygmy rattlesnake. To the authors' knowledge, this is the first report of cytohistological correlation of a fibrosarcoma in a specimen of the order Squamata, suborder Serpentes.

SP-45: URORECTAL SEPTUM MALFORMATION SEQUENCE IN A STANDARDBRED FOAL: A CASE STUDY AND COMPARATIVE REVIEW OF THE LITERATURE

Samantha L. Darling, Carla Carleton, Dalen Agnew

The body of a euthanized Standardbred foal exhibiting atresia ani and ambiguous genitalia was presented to Michigan State University Veterinary Diagnostic Laboratory for necropsy. Gross examination revealed that the rectum, vagina, and bladder all opened into a common cloacal cavity. The cloaca had no communication to the external environment but terminated in a blind-ended slit within the superficial perineal vestibular tissue which extended anteriorly along the ventral midline 20 cm. Histopathology showed that at the colonic-cloacal junction an abrupt transition from colonic epithelium to stratified squamous epithelium occurred. Distally, the cloaca was lined by fissures which extended 1.6 mm into the mucosal surface and was lined by stratified squamous to transitional epithelium. The pattern of congenital abnormalities described in this foal resembles the condition "urorectal septum malformation sequence" (URSMS). This condition is lethal and, if born alive, the neonate usually succumbs shortly after birth. Although rare, several other cases of foals with persistent cloaca and atresia ani have previously been reported. Since this condition was first described, a variety of theories on the etiology and pathogenesis of this syndrome have been proposed; however, much controversy still surrounds this topic. While there is no substantial evidence to support that URSMS is heritable in horses, research in other species has shown that a genetic component may contribute to these congenital defects.

SP-46: BLACKLEG: A RETROSPECTIVE STUDY OF THE FREQUENCY OF CARDIAC LESIONS IN FATAL CASES OF CLOSTRIDIAL MYOSITIS

Caitlin Maureen Culligan, Kim Newkirk

Clostridial myositis (blackleg) is a common cause of death in cattle and is classically caused by Clostridium chauvoei. The characteristic lesions include hemorrhage, necrosis, edema, and emphysema within skeletal muscle, although in some cases, there may also be cardiac involvement. This retrospective study aimed to determine the frequency of cardiac lesions in cases of Clostridial myositis diagnosed at the University of Tennessee. A search of the pathology records and review of the case materials (as available) identified 48 cases diagnosed based on gross findings and/or histopathology and culture results. Two of the 48 cases (4%) had only cardiac involvement without reported lesions in skeletal muscles; one had necrotizing myocarditis and one had only fibrinosuppurative endocarditis, myocarditis and epicarditis from which DNA sequencing identified Clostridium chauvoei. Of the 46 cases with skeletal muscle lesions, 27 (58.7%) had associated cardiac lesions, which included fibrinous to fibrinosuppurative peri-, epi-, or endo-carditis (n=12), necrotizing myocarditis and fibrinous to fibrinosuppurative peri-, epi-, or endo-carditis (n=9), or necrotizing myocarditis (n=6). These data demonstrate that cardiac lesions in cases of Clostridial myositis are common.

SP-47: VEGETATIVE ENDOCARDITIS AS A CAUSE OF SEPTIC POLYARTHRITIS IN A DOG

Zachary S. Croslin, James H. Meinkoth, Laura A. Nafe, Rebecca S. Tims

Anamnesis: Leo, a ten year-old, castrated male, mixed breed, dog was presented for fever, severe lethargy and joint pain. A tentative diagnosis of Rocky Mountain Spotted Fever was based on a single titer (1:128). Treatment consisted of doxycycline, dexamethasone SP, fluids and prednisone. Two days later, fever persisted and joint pain worsened prompting referral.

Clinical Findings: Leo presented non-ambulatory with a temperature of 103.8F. Both elbows were severely effusive, hot and painful. Tachycardia with a sinus arrhythmia and grade I-II/VI left systolic heart murmur were noted.

Diagnostic Procedures: Significant CBC findings included thrombocytopenia ($37x10^3$/ul, RI 170-400 $x10^3$/ul) and neutrophilia (35,670/ul, RI 2,060-10,600/ul) with moderate toxic change. Serum chemistries showed hyperbilirubinemia (10 mg/dL, RI 0.1-0.3 mg/dL), increased ALT (226 IU/L, RI 12-118 IU/L) and ALP (1,908 IU/L, RI 5-131 IU/L) activities. Thoracic radiographs revealed possible focal pneumonia. Elbow arthrocentesis and cytology showed marked pyogranulomatous inflammation and many bacterial cocci within neutrophils, indicating septic arthritis. Subsequently, echocardiography showed evidence of vegetative endocarditis on the anterior leaflet of the mitral valve. Culture of the synovial fluid grew *Streptococcus canis.* Following long-term antimicrobials, pain management and physical therapy, Leo's condition has greatly improved.

Clinical Relevance: Septic arthritis is uncommon in adult dogs, but can result from either direct inoculation or hematogenous spread. With evidence of polyarticular disease and no insult to synovial membranes, hematogenous spread was suspected. Attempting to identify a source of sepsis, an echocardiogram was performed leading to diagnosis of vegetative endocarditis. Endocarditis should be considered as a cause of septic arthritis.

SP-48: NEUTROPHIL LOCALIZATION AND INTERACTIONS IN THE SPLEEN OF A MOUSE MODEL OF SYSTEMIC LUPUS ERYTHEMATOSUS

Catharine Cowan, Rujuan Dai, Michael Edwards, Bettina Heid, S. Ansar Ahmed

Systemic Lupus Erythematosus (SLE) is a complex autoimmune disorder that affects humans, canines, and rarely felines and equines. SLE manifests with a constellation of clinical signs and is driven by antigen-autoantibody complex deposition. It is generally thought of as an adaptive immune system disease, however recent work has elucidated dysregulation of the innate immune system as well. Neutrophils are the primary innate immune defenders against pathogens and are commonly short-lived and rapidly recruited to sites of inflammation. Recent studies have demonstrated significant plasticity in their behavior, particularly in SLE. Neutrophil numbers are increased in the spleen of SLE patients and SLE mouse models, and these neutrophils secrete cytokines such as BAFF and IL-17 that can regulate B and T cell function as well as act in an autocrine fashion. Neutrophils are found in both the marginal zone and T cell zone of the white pulp, with variable localization depending on the model and disease course. We are investigating potential alterations in the phenotype and function of splenic neutrophils in the MRL/MpJ-*Fas*^{lpr} SLE mouse model. We have found increased numbers of neutrophils that localize to the T cell zone during active disease. Flow cytometric and image analysis of this neutrophil population shows increased cytokine expression, MHC-II expression, and direct interaction with T cells, while other canonical neutrophil functions (ROS production, phagocytosis) are not different from controls. Analysis of chemotactic migration behavior is ongoing. Further work is necessary to elucidate the role this unusual neutrophil population plays in the pathogenesis of SLE.

SP-49: FELINE LEUKEMIA VIRUS REPLICATES FASTER IN PANTHER CELLS

Elliott S. Chiu, Melody Roelke, Sue VandeWoude

Feline leukemia virus (FeLV) is a common domestic cat disease that has been documented in a range of wild felids. While the progression of FeLV infection has been well documented in the domestic cats, infection has not been thoroughly investigated in atypical hosts. In domestic cats, four types of infection (abortive, latent, regressive, and progressive) have been documented. It has been hypothesized that presence and activity of the endogenous form of the virus (enFeLV) account for some differences in disease phenotypes. Since the early 2000's, Florida panthers (*Puma concolor coryi*) have experienced two FeLV outbreaks resulting in at least 29 unique infections. As Florida panthers lack enFeLV and that viruses in naïve hosts may often be more virulent that in native hosts, we hypothesize that FeLV viral load, proviral load, and viral replication will be greater in panthers compared to domestic cats. We measured viral and proviral tissue loads by qPCR in domestic cats and panthers following natural and experimental infections *in vivo* as well as experimental infections *in vitro*. Fibroblasts derived from panthers showed accelerated FeLV viral replication rates compared to domestic cat fibroblasts, resulting in higher proviral load and productive viremia. Endpoint naturally occurring FeLV-A infection in panthers had similar tissue viral loads as experimentally infected cats in a small set of animals. Our research documents that FeLV-A can replicate with accelerated kinetics in panther cells in vitro, but endpoint viral loads of panthers may be similar to that observed in domestic cats.

SP-50: SYSTEMIC BLASTOMYCOSIS IN AN ADULT FEMALE DOG

Mitchell T. Caudill, Katie M. Boes, Stefanie M. DeMonaco, Vanessa Wallace, D. Phillip Sponenberg

A 5-year-old female spayed Golden Retriever presented to the to the Virginia-Maryland College of Veterinary Medicine Veterinary Teaching Hospital for progressive respiratory distress, facial nerve paralysis and a wound on right foot pad. A previous diagnosis of blastomcycosis had been made by the referring veterinarian, and disease progression was confirmed at presentation via cytology of the footpad that revealed ~15 um round yeast organisms with thick refractile double-contoured walls and broad-based budding, variably degenerate neutrophils, macrophages and lymphocytes, and a mixed population of presumed secondary bacteria. The dog died spontaneously from respiratory arrest despite treatment, and a necropsy was performed. The right footpad was diffusely ulcerated, and all lung lobes displayed multiple, tan, firm, miliary coalescing nodules that extended throughout the parenchyma. On histopathology, the lung was effaced by necrotizing pyogranulomatous inflammation consisting of macrophages, multinucleated giant cells, and degenerate neutrophils. Numerous yeasts with thick refractile double contoured wall and broad based budding were present, along with occasional lymphoplasmacytic infiltrates and fibrin. The footpad showed complete ulceration and similar pyogranulomatous inflammation and *Blastomyces* yeasts.

SP-51: HARD YELLOW LIVER DISEASE IN WEST TEXAS CATTLE

Natalie J. Castell, Meredyth L. Jones, Eric R. Snook, Barbara C. Lewis, Brain F. Porter

Six adult cattle from a single farm were euthanized to further diagnose a chronic herd health problem. All cattle were losing weight and in poor body condition, and recently slaughtered animals had been condemned for liver abnormalities. The livers of all six were diffusely pale tan to yellow and firm. Histologically, the hepatic parenchyma was effaced and replaced by hepatocellular fatty cysts and lipogranulomas, and there was marked bridging portal fibrosis and biliary hyperplasia. Macrophages variably contained a yellow-brown pigment, and some areas also showed necrosis and neutrophilic and lymphoplasmacytic inflammation. Surviving hepatocytes often contained fine lipid vacuoles. Cytokeratin and Iba-1 immunostains demonstrated that hepatocytes were adjacent to areas of biliary hyperplasia, and lipid effaced regions were composed primarily of macrophages. Masson trichrome stain showed extensive fibrosis in these lipid filled areas. The cortex of mediastinal lymph nodes also contained multiple lipogranulomas with similar pigment. The nodes also showed reactive lymphoid hyperplasia and sinus histiocytosis. There was moderate to marked hemosiderosis within the spleen. The findings are consistent with hepatic fatty cirrhosis, or hard yellow liver disease, a lethal condition that affects ruminants in West Texas. The disease has been reported in cattle, sheep, and goats, and it has a seasonal, though sporadic, pattern of occurrence. Though likely caused by a phytotoxin or mycotoxin, no specific etiologic agent has been identified.

SP-52: EVALUATION OF CANINE AND FELINE PLATELET PARAMETERS DURING STORAGE IN MgSO4 AND EDTA ANTICOAGULANTS

Michelle Kathryn Bourgeois, Shannon Dehghanpir, Kelsey Legendre, Stephen Gaunt

Background: Magnesium sulfate ($MgSO_4$) is an alternative anticoagulant for human blood samples with platelet clumping induced by EDTA. Veterinarians often use regional laboratories for CBCs, which can delay sample evaluation for 24 to 48 hours and induce artifactual changes in platelets. This study evaluates the usefulness of $MgSO_4$ as an anticoagulant in veterinary medicine and its effect on platelet parameters over 48 hours.

Methods: Whole blood from 9 dogs and 10 cats was collected directly into ThromboExact (Mg2+ at 0.82mg/ mL; S-Monovette®, Sarstedt) and K_3EDTA tubes (S-Monovette®, Sarstedt) in alternating order. The $MgSO_4$- and EDTA-anticoagulated blood samples were analyzed with Advia 120 hematology analyzer at 0, 12, 24, 36, and 48 hours. Parameters analyzed were platelet concentration, mean platelet volume (MPV), mean platelet component (MPC), plateletcrit (PCT), platelet distribution width (PDW), and mean platelet mass (MPM). Blood smears were stained and examined at each time interval and graded for platelet clumping.

Results: In dogs, $MgSO_4$ induced significant platelet clumping and caused decreased platelet concentration, MPC, PCT, and MPM, as well as increased MPV and PDW. Feline $MgSO_4$-anticoagulated blood had similar results, except MPV was decreased. EDTA maintained stable platelet values for 48 hours in dogs; however, in cats, EDTA-induced changes were observed in platelet concentration, PCT, MPV, and MPC.

Conclusions: $MgSO_4$ is not a suitable EDTA substitute for anticoagulation of canine and feline blood, as platelet parameters were significantly altered during storage. The results of this research contrast those of human literature, suggesting species-specific differences in platelet inhibition with $MgSO_4$

SP-53: PROTEOMIC ANALYSIS OF PRIMARY AND METASTATIC CANINE HEMANGIOSARCOMA TISSUES

Miranda N. Frohlich, Erin E. Trageser, Christopher R. Olson, Tony Tullot, Valerie M. Wong, Katherine A. Blincoe

Hemangiosarcoma is a common and highly aggressive neoplasm in dogs, and metastasis is considered a major contributing factor to mortality. Despite the clinical significance of metastasis, its mechanisms in canine hemangiosarcoma are largely unknown. The goal of this study was to compare the proteomic profiles in paired primary and metastatic tumor tissues. Dogs suspected to have splenic hemangiosarcoma were recruited into this study from the Phoenix Metropolitan Area, AZ. At necropsy, suspected tumor tissues were fixed in formalin for routine histopathology and flash-frozen in OCT medium for cryosectioning. Inclusion criteria for final proteomic analysis were: (1) presence of largest tumor burden in the spleen grossly; (2) presence of small, multifocal, blood-filled, cavernous (i.e. presumed metastatic lesions) lesions in liver; and

(3) histopathologic findings of the splenic and liver lesions consistent with hemangiosarcoma. To obtain pure tumor tissues, both primary and liver metastatic tissues (n=5 pairs) were isolated by laser microdissection. The proteomic profiles of paired primary and metastatic tissues were compared with 2-dimensional fluorescence difference gel electrophoresis (2-D DIGE), which identified seventy-four differentially expressed proteins in a cross-gel analysis. In particular, prostaglandin reductase 2 (PTGR2), an enzyme which suppresses PPAR gamma-mediated adipogenesis, was found to be upregulated in metastatic tissues compared to primary tissues. This finding further supports previously reported links between hemangiosarcoma and adipogenesis, but the exact role of PTGR2 in the pathogenesis of hemanigosarcoma remains to be elucidated.

SP-54: SPINAL GANGLIONEUROMA IN A BEAGLE

Jillian M. Athey, Paula R. Giaretta, Melissa L. Blazevich, Brian F. Porter

A six-year-old, tri-colored, spayed female Beagle presented to Texas A&M Small Animal Emergency with a three-week history of progressive ataxia. Proprioceptive deficits were found in the left and right forelimb and the left hind limb. At necropsy, a 1x0.4x0.5 cm, tan, soft, subdural mass was found in the cervical segment of the spinal cord (C3-C4). The mass involved the right dorsal and ventral nerve roots and compressed the right lateral aspect of the spinal cord. Histologically, the mass was poorly demarcated, unencapsulated, expansile, and moderately cellular. It infiltrated the spinal nerve roots, the dorsal root ganglion, and the meninges. The neoplastic cells were arranged in streams and contained large ganglion cells intermixed with Schwann cells and axons. Anti-neurofilament immunohistochemistry and a Holmes stain confirmed the presence of axons within the neoplasm. The mass was diagnosed as a ganglioneuroma, presumably arising from the dorsal root ganglion. Ganglioneuromas are rare peripheral nervous system tumors. In dogs, ganglioneuromas have been documented in the jejunum, rectum, nasal cavity, thoracic area, oral mucosa, and urinary bladder. There is only one previous report of a canine ganglioneuroma arising within the vertebral canal.

ACVP Late-Breaking Poster

LB-01: PARALLEL IN TRANSCRIPTION PROFILE, MORPHOLOGIC AND DENSITROMETRIC FINDINGS IN BONE TISSUES AFFECTED BY EQUINE SILICATE ASSOCIATED OSTEOPOROSIS (SAO)

Regina Zavodovskaya, Carrie J. Finno, Tanya Garcia-Nolen, Susan M. Stover

Background: Skeletal deformities and fatal pathological fractures are the major sequelae of progressive osteoporosis in horses that have pulmonary silicosis from inhalation of soil-derived cytotoxic silicate crystals. The putative link between the skeletal and respiratory conditions and the mechanism of the accelerated osteolysis are unknown.

Objective: We hypothesized *that the systemic effect of silicosis is disruption of bone remodeling resulting in bone fragility.* We predicted that differential gene expression and variation in µCT parameters between SAO^+ and unaffected rib bone specimens would be consistent with enhanced osteolysis.

Methods: Rib bone, which consistently contains SAO lesions and cells involved in bone remodeling, was sampled for molecular, imaging and histology analyses. Differential gene expression analysis was performed on mRNA extracted from medullary cores (8 SAO^+ and 8 unaffected). Rib specimens (24 SAO^+ and 20 unaffected) were scanned with µCT.

Results: Ten of the 17 genes differentially expressed between SAO^+ and unaffected horses were consistent with mediators of bone formation, none indicated osteoclast activation. SAO^+ ribs medullary bone tissue had lower bone mineral density and altered structural parameters, while bone volume fraction was not different and trabecular numbers were high. Edema, congestion and loss of the hematopoietic bone marrow co-localized with neovascularization, and high osteoclastic and osteoblastic activity. Transcription profile and µCT data correlated with morphologically high numbers of active osteoblasts on bone surfaces.

Conclusions: Study reveals a consistent trend of either ***compensatory*** or ***pathologic*** increase in **osteoblast activation**, and provides specific gene targets to explore as SAO molecular markers.

LB-02: METAPHYEAL AND DIAPHYSEAL DYSPLASIA OF THE THIRD CERVICAL VERTEBRA (C3) SECONDARY TO PHYSEAL NECROSIS IN A QUARTER HORSE FOAL

Ching Yang, Steven E. Weisbrode, Christopher Premanandan

A 2-month-old Quarter horse foal was presented for an acute onset of ataxia and a one week history of diarrhea. A vertebral fracture was suspected. The foal failed to improve after a month of medical treatment and was humanely euthanized due to the poor prognosis.

There was severe focal narrowing of the spinal canal due to ventral compression by the clockwise rotation of the third cervical vertebra (C3) in sagittal sections of the vertebral column. The metaphysis and diaphysis of C3 were markedly shortened than the adjacent vertebrae and were tan-white. The epiphyses appeared normal.

Histologically, there was a complete loss of dorsal compact bone of C3 with replacement by fibrous tissue. More than 80% of the transverse growth plate was replaced by thickened viable trabecular bone and infrequent foci of retained cartilaginous matrix. Both physes of C3 showed widespread linear coagulative necrosis of the zone of hypertrophy with multifocal transphyseal fibrovascular bridges. There was marked transverse trabeculation and fibrosis in the metaphyses. Severe bone marrow hypoplasia with slight fibrosis was observed. There was no evidence of fracture. The bone, cartilage and marrow and of the epiphyses were microscopically normal.

It was hypothesized that a regional transient incomplete and possibly multiphasic ischemia caused necrosis of the transverse growth plate and physes resulting in dysplasia of the bone. Ischemic injury to growth plates should be considered in the pathogenesis of focal bone dysplasia in horses.

LB-03: NEUROAXONAL DYSTROPHY IN A 4-YEAR-OLD DROMEDARY CAMEL

Andrea L. Vanderpool, Yava L. Jones-Hall, Sandra D. Taylor

Neuroaxonal dystrophy is a neurodegenerative disorder characterized by degeneration of nerve cell bodies and axonal swellings (spheroids) in the central nervous system. A specific neuroaxonal dystrophy, equine degenerative myeloencephalopathy (EDM), is common in horses and manifests as slowly progressive limb hypermetria with a suspected genetic basis; some cases have been linked to low serum concentrations of Vitamin E. Here, we report a case of rapidly progressive neuroaxonal dystrophy in a 4-year-old Dromedary camel. Following a 2-day history of hind limb ataxia, the camel became nonambulatory. Complete blood count, chemistry panel, spinal radiographs, CSF analysis, EHV-1 serology and nasal swab PCR, and West Nile Virus serology were all unremarkable. Despite hospitalization and supportive treatment, including oral vitamin E administration, neurologic signs progressed and the camel died. Gross lesions were not present at the time of autopsy. Histopathology revealed neuron cell body degeneration and numerous spheroids confined to dorsal and ventral funiculi and adjacent gray matter horns of cervical and thoracic spinal cord segments. Liver vitamin E concentrations measured 7.47 ppm; well below reference intervals for adult cattle and horses (camel ranges have not been established). This represents a unique case of spinal cord neuroaxonal dystrophy in a camel. Although neuronal changes and vitamin E concentrations were similar to EDM, the rapidly progressive nature of the disease and absence of typical brainstem lesions may suggest an alternate underlying pathogenesis.

LB-04: MORPHOLOGICAL AND IMMUNOHISTOCHEMICAL STUDY OF INTESTINAL TUMORS INDUCED BY HETERAKIS GALLINARUM INFECTION IN PHASIANIDAE

Alexandru Flaviu Tabaran, Gerard Michael O'Sullivan

Heterakiasis is a parasitic helminthic disease that affects domestic and wild *Galinaceae* birds. Infection with *Heterakis gallinarum* and *Heterakis isolonche* in pheasants and quails is occasionally associated with cecal proliferative lesions that are manifested as exuberant granulomas or occasionally as intestinal tumors. The oncogenic mechanism of such parasite-induced tumors is largely unknown, nevertheless, chronic inflammation induced by long term intramural parasite persistence is generally accepted to be an important factor in transformation.

The objectives of this study were to determine the morphological and immunohistochemical features of intestinal tumors produced by *Heterakis spp.* in Phasianidae. Seven cases (five common pheasants- *Phasianus colchicus* and two golden pheasants *Chrysolophus pictus*) previously diagnosed as *Heterakis spp.*-induced intestinal spindle-shaped cell tumors were evaluated for immunohistochemical expression of vimentin, CD34, α-SMA, S100 protein, and desmin. All tumors were diagnosed as leiomyomas, and, based on its consistent presence within the tumors, *Heterakis gallinarum* was interpreted to be the causative agent. The tumors had variable amounts of associated inflammation, with large tumors (up to 1 cm diameter) eliciting minimal inflammation. The tumors contain spindle-shaped or stellate cells, that exhibit strong immunoreactivity for α-SMA and vimentin but are negative for CD34 and S100 protein. Interestingly, and in contrast to the desmin immunoreactivity reported for leiomyomas, all tumors were negative for desmin, indicating a "partial" muscle immunophenotype and suggesting a myofibroblast origin.

LB-05: HISTOPATHOLOGICAL APPLICATION OF HIGH RESOLUTION DARK-FIELD MICROSCOPY AND HYPERSPECTRAL IMAGING IN GOLD NANOPARTICLE TOXICITY RESEARCH

Alexandru Flaviu Tabaran, Gerard Michael O'Sullivan

The developing interest for nanoparticle (NP) research brought new challenges for the pathologist, especially from the perspective of toxicological analysis of biological samples. Due to the compartmentalized distribution of nanoparticles, a comprehensive toxicological approach implies correlation between NP tissue presence and histological changes.

The main limitation of current physico-chemical techniques used for the assessment of NP biodistribution is their inability to detect the sub-organ and cellular localization of NP. For two studies on C1 mice employing intraperitoneal administration of gold nanoparticles (GNP) (8 mg/kg), these challenges have been addressed by using high-resolution microscopy coupled with hyperspectral imaging (HRDF-HI).

The routinely processed histopathological slides used for the assessment of toxic effects were further evaluated by HRDF-HI for nanoparticle identification and tissue mapping. The biodistribution data were compared with those obtained on the same samples by a previously standardized confocal laser scanning microscopy (CLSM) technique, and further confirmed by transmission electron microscopy (TEM) analysis. Samples were additionally investigated for immunohistochemical expression for TNF-α, IL6 and cleaved caspase-3.

Based on the light scattering and surface plasmon resonance properties of GNP, the multispectral analysis of the unique optical signatures of GNP allowed the tissue mapping of nanoparticles on the routinely processed histopathological slides. Interestingly, the data achieved by HRDF-HI have a better correlation with TEM compared with those obtained by CLSM. The tissue expression for cleaved caspase-3, TNF-α and IL6 was significantly increased in organs determined to concentrate the nanoparticles (spleen, liver and lymph nodes), but without the presence of any significant histopathological findings.

LB-06: UTILIZING GÖTTINGEN MINIPIGS IN OCULAR RESEARCH – REVIEW AND CURRENT PERSPECTIVES

Stephanie M. Shrader, William F. Greentree

Göttingen minipigs were developed in the 1960's at the University of Göttingen in Germany for use in biomedical research. Since then, they have become a popular alternative to other non-rodent animal models, especially for ocular studies, because of their ease of handling, smaller size (compared to other miniature swine breeds), rigorous genetic maintenance, and ocular anatomy that is similar to humans. Although Göttingen minipigs are commonly used in ocular research, there is a paucity of literature available to researchers, ophthalmologists, and pathologists regarding their overall use in such research endeavors. To date, the Göttingen minipig has been used for a variety of ocular studies, including those evaluating the safety and efficacy of ocular therapeutics (delivered via various routes of administration), glaucoma etiopathogenesis and treatment, novel biomaterials and implantable devices, and novel surgical procedures. Additional diagnostics tools that are often utilized during the course of ocular studies include in vivo procedures (e.g. electroretinography, optical coherence tomography, fundoscopic imaging, and fluorescein angiography) and histopathologic evaluation of enucleated globes. This review provides an understanding of porcine ocular anatomy, human ocular therapeutic needs, surgical procedures, feasibility of implantable biomaterials, useful adjunctive diagnostic tools, and demonstrated uses of this animal model in ocular research which are critical for the appropriate design of future studies.

LB-07: VALIDATION OF VESICULAR STOMATITIS VIRUS NUCLEOPROTEIN IMMUNOHISTOCHEMISTRY FOR VESICULAR STOMATITIS-INDIANA SEROTYPE VIRUSES

Leon Daniel Schermerhorn, Mya Masterson, Jessie D. Trujillo, Igor Morozov, Brian Martin, Tom Monath, Juergen Richt, Sally Davis

Vesicular stomatitis virus (VSV) is an enveloped, negative sense, single-stranded RNA virus within the *Rhabdoviridae* family, genus *Vesiculovirus*. It can cause lesions in swine and cattle indistinguishable from Foot and Mouth Disease. The VSV genome can be readily manipulated, facilitating its use as a vaccine platform. To examine viral antigen distribution in tissue for a project investigating the potential for agricultural spillover of a replication-competent, recombinant VSV-IN virus from which the VSV-I glycoprotein (G) is completely deleted and replaced with the Ebola vaccine glycoprotein (rVSV-ZEBOV), we developed an immunohistochemical (IHC) assay specific for VSV nucleocapsid (N). Since both the wild-type and recombinant vector have the N gene, the IHC assay will detect both viruses. First, we validated the anti-VSV-N antibody via Western blot against lysates of Vero cells infected with wild type VSV-IN viruses L2-83, L134-85, or 97-31247. Second, sections of VSV-IN infected and uninfected swine nasal planum were used to test a variety of antigen retrieval (AR), primary concentrations and detection methods until specific results with minimal background were achieved. Additionally, we validated the antibody against formalin-fixed, paraffin-embedded VSV-IN infected Vero cell pellets. The antibody specifically labeled an ~47kD protein consistent with N in all infected cell lysates. VSV-IN infected nasal planum and cell pellets labeled best with a 40 minute, low pH AR, anti-VSV-N diluted 1:500, and a polymer-based, alkaline phosphatase detection system. Our results show specific labeling of VSV nucleocapsid protein. These tools can be used to investigate viral distribution and its tissue tropism in future studies.

LB-08: DISSEMINATED MYCOBACTERIOSIS CAUSED BY MYCOBACTERIUM KANSASII IN A POT-BELLIED PIG

Ryan M. Schafbuch, Stacy H. Tinkler, Chee K. Lim, Rebecca M. Wolking, José A. Ramos-Vara

A 1.5 year-old female spayed Juliana pot-bellied pig presented to the Purdue University Veterinary Teaching Hospital with a history of wasting and anorexia and was diagnosed with disseminated mycobacteriosis. Enlarged and partially mineralized lymph nodes were identified on radiographs and computed tomography scan. On post-mortem examination, there was generalized peripheral, thoracic, and abdominal lymphadenomegaly; additionally, disseminated pale tan nodules in the lungs, liver, spleen, and kidneys were identified. Histologic examination of nodules and lymph nodes revealed caseonecrotic granulomas with numerous intracellular, acid-fast, beaded bacilli. *Mycobacterium kansasii* was identified as the etiologic agent by PCR amplification using universal *Mycobacterium* primers, direct sequencing of the PCR amplicon, and comparison to sequences in GenBank. *Mycobacterium kansasii* is a non-tuberculous mycobacteria most known as an opportunistic pathogen that causes pulmonary disease in immunocompromised humans. Disseminated mycobacteriosis in pigs is most commonly caused by *Mycobacterium avium/intracellulare* complex, and less frequently associated with *Mycobacterium bovis* or *Mycobacterium tuberculosis*. To our knowledge, this is the first report of disseminated mycobacteriosis in a pot-bellied pig caused by *Mycobacterium kansasii*. Diagnostic testing in cases of mycobacteriosis is critical to differentiate those caused by atypical mycobacterial pathogens from the more serious tuberculous species.

LB-09: REDUCED LEVELS OF TISSUE MYCOBACTERIUM AVIUM SUBSPECIES PARATUBERCULOSIS AFTER PEYERS PATCH INOCULATION IN CALVES

Kevin J. Stinson, Monica M. Baquero, Brandon L. Plattner

Mycobacterium avium subspecies *paratuberculosis* (*Map*) is the cause of Johne's disease (JD), a chronic progressive enteritis of cattle. Animal prevalence remains low in North American dairy herds despite high herd prevalence, though the reasons for this remain unclear. A major knowledge gap is in understanding early host-pathogen interactions, and how early defense events influence progression of JD. To address this, we modeled early *Map* infection in calves by direct ileocecal Peyer's patch injection. The objective was to understand development and progression of early *Map* infection in the distal small intestine and draining lymph node by characterizing persistence of *Map* in tissues, *Map*-induced intestinal lesions, fecal *Map* shedding, and *Map*-specific serum antibodies, through the first 28-weeks post inoculation (WPI). We hypothesized that *Map* persists locally within the intestine, and that lesion scores, fecal

Map shedding and *Map*-specific serum antibodies increase over time. Instead, our data show that persistence of *Map* in the intestine and draining node significantly decrease over time. 100% of calves had evidence for *Map* in multiple tissues at 4 and 12 WPI, but by 20 WPI, most retained *Map* in nodes but not intestine. By 28 WPI, only one calf had detectable *Map* in multiple tissues while 50% had no detectable *Map*. These data suggest that most calves rapidly achieve localized clearance of *Map* following intestinal infection. This may explain why low individual prevalence is observed despite high herd prevalence, and provides a platform for elucidating mechanisms important during early mucosal host defense.

LB-10: HERMAPHRODITISM, THE GENDER PRETENDER: A CASE REPORT OF OVOTESTICULAR DISORDER OF SEX DEVELOPMENT (OT-DSD) IN A BABOON

Ekaterina Perminov, Sara Mangosing, Alexandra Confer, Ana Alcaraz, Jason R. Crawford, Olga Gonzalez, Shyamesh Kumar, Edward Dick

Background: Disorders of sexual development are rare in the non-human primate literature. We report a case of true hermaphroditism in a phenotypically female baboon. A 19-year-old, nulliparous, female, baboon (*Papio spp.*) was presented for necropsy. The animal was in good body condition with no major medical history.

Methods: A complete necropsy with histology was performed.

Results: At necropsy, the animal was obese with adequate muscle mass and hydration. The myometrium contained two, 1-1.5 cm diameter, firm nodules. The cervical mucosa was thickened and dark. Both gonads appeared to be grossly unremarkable ovaries; histologically both were ovotestes, surrounded by germinal epithelium and containing discrete areas of ovarian and testicular tissue. There were follicles in various stages of development surrounded by ovarian stroma. Other areas contained hypoplastic seminiferous tubules that were lined by Sertoli cells, but lacked germ cells and spermatozoa. The uterine lesions were consistent with adenomyosis and cystic endometrial hyperplasia. The cervical lesion was consistent with atypical glandular hyperplasia with squamous metaplasia.

Conclusions: Ovotesticular disorder of sexual development (DSD) or true hermaphrodite, is rare in animals and non-human primates and to our knowledge has not been reported in *Papio spp.*

LB-11: ALLOGRAFT INFLAMMATORY FACTOR-1 IS A USEFUL MARKER FOR MACROPHAGES IN MULTIPLE SPECIES

Kathleen M. Donovan, Logan McQuillen, Mariah Leidinger, J. Adam Geoken, Christine M. Hogan, Heather A. Flaherty, David K. Meyerholz

Increasing the relevance and reproducibility of scientific data are growing areas of emphasis in translational scientific studies. For pathologists, the ability to use an immunohistochemical marker to study animal model as well as human tissues is a valuable resource. Allograft inflammatory factor-1 (AIF1) has been traditionally a microglia marker in the brain, but has recently been shown to have utility in other tissues including lung in humans. We studied tissues (male and female) from mice (n=20, several strains), rats (n=15, multiple strains), ferret (n=4), pig (n=4) and human (n=4, lungs only). We evaluated AIF1 immunostaining to study its cellular localization and potential utility in translational research. In the spleen, AIF1 immunostaining was most intense and abundant in red pulp with some scattered immunostaining in white pulp for all species. In the liver of all species, immunostaining was moderate in scattered interstitial cells consistent with Kupffer cells. In the lungs, alveolar macrophages generally had weak immunostaining. In incidental disease conditions, (e.g. tumors or inflammation), AIF1 immunostaining was robust within activated macrophages independent of organ. Our study suggests that 1) AIF1 immunostaining is consistent with known cellular sites for macrophages in various organs, and 2) AIF1 may be a useful marker for tissue macrophages in multiple tissues and across several species (including humans), making it attractive for translational studies.

LB-12: SQUAMOUS METAPLASIA IN THE NEWBORN TRACHEA

Logan McQuillen, David K. Meyerholz

Squamous metaplasia has been reported as a benign, noncancerous change in response to various injurious or irritating stimuli (e.g. chronic cigarette smoking) and can be associated with Vitamin A deficiency. In young children, squamous metaplasia has been recorded and these cases have been linked to injury associated with intubation during hospitalization. Interestingly, controls for these studies were often lacking. We investigated whether squamous metaplasia could be detected in tracheas from archival blocks from newborn (<2 days of age) pigs without a history of respiratory disease or intubation. Squamous metaplasia was morphologically distinguished by loss of pseudostratified ciliated epithelium to a squamous epithelium that had flattened nuclei extending all the way to the epithelial apical surface. We identified 41 cases from the archival blocks and we found focal squamous change in 13 samples. The detection squamous metaplasia in newborn pigs prior to development of disease and lack of injury (e.g. intubation) would suggest a congenital origin. The focal nature of this small lesion may also suggest that it is an under-reported by most studies and that it may represent a “normal” developmental aberration.

LB-13: AVOIDING PITFALLS IN STATISTICAL ANALYSIS OF TISSUE SCORES

David K. Meyerholz, Amanda P. Beck

There is a growing initiative to increase the reproducibility of data in scientific studies. For pathologists, increasing the reproducibility of studies often includes application of semi-quantitative or quantitative scoring methods to enhance and validate the interpretation. Effective tissue scoring requires suitable statistical analysis to validate the group comparisons. Choice of appropriate statistical test is greatly influenced by experimental design (e.g. paired, unpaired, parametric, nonparametric, longitudinal, repeated measures, multiple factorial). We evaluated whether the type of statistical analysis could impart skewed interpretations of the data. We evaluated mock tissue scores for inflammation representing two groups of animals. For these semi-quantitative scores, we analyzed the data using 3 common statistical tests for comparing two groups: paired T-test, unpaired T-test and nonparametric Mann-Whitney test. From these tests, the data produce P values of 0.015, 0.067, and 0.174, respectively. The data was further analyzed as to what type of test should have been run. The data failed a normality test ($P<0.05$, Shapiro-Wilk test) and given the semi-quantitative scoring approach, we had to exclude use of the parametric tests because the foundational assumptions of normality were violated. While it is tempting for investigators to use a statistical test that shows significance, those statistical tests that give the best "confidence" in the data analysis should ideally be selected. Our study highlights how selection of statistical tests is an important step towards proper and reproducible interpretations.

LB-14: BABESIA FELIS INFECTION IN A YOUNG CAT FROM SOUTH AFRICA

Anna M Meredith, Dorothee Bienzle, Janet Beeler-Marfisi

A 2-year-old female spayed domestic shorthair cat was referred to the Ontario Veterinary College Health Sciences Centre for evaluation of vomiting, lethargy, anorexia, and icterus of approximately one week's duration. Serum biochemical abnormalities prior to referral included increased ALT of 1533 U/L (31-105 U/L) and total bilirubin of 38.7 µmol/L (0-4 µmol/L). The cat had been adopted from South Africa one month prior to presentation, was seronegative for feline leukemia virus antigen and feline immunodeficiency virus antibody, and had up-to-date vaccinations. On physical examination, the cat was dull but responsive, tachycardic, tachypneic, pyrexic (40.0°C), and had pale, icteric mucus membranes. Admission CBC abnormalities included: HCT 19 L/L (0.28-0.49 L/L), MCV 75 fL (39-52 fL), MCHC 309 g/L (317-350 g/L), reticulocyte count 235 x 10^9/L (0-60 x 10^9/L), and rubricytes 4.9 x 10^9/L. Review of the blood smear revealed inclusions in approximately 20% of erythrocytes. PCR results for *Mycoplasma* spp. were negative. A presumptive diagnosis of *B. felis* was made, and treatment with primaquine phosphate (0.5 mg/kg per os once daily) was initiated. Whole blood samples were submitted to the Vector Borne Disease Diagnostic Laboratory at North Carolina State University for PCR testing of *Babesia* spp. and *Cytauxzoon* spp. The PCR results were positive for *B. felis* and negative for *Cytauxzoon* spp. The cat improved with primaquine phosphate therapy, and the parasitemia and anemia resolved 3 and 5 days after initiation of treatment, respectively. To the authors' knowledge this is the first documented case of *B. felis* infection in North America.

LB-15: UTERINE TUMORS RESEMBLING OVARIAN SEX CORD TUMOR (UTROSCT) IN TWO BABOONS (PAPIO SPP.)

Sara Jane Mangosing, Ekaterina Perminov, Olga Gonzalez, Ana Alcaraz, Erica K. Barkei, Emily M. Corbin, Shyamesh Kumar, Edward J. Dick Jr.

Background: Uterine tumors resembling ovarian sex-cord tumors (UTROSCTs) are rarely reported neoplasms in humans that resemble ovarian sex cord stromal tumors and exhibit prominent sex cord-like differentiation. They have not been reported in the veterinary literature. We report UTROSCTs in a nulliparous 12-year-old and a multiparous 23-year-old female baboon (*Papio* spp.) housed at the Southwest National Primate Research Center.

Methods: A complete necropsy with histology was performed.

Results: Both baboons were in good body condition, and did not have any gross lesions in the uterus. Histologically, there were multiple, well-demarcated, unencapsulated, expansile masses in the myometrium. The neoplastic cells were cuboidal to columnar, arranged in anastomosing cords, trabeculae, small nests, and packets, and supported by a fibrovascular stroma. Occasionally, the cells surrounded a central space filled with eosinophilic fluid (Call-Exner bodies), resembling granulosa cell tumors. In both cases, the neoplastic cells were diffusely and strongly positive for WT-1 and diffusely negative for inhibin, calretinin, CD10, and desmin. The location, histologic appearance, and immunohistochemical properties are consistent with a diagnosis of uterine tumor resembling ovarian sex-cord tumor (UTROSCT).

Conclusions: To the best of our knowledge, this is the first report of this neoplasm in non-human primates.

LB-16: FEMALE CD-1 MICE ARE MORE SUSCEPTIBLE TO MICROCYSTIN-LR-RELATED MORTALITY AND TOXICITY THAN MALE CD-1 MICE

Shambhunath Choudhary, Bridget Lewis

Microcystin-LR (MCLR) is a potent hepatotoxin produced during blooms of *Microcystis aeruginosa* cyanobacteria in freshwater resources. Toxicity of MCLR has been well characterized in mice, but most of the earlier studies were conducted in a single sex, with strains that varied from study to study and, in all cases, the purity of the microcystin LR was likely questionable. In this study male and female CD-1 mice were given MCLR orally at dose levels of 0, 3000, and 5000 µg/kg/day for up to 7 days. Mortality reached 30% (3/10) in females as compared to 10 % (1/10) in males at 5000 µg/kg/day. Histopathological findings in early deaths included marked hemorrhage and widespread foci of necrosis in the liver. Histopathological findings in terminally euthanized animals included dose-dependent centrilobular hepatic necrosis, degeneration, and hypertrophy. MCLR-related clinical chemistry parameters were also more prominent in females and included significant elevations in AST (up to 4.1X in males and 13.0X in females) and ALT (up to 10.7X in males and 24.8X in females) compared to the respective controls. Glutathione (GSH) and glutathione-S-transferases (GSTs) are two primary lines of defense against MCLR toxicity. Female CD-1 mice have been shown to have both lower constitutive GST activity and lower levels of GSH than male CD-1 mice, which likely predisposed CD-1 female mice to microcystin LR-related hepatotoxicity and death in this study. To the author's knowledge, this is the first published report of a sex difference in mortality and toxicity in mice following MCLR exposure.

LB-17: DETERMINING CLINICAL SIGNS AND SICKNESS BEHAVIOR IN A HUMAN RSV NEONATAL INFECTION LAMB MODEL

Alejandro Larios Mora, Jack Michael Gallup, Mark R. Ackermann, Sarhad Alnajjar, Chong Wang, Rebecca L. Parsons, Suzanne T. Millman

Rationale: The aim of this study was to determine the clinical symptomology and sickness behavior of RSV- infected lambs.

Methods: Colostrum-deprived newborn lambs were nebulized with M37 human RSV (infected) or cell-conditioned growth medium (control) at day 0, and were observed daily for general clinical signs of illness, monitored for clinical data until day 8 p.i. Behavior data was collected by 10-minute instantaneous scan sampling from video recordings using Noldus Observer software with researcher blind to treatment group, lamb and trial day. Data was analysed using generalized mixed models (GLIMMEX) to compare changes in response across trial days and between control and inoculated lambs.

Results: General illness signs were sparse for infected and control groups. Mean respiratory rates (RR) and body temperatures (BT) differed on days 1 p.i. (RR and BT) and 5 p.i. (BT). Expiratory effort was only observed for 2 lambs in the infected group on days 4 (n=2) and 7 (n=1) p.i. Contrary to our hypothesis, infected lambs spent significantly less time lying and more time standing than control lambs. Infected lambs did not spend more time near the heat lamp than controls nor did this behavior differ by trial day. However, both treatment groups displayed tremoring and lameness, and bacterial pneumonias were observed at necropsy in the control group, confounding clinical signs, clinical data, and lamb behaviors due solely to RSV infection.

Conclusion: As long as bacterial infections are present, interpretation of clinical data and sickness behavior should be performed with caution when drawing conclusions.

LB-18: PIGMENTED THYMOMA IN A DOG

Misa Komine

Pigmented thymomas are a not well-known entity of canine thymomas, which are rarely described and only mentioned shortly in few previous reports and literature without pictures. Here we present a case of pigmented thymoma in a dog with detailed case information, as well as gross and histological images.

A ten-year-old, castrated male Labrador retriever presented to the Daktari Animal Hospital Kyoto Medical Center for a thoracic mass. Computed topography (CT) scans showed a 4.4 x 3.8 x 2.6 cm cranial mediastinal mass in the thorax. There were no other significant findings in the abdominal cavity, oral cavity, eyes, or digits. Needle aspirates were taken from the mass and showed clusters of mature lymphocytes as well as pigmented cells. The two primary differentials were metastatic melanoma and thymoma. The mass was removed and submitted to IDEXX Laboratories for histopathology. Grossly the mass was dark red to brown and firm.

Histologically, the mass was mostly covered with a fibrous capsule and composed of mixture of polygonal cells and mature lymphocytes. Polygonal cells had a moderate amount of eosinophilic cytoplasm and an oval nucleus with a single nucleolus. Some of these cells contained intracytoplasmic brown pigment. Anisocytosis and anisokaryosis were mild, and mitotic figures were rarely observed.

On immunohistochemical examination, the majority of the polygonal cells were positively stained with pan-cytokeratin. Fewer numbers of polygonal cells, as well as pigmented spindle cells, were positive for PNL2 and Melan A.

These histological findings, combined with the clinical information, support a diagnosis of pigmented thymoma.

LB-19: BIOMARKERS OF HEPATOFIBROSIS FOR SUBACUTE ORAL EXPOSURE TO PROTOTYPE LIVER TOXINS

Keith Koistinen, Jason Koontz, Erica Carroll, Danielle Ippolito

Background: More than 80,000 chemicals are in commercial use worldwide, with hepatic metabolism of xenobiotics to toxic intermediates increasing risk of injury after exposure. Estimating hepatotoxicity with minimally invasive assays is critical to early diagnosis of liver injury.

Objective: Our objective was to correlate serum miR-122, a liver-specific biomarker of injury, to histopathologic and clinical chemistry evidence of liver injury in order to facilitate early prediction of hepatic fibrotic injury without biopsy.

Methods: Male Sprague-Dawley rats were orally administered the archetypal hepatotoxicants 4,4'-methylenedianiline (4,4'-MDA), allyl alcohol (AA), carbon tetrachloride (CCL4), or bromobenzene for 1, 2, 3, 4, 5, 14, and/or 28 days. Sections of liver and kidney were routinely processed and stained with hematoxylin and eosin, periodic acid-Schiff, Masson's Trichrome, and Oil-Red-O. Tissue sections were scored for inflammation, fibrosis, necrosis, and lipid accumulation. Serum miR-122 concentrations were assessed by qPCR.

Results: Hepatoxicant administration resulted in elevated serum miR122, liver enzymes, and bilirubin which correlated with dose, time, and lesion severity. Hepatocellular necrosis and vacuolation was the most consistently observed histopathologic response to hepatotoxicant administration. This necrosis was variably accompanied by ductular reaction (biliary hyperplasia), and fibrosis especially after longer exposure and in animals that had a recovery period. Renal histopathologic findings were equivocal.

Conclusions: Oral administration of archetypical hepatotoxicants, reliably produced liver injury that resulted in the ability to assess the effectiveness of the biomarker miR122. miR122 could potentially be used in conjunction with other markers to facilitate early prediction of hepatic fibrotic injury without biopsy.

LB-20: CRIMEAN-CONGO HEMORRHAGIC FEVER IN HU-NSG-SGM3 MICE

M. Kelly Keating, Jessica R. Spengler, Anita K McElroy, Marko Zivcec, JoAnn Coleman-McCray, Jessica R. Harmon, Brigid Bollweg, Cynthia S. Goldsmith, Eric Bergeron, James G. Keck, Sherif R. Zaki, Stuart T. Nichol, Christina F. Spiropoulou

Crimean-Congo hemorrhagic fever (CCHF) is a tick-borne disease caused by Crimean-Congo hemorrhagic fever virus (CCHFV), a negative-strand RNA virus in the family Nairoviridae. The disease manifestations range from mild, febrile illness to a severe hemorrhagic diathesis that is seen exclusively in humans with the absence of overt disease in animals. To evaluate a novel mouse model of CCHF, humanized mice (hu-NSG-SGM3) were infected by intraperitoneal injections with sterile Dulbecco's modified Eagle's medium (control) or CCHFV isolated from a human patient in Oman or Turkey. Samples were collected for clinical chemistry, viral RNA quantification, flow cytometry, histochemical staining, immunohistochemistry (IHC) and electron microscopy. At 4 days post-infection (dpi), a subset of mice was euthanized and viral RNA was detected in all blood/tissue sampled in both CCHFV-TR and CCHFV-OM mice. However, all remaining CCHFV-OM infected mice survived until study completion, while all CCHFV-TR mice had progressive weight loss ultimately requiring euthanasia. Notably high levels of viral RNA were detected in the brains of CCHFV-TR mice that succumbed to disease. Histological evaluation in CCHFV-TR mice revealed increased hepatocellular cell death and/or vacuolation. In addition, mild to moderate meningitis, meningoencephalitis, edema, vascular congestion and early glial nodules was observed in CCHFV-TR mice, and correlated to regions of CCHFV-immunostaining in the meninges, astrocytes including subpial foot processes and cells within the cerebellar cortex. These data support Hu-NSG-SGM3 mice as a new lethal mouse model of neurologic and hepatic CCHF. In addition, neuropathological changes, suggests astrocyte damage and/or dysfunction as a contributing factor of disease.

LB-21: CANINE AND FELINE BLADDER DISEASE – PATHOLOGICAL FEATURES AND A NOVEL QUALITATIVE ANALYSIS OF PATHOLOGY REPORTS

Emily Jones, John Al-Alawneh, Chiara Palmieri, Karen Jackson, Mary Thompson, Rachel Allavena

Background: Bladder diseases are a cause of high clinical morbidity and occasional mortality in dogs and cats. Despite this their pathologic features are understudied. Bladder diseases can be grouped into inflammatory, infectious and neoplastic disorders.

Objectives: Produce case definitions for bladder diseases in dogs and cats, produce standardized morbidity ratios of bladder diseases relative to signalment and histological features, and perform thematic and concept analysis of histology reports.

Method: Canine and feline bladder biopsies submitted to The University of Queensland School of Veterinary Science Veterinary Laboratory Service or taken during routine necropsy (1995-2016, n=437) were included in this study by searching the pathology database. Proportionate morbidity and standardized morbidity ratios were calculated relative to signalment (species, breed, age, sex, neuter status). Thematic and concept analysis of bladder histopathology reports was performed using Leximancer; an analytics software for unstructured textual data, which has not previously been applied to veterinary medicine. Concepts identified from the reports were analysed to identify associations between keywords used and the final diagnosis.

Results: Proportionate morbidity of cystitis in cats and dogs was similar (29% and 28% respectively, p=0.91), and was 3% for both species for urolithiasis (p=0.93). Differences were observed in the occurrence of neoplasia – cat 8%, dog 20% (p<0.01) and normal – cat 50%, dog 37% (p=0.01).

Conclusions: This work has quantified proportionate morbidity of bladder diseases in cats and dogs, and thematic/concept analysis is a promising novel method for defining keywords and diagnostic concepts in the development of case definitions and disease classification.

LB-22: CREATION OF AN OPEN ACCESS TEXTBOOK FOR VETERINARY CLINICAL PATHOLOGY

Marion L. Jackson, Beverly A. Kidney, Nicole J. Fernandez, Brianne Y. Cheng

Background: Veterinary textbooks are increasingly costly and may be unaffordable for many students. In 2007, Veterinary Clinical Pathology – An Introduction was published as a traditional textbook (ISBN-13: 978-0-8138-2140-5) and intended as an aid for teaching undergraduate veterinary students upon their first exposure to clinical pathology. Later, the textbook required revision and co-authors were attracted to the project. We became intrigued with the idea of creating a more comprehensive teaching resource and one that would be available free to anyone with internet access.

Objective: The objectives were: to revise and update an introductory textbook of Veterinary Clinical Pathology; to enrich the project by adding a Protocol Manual for sample submission, a Laboratory Manual, and practical videos; and to create the work only in electronic format and distribute it globally for free.

Methods: Funding was obtained from the Open Textbook Fund, University of Saskatchewan, and other sources. A Creative Commons "attribution-noncommercial-noderivatives" license was chosen, meaning these resources can be downloaded and shared but the authors must be credited and the material cannot be altered or used commercially.

Results: The electronic textbook was used by veterinary students in Saskatoon and Calgary in 2016-2017 and now the entire work is available globally, for free.

Conclusions: This project has allowed us to assist financially challenged students anywhere in the world and to provide ongoing updates compared to a traditional textbook.

LB-23: EMBRYONIC ATRAZINE EXPOSURE ELICITS AGE AND SEX-SPECIFIC ALTERATIONS IN ZEBRAFISH BEHAVIOR, BRAIN TRANSCRIPTOME, AND BODY AND BRAIN WEIGHTS

Katharine A. Horzmann, Boghos Taslakjian, Jennifer L. Freeman

Background: Atrazine (ATZ) is a commonly used pesticide that frequently contaminates rural and urban water sources. Exposure to ATZ is linked to cancer, endocrine disruption, and alterations in neurochemistry and behavior.

Objective: This study explores the hypothesis that embryonic exposure to ATZ results in age and sex-specific changes in behavior, body and brain size, and the adult brain transcriptome.

Methods: Zebrafish (*Danio rerio*) embryos were exposed to 0, 0.3, 3, or 30 parts per billion (ppb; mg/L) ATZ immediately after fertilization through 72 hours post fertilization (hpf). At 120 hpf, a visual motor response assay examined larval behavior. Larvae were also grown to 9 months post fertilization (mpf) or 14 mpf. At 9 mpf, a novel tank test, a light-dark box, and an open field test evaluated adult behavior. Microarray analysis investigated ATZ related differences in gene expression. At 14 mpf, the body length, weight, and brain weight was measured to evaluate effects of ATZ on mature body and brain size.

Results: Only the 30 ppb treated larvae were hypoactive; however, the adult tests found non-monotonic, sex-specific behavior changes in all treatments. Microarray analysis identified sex-specific transcriptomic alterations, with females having altered expression of organismal injury, neurological disease, and endocrine system disorder pathways and males having altered expression of endocrine and reproductive system disorder and nervous system development and function pathways. Adult zebrafish also had non-monotonic, sex-specific alterations in body length, body weight, and brain weight.

Conclusion: Developmental ATZ exposure elicits sex-specific alterations in adult size, behavior, and gene expression.

LB-24: INTRAOCULAR COLLISION TUMOR IN A CAT: PLEOMORPHIC IRIDOCILIARY CARCINOMA AND POST-TRAUMATIC LYMPHOMA

Megan E. Climans, Gillian C. Shaw, Leandro B.C. Teixeira

The formalin-fixed right globe of a 9-year-old neutered male domestic shorthaired cat was submitted to the Comparative Ocular Pathology Laboratory of Wisconsin following a 2-year history of an enlarging iris mass. Gross examination revealed a tan-to-white mass expanding the anterior uvea, equatorial sclera and episclera. Histopathologically two cell populations comprised the mass. Pleomorphic polygonal-to-spindle-shaped cells with osseocartilaginous metaplasia formed the bulk of the mass and were surrounded by sheets of monomorphic round cells. Initial differentials were poorly-differentiated iridociliary carcinoma or metastatic carcinoma with possible additional round cell tumor. Immunohistochemical/histochemical stains revealed that the polygonal-to-spindle-shaped population expressed cytokeratin, vimentin and NSE with PAS-positive basement membranes, confirming neuroepithelial origin (iridociliary). The round cells were CD20-positive and CD3-negative, confirming B-lymphocyte origin. The IHC expression, epithelial differentiation and thick PAS-positive basement membranes of the main neoplastic component in this case are typical of iridociliary epithelial tumors, but the invasiveness, pleomorphism and osseocartilaginous differentiation indicate malignancy, suggesting pleomorphic iridociliary carcinoma. The distribution and IHC-profile of the neoplastic lymphocytes are consistent with feline post-traumatic lymphoma (sarcoma, round-cell variant). To the authors' knowledge, this is the first reported case of an aggressive iridociliary tumor in a cat with features of pleomorphic iridociliary carcinoma, a rare truly malignant ocular tumor described in humans and dogs, that is associated with trauma and/or long-standing eye disease. The simultaneous post-traumatic ocular lymphoma suggests that chronic ocular disease and trauma, evidenced by the clinical history and hypermature cataract, contributed to the development of these two rare neoplasms in this case.

LB-25: URINARY BLADDER STASIS IN BEEF CATTLE ASSOCIATED WITH SUBCUTANEOUS INJECTIONS IN THE ISCHIORECTAL FOSSA

Fernanda Castillo-Alcala, Susan Brown, Tim Crawshaw, Richard Laven, Garth Riddle, John Munday

An outbreak characterized by urinary retention and urinary bladder rupture in beef cattle is described. The herd consisted of 39, raising 2-year-old Friesian bulls grazed on rye-grass/white clover pasture. Overall herd mortality rate was 26% (10/39). Clinical history indicated that 10 days prior to the outbreak, the bulls had been treated with anthelmintic via subcutaneous injection in the ischiorectal fossa. Affected bulls presented with anorexia, abdominal pain and urine dribbling. Ultrasound examination revealed urinary bladder distention and peritoneal fluid accumulation. Necropsies performed on the farm revealed urinary bladder rupture with no evidence of urinary tract obstruction. One affected live bull was submitted to Massey University for clinical examination and subsequent necropsy. Gross examination of the abdomen revealed a markedly distended urinary bladder, massive subcapsular and pericapsular renal edema with retroperitoneal fluid accumulation, minimal hydronephrosis and no gross evidence of urinary outflow obstruction. The right ischiorectal fossa revealed multifocal areas of tissue fibrosis that extended into areas innervated by the distal cutaneous branch of the pudendal nerve and the pelvic nerve. Histologically, there was extensive fibrosis, myonecrosis and neurodegeneration localized to the injection site. Lesions in the urinary bladder and the kidneys were consistent with chronic urinary stasis and secondary bacterial overgrowth. No lesions were present on other organs examined, including the central nervous system. On the basis of epidemiology, clinical signs, clinical and pathological findings, we suggest that a local inflammatory reaction at the injection site led to peripheral nerve damage and dysfunction, resulting in urinary bladder stasis.

LB-26: HEMORRHAGIC ENTEROPATHY ASSOCIATED WITH CLOSTRIDIUM CHAUVOEI IN A FARM RAISED CATFISH

Wes A Baumgartner, Larry Hanson, John Brooks

Reports of anaerobic bacterial infections in fish are uncommon. An adult catfish from a farm pond mortality event had severe gastrointestinal hemorrhage, hemorrhagic ascites, hepatic and splenic hyperemia, and fin hemorrhages. The intestine had severe transmural hemorrhage with mucosal necrosis and large numbers of Gram positive rod bacteria that were positive by immunohistochemistry for *C. chauvoei*. A light growth of *Aeromonas sobria* was cultured from the mesonephros. A single anaerobic colony type was cultured from the liver and was screened by PCR using targeted primers for *Clostridium perfringens* alpha toxin gene, hemolysin, fliC, *Clostridium septicum* alpha toxin, spoOA sporulation gene, and the16S-23S intergenic space. The complete 16S rRNA gene was also sequenceding was also conducted.

Morphologic and molecular data is consistent with *C. chauvoei*, and the it shared 99% sequence homology 16S sequence was 99% identical. The gross and histopathological findings in this case are far more severe than is seen in cases of simple *A. sobria* infection; it is an opportunistic bacteria that is typically seen in conjunction with more virulent pathogens. The microbial and molecular data indicate that *C. chauvoei* contributed significantly to disease in this fish. Based on knowledge of *C. chauvoei*, it seems likely that the *Aeromonas* infection in conjunction with spore ingestion led to clostridial overgrowth, toxin elaboration, and death. This is the first report of *C. chauvoei* infection causing disease in a fish.

LB-27: A NOVEL AEROSOLIZED MANNHEIMIA HAEMOLYTICA CHALLENGE MODEL IN COLOSTRUM-DEPRIVED CALVES

Laura Lynn Bassel, Sarah-Nicole A. Alsop, Emily I. Kaufman, Kevin J. Stinson, Ksenia Vulikh, Laura R. Siracusa, Linda Groocock, Veronique Carson, Tony Cengija, Jeff L. Caswell

Mannheimia haemolytica causes severe respiratory disease in susceptible cattle. It can be isolated from the nasopharynx of healthy animals but additional factors such as stress and viral infection are necessary for the development of pneumonia. As such, experimental models have relied on a combination of viral infection (e.g. bovine herpesvirus 1) followed by aerosolized *M. haemolytica* challenge or inoculation of *M. haemolytica* in a relatively large volume of saline directly into the trachea or bronchi. Although these models produce disease that is like that seen with natural infection, they inhibit or bypass and dilute many of the innate immune defenses that are important in the initial establishment of infection. In pilot studies, we investigated the effect of aerosolization of 3 different strains of *M. haemolytica* on the development of pneumonia in colostrum-replete or colostrum-deficient calves. Colostrum-deprived calves consistently established infection and pneumonia following aerosol challenge with one of the two tested field strains of *M. haemolytica. Mannheimia haemolytica* was isolated from the nasal cavity from day 1 to day 3 following challenge, when calves were euthanized. Clinical signs; elevated blood neutrophils, haptoglobin and fibrinogen; and ultrasonographic, gross and histologic lesions of variable severity developed in all challenged calves. *Mannheimia haemolytica* was isolated from lungs at post mortem examination. This experimental model may be useful in testing interventions and in investigating the initial events occurring during the development of *M. haemolytica* infection independent of viral infection.

LB-28: THE TRIUMPH OF DEATH: 1,090 CATTLE DIED IN ONE OUTBREAK OF BOTULISM IN MIDWESTERN BRAZIL

Ricardo A.A. Lemos, Carolina Guizelini, Juliana L.P. de Paula, Rayane C. Pupin, Claudio S.L. Barros, Daniela A. Neves, Lilian O.B. Alcântara, Tessie B. Martins

Background: Bovine botulism is caused by *Clostridium botulinum* toxins C and D, which prevent the release of acetylcholine from neuromuscular synapses, thus causing flaccid paralysis, but not central nervous system signs.

Objective: To describe an outbreak of bovine botulism in Brazil.

Results: On August 2017, in a feedlot in Midwestern Brazil (S 2026'34" W 54°38'47") 1,700 18-month-old cattle were held in 10 contiguous lots. Cattle were fed 5 times a day with a ration consisting of moist corn grains, guinea grass, hulls of soy bean grain, cotton seeds, minerals, urea, vitamins, monensin sodium, and limestone. Water was available *ad libitum*. Roughly 1,035 cattle died within the first 48 hours in the feedlot and other 55 died within 3 to 16 days after the onset of the outbreak. Main clinical signs were flaccid paralysis, protruded flaccid tongue, and abdominal breathing. Recumbent cattle were bright and alert. No gross or microscopic lesions were found in 21 necropsied cattle. The source of toxin was traced to the corn in the feedlot silage, which was later fed to a sheep that developed similar clinical signs and recovered after treated with antibotulin serum (Botulin C-D). The bioassay in mice done with serum and liver from two affected cattle resulted positive for toxins C and D.

Conclusions: The diagnosis of botulism in the current outbreak was based on epidemiological data (high mortality within 48 hours), characteristic clinical signs, reproduction of the disease in one sheep, and bioassay results.

LB-29: IMMUNOHISTOCHEMISTRY PROFILE TO STUDY A NON-INFECTIVE MENINGOENCEPHALITIS IN CANINE: CASE REPORT

Cristhian Camilo Rozo, Ricardo Piñeros, Paola Barato

Non-infectious meningoencephalitis (NIME) is an idiopathic inflammatory disease of the central nervous system (CNS) of canines. Granulomatous meningoencephalitis (GME), necrotizing meningoencephalitis (NME) and necrotizing leukoencephalitis (NLE) are three of the NIME forms of presentation and the final diagnosis is defined by the histopathological and immunohistochemical findings.

We present a case in a canine, pinscher, female, 8 years old, with a neurological signs and presumptive diagnosis of Erhlichiosis, who arrived at the veterinary clinic of the University of La Salle, Bogota, Colombia in 2016.

The most relevant macroscopic findings were in the Central Nervous System (CNS) with moderate congestive changes on meninges, cerebrospinal fluid increased and irregular brownish spots in white and gray matter. Histopathological findings revealed multifocal areas of granulomatous mononuclear inflammatory reaction in perivascular areas, leptomeninges and part of the neuropil, and without evidence of microorganisms with special colorations (Giemsa and Gram). Moderate positive immunostaining with T-cell (CD3), B-lymphocyte (CD79a), IgM and macrophages / histiocytes (CD68) lymphocytes were observed. GFAP and IgG were negative. Gross, histopathological and immunohistochemistry pattern suggest granulomatous meningoencephalitis (GME).

LB-30: EVALUACION OF BUTYRICICOCCUS PULLICAECORUM AS GROWTH PROMOTER AND ITS IMPACT IN INTESTINAL HEALTH, IMMUNE RESPONSE AND PRODUCTIVE PARAMETERS IN BROILER CHICKENS

Ricardo Piñeros, Andrés Rodríguez-Ávila, Alvaro Pedroche, Eduardo Castro, Angélica López, Paola Barato

Butyricicoccus pullicaecorum was evaluated as replacement for antibiotic growth promoters. Broilers Ross 308 were housed in 5 groups each one of 500 birds with following treatments: T1: Anticoccidial and antibiotic growth promoter. T2: Without growth promoter or anticoccidial. T3: Day 1 to 7 of age with B. pullicaecorum 500 g/Ton of food and from day 8 to 42 witgh250 g/Ton. T4 = T3 plus anticoccidial, and T5 =T4 decreasing 50 Kcal of total metabolizable energy and 2% essential amino acids. On days 14, 21, 28, 35 and 42, duodenum, jejunum, ileum and cecum of 10 birds from each group were taken for evaluation of villus height, crypt depth and inflammatory lesions. On days 1, 21 and 42 serological samples were taken for IBV, IBDV, NDV, CAV and ARV from 20 chickens of each group. The productive parameters were evaluated weekly. Treatment 3 had significant differences in villus height compared to control groups (P <0.05) and showed significant differences in crypt depth in comparison with the control groups, T4 and T5 (P<0.005). The weight gain was statistically significant higher in T3 (P <0.005) as well as lower feed conversion in the same group (P <0.05) compared to controls, but not with the T4 and T5 groups. No significant differences were found between groups for the immune response. In conclusion, T3 showed better productive and intestinal health parameters compared to the control groups, suggesting that it can be used in the field by replacing antibiotic as growth promoters.

LB-31: NEW GENOTYPE OF CANINE DISTEMPER VIRUS AFFECTING DOGS IN COLOMBIA: IMMUNOHISTOCHEMICAL, MOLECULAR AND PHYLOGENETIC STUDY

Paola Barato, Ricardo Piñeros, Norma Forero, Alisson Bello, Maria Paula Santos, Miguel Montufar, Cesar Diaz, Carlos Venegas, Germán Rodriguez, Henry Benavides, Elisabete Martins

We evaluated by immunohistochemistry (IHC) 36 cases from 2007 to 2013 and by RT-PCR for *N* gene 201 samples between January 2014 to January 2015, with previous clinical or histopathological diagnosis of CDV infection in Bogotá, Colombia. Twenty-eight samples from seven epidemiological events of canine distemper and two commercial vaccines were also amplified and sequenced to *fsp* gene for phylogenetic study. For immunohistochemistry study with a monoclonal antibody against nucleocapsid of CDV was selected respiratory, nervous and lymphoid tissues (lymphoid nodes, tonsils and spleen). Thirty nine percent of cases (n=14) had IHC positive to CDV, and twenty-two cases (61%) were negative. The histopathological analysis found lesions compatible with other viral infections that were previously diagnosed clinically as distemper infection (n=14, 32%) and was seen interstitial pneumonia in two cases without IHC positive. For RT-PCR of *N* gene 28% of samples were positive and 72% negative, and the phylogenetic analysis of *fsp* gene detected a new linage of CDV in field very distant of vaccines. All results suggest: a. Over diagnosis of CDV infection based on clinical signs emphasising to use molecular analysis, histopathology and immunohistochemistry to find the final diagnosis; b. To consider differential diagnosis to respiratory symptoms such as canine adenovirus type 2, canine herpesvirus and canine influenza; c. New lineage of CDV in Colombia, it could be called SA4 (South America 4) as its counterpart with *H* gene; and d. Further studies are needed to evaluate if genotypic difference between field strains and vaccines affect immune response

LB-32: TO FREEZE OR NOT TO FREEZE: MINIMIZING HISTOLOGY FREEZING AND TISSUE PRESERVATION ARTEFACTS IN WHOLE CADAVERS, A PRACTICAL GUIDE

Megan Du Toit, John Al-Alawneh, Emily Sheehan, Taylor Argent, Rachel Allavena

Background: It is well known that the process of freezing and thawing cadavers impedes histological examination, however the degree of damage observed in frozen tissues can vary remarkably and unpredictably between organs and cadavers.

Objectives: This study aims to provide a guide to minimize autolytic change and freezing artefact due to storage temperatures of tissues for histological analysis.

Method: Twenty-eight cat cadavers were stored for varying lengths of time at 18 to 20C, 4C, -20C and -80C, and freezing and thawing time was rapid at room temperature, or gradual at 0-4C. The tissue specimens were then examined grossly and histologically and graded on freezing artefact and autolytic change. Pilot studies were conducted on cold or room temperature formalin, repeated thawing and re-freezing, and prolonged storage.

Results: Autolysis was greatly accelerated in cadavers stored at ambient temperature, with gross changes such as bloating, hypostasis, hemoglobin inhibition and biliary inhibition. Histologically, autolysis was more severe in organs with high enzymatic activity, the worst being the liver. Logically refrigeration at 4C had a less detrimental effect on freezing artefact, than storage at -20C or -80C. However freezing of specimens was beneficial to prevent autolysis if cadavers had to be stored for longer than a week at 4C. Repeat freeze-thaw was the most detrimental protocol for the histological integrity of tissue, and cold formalin exacerbated artefacts compared to room temperature fixation.

Conclusions: Thus, to maximize diagnostic findings from histology freezing should be avoided unless the postmortem interval is greater than one week.

AUTHOR INDEX

AUTHOR INDEX

AUTHOR INDEX

AUTHOR INDEX

AUTHOR INDEX

AUTHOR INDEX

AUTHOR INDEX

AUTHOR INDEX

AUTHOR INDEX

AUTHOR INDEX

AUTHOR INDEX

AUTHOR INDEX

AUTHOR INDEX